ADVANCES IN CHEMICAL PHYSICS

VOLUME LXVIII

EDITORIAL BOARD

Advances in
CHEMICAL PHYSICS

EDITED BY

I. PRIGOGINE

University of Brussels
Brussels, Belgium
and
University of Texas
Austin, Texas

AND

STUART A. RICE

Department of Chemistry
and
The James Franck Institute
The University of Chicago
Chicago, Illinois

VOLUME LXVIII

AN INTERSCIENCE® PUBLICATION
JOHN WILEY & SONS
NEW YORK · CHICHESTER · BRISBANE · TORONTO · SINGAPORE

An Interscience® Publication

Library of Congress Cataloging Number: 58-9935

ISBN 0-471-84901-4

Printed in the United States of America

10 9 8 7 6 5 4 3 2 1

CONTRIBUTORS

VLASTA BONAČIĆ-KOUTECKÝ, Institut für Physikalische und Theoretische Chemie, Freie Universität Berlin, D-1000 Berlin (West) 33, Takustrasse 3

F. HEISEL, Groupe de Photophysique Moléculaire, Centre de Recherches Nucléaires, F-67037 Strasbourg, B.P. 20, France

I. G. KAPLAN, L. Ya. Karpov Institute of Physical Chemistry, ul. Obukha, 10, 107120, Moscow B-120, USSR

YU. L. KLIMONTOVICH, Moscow State Lomonossov University, Faculty of Physics, Moscow, 117234, USSR

W. D. KRAEFT, Ernst-Moritz-Arndt-Universität Greifswald, Sektion Physik/Elektronik, DDR

D. KREMP, Wilhelm-Pieck-Universität Rostock, Sektion Physik, Rostock 2500, DDR

E. LIPPERT, Iwan-N. Stranski-Institut für Physikalische und Theoretische Chemie, Technische Universität Berlin, D-1000 Berlin (West) 12, Strasse des 17.Juni 112

J. A. MIEHÉ, Groupe de Photophysique Moléculaire, Centre de Recherches Nucléaires, F-67037 Strasbourg, B.P. 20, France

A. M. MITEREV, L. Ya. Karpov Institute of Physical Chemistry, ul. Obukha, 10, 107120, Moscow B-120, USSR

W. RETTIG, Iwan-N. Stranski-Institut für Physikalische und Theoretische Chemie, Technische Universität Berlin, D-1000 Berlin (West) 12, Strasse des 17.Juni 112

INTRODUCTION

Few of us can any longer keep up with the flood of scientific literature, even in specialized subfields. Any attempt to do more and be broadly educated with respect to a large domain of science has the appearance of tilting at windmills. Yet the synthesis of ideas drawn from different subjects into new, powerful, general concepts is as valuable as ever, and the desire to remain educated persists in all scientists. This series, *Advances in Chemical Physics*, is devoted to helping the reader obtain general information about a wide variety of topics in chemical physics, which field we interpret very broadly. Our intent is to have experts present comprehensive analyses of subjects of interest and to encourage the expression of individual points of view. We hope that this approach to the presentation of an overview of a subject will both stimulate new research and serve as a personalized learning text for beginners in a field.

I. PRIGOGINE

STUART A. RICE

CONTENTS

ADVANCES IN CHEMICAL PHYSICS

VOLUME LXVIII

PHOTOPHYSICS OF INTERNAL TWISTING

E. LIPPERT* AND W. RETTIG

Iwan N. Stranski-Institut für Physikalische und Theoretische Chemie
Technische Universität Berlin
D-1000 Berlin (West) 12, Strasse des 17. Juni 112

V. BONAČIĆ-KOUTECKÝ

Institut für Physikalische und Theoretische Chemie
Freie Universität Berlin
D-1000 Berlin (West) 33, Takustrasse 3

F. HEISEL AND J.A. MIEHÉ

Groupe de Photophysique Moléculaire
Centre de Recherches Nucléaires
F-67037 Strasbourg Cedex

CONTENTS

* Adjunct Professor at the Institute for Molecular Science, Okazaki, Japan, from January to July, 1985, where the basic conceptions were developed.

1. INTRODUCTION

A. Adiabatic Photoreactions Involving Charge Transfer

Recently, some statistical approaches have been reported concerning dynamics in molecular liquids, and this especially in the deduction of (1) equations of state, for hard-core convex molecules; (2) equations of the conformational equilibrium of small-chain molecules, and (3) by generalizations of the linear response theory and the fluctuation dissipation theorem, equations of the entropy production due to dissipation.[1] This contribution deals with another problem of the interaction of light with fluid molecular solutions, namely, (1) with the photochemical reorientation of different functional groups against each other while the dissolved molecule as a whole remains in electronically excited states, and (2) with the photophysical reorientation of the surrounding solvent molecules as induced by alterations in the charge distribution of the solute when it proceeds along its reaction pathway.[2]

If such reorientational relaxation processes lead to other minima on a single excited-state energy surface of a given multiplicity, those horizontal radiationless transitions are called adiabatic photoreactions.[3] In principle, they can be observed by their ability to emit fluorescence light from both the initial and the final excited state, and the emissions are situated at shorter and longer wavelengths, respectively.[4]

Dual fluorescence due to this type of adiabatic photoreaction was first observed in 1961[5] in dilute solutions of *para*-cyano-*N*,*N*-dialkylanilines (e.g., DMABN, Fig. 2.1[6]) and first interpreted in 1972[7] by a "twisted intramolecular charge-transfer" (TICT) mechanism.

The term *twist* is selected instead of *rotation* (like a propeller) or *torsion* (like a vibrational mode) because the molecular geometry changes into a new stable excited state, for example, in the case of DMABN from an almost planar to the orthogonal configuration, with the torsional angle θ between the planes of the two functional groups of the molecule (dialkylamino group and aromatic ring).

This twisting from the initial (Franck–Condon) to the final (TICT) configuration in DMABN is connected with the transfer of one electronic charge from the amino to the cyano group and by a decrease to zero of the $S_1 \rightarrow S_0$ transition moment[8]; fluorescence from the TICT state is symmetry forbidden, and its ground-state-recovery (GSR) rate deviates from zero only due to some vibrational modes, for example, the amplitude of torsion from orthogonality. Therefore, GSR increases with temperature, and with further increasing temperature the back reaction to the initially excited state eventually becomes possible, so that at sufficiently high temperatures thermal equilibrium could be achieved[9,10] (see also Fig. 2.9).

In this chapter we shall use Grabowski's proposal,[7,11] which has been generalized in literature and which defines the state with maximum charge separation as the "charge-transfer" excited state even though in *intra*molecular charge transfer no full charge separation into independent units occurs as compared with most cases of *inter*molecular charge transfer reactions. Intramolecular maximum charge separation in the excited state is achieved by differences in electronegativity of different functional groups of the free molecule alone[12–14] (in the gas phase), but in fluid solutions this effect is enhanced by solvent polarity.

There are large families of chemical compounds between those of DMABN and ethylene that are either polar or nonpolar in S_0, but after photoexcitation the twisted charge-separated biradical excited states develop via adiabatic pathways (Section III). This has been displayed by experimental[2] and theoretical[14] studies.

Let us consider a liquid solution in its thermodynamic equilibrium state. Starting at time $t = 0$ it is irradiated by light for which the solvent is transparent but the solute is not. After absorption of a photon by a solute molecule, the equilibrium is disturbed at least in a certain volume element, which is assumed to consist of the excited molecule and its nearest neighborhood, that is, the cage of surrounding solvent molecules, while the rest of the solution is considered as a heat bath. Instantaneously after absorption a relaxation process starts from the Franck–Condon state, proceeding to the TICT state, and that can be studied by the time dependences of the decay and rise of the fluorescence intensities which are emitted from the initial state B and the final state A (Fig. 2.4). Connected with the radiative or nonradiative GSR there will occur specific reorientational reactions in the cage owing to changes in the geometric and electronic structure of the solute along its GSR pathway, and this is in addition to the dissipated fraction Q of the excitation energy,[15] which increases the entropy of the heat bath.[1] This program implies that these volume elements are small compared to their distances, that is, the solute concentration is sufficiently small.

The situation where the dissipation takes place on a time scale characterizing the interactions between the solute and its cage can be described by a generalized Langevin equation (GLE) in which the time-dependent friction and the Gaussian random force are related by an expression of the fluctuation-dissipation theorem (FDT). The GLE implies that the process is non-Markovian, and reduces to the conventional Langevin equation only if the friction decays on a time scale much shorter than the time development of the velocity.

Kramers' theory[16] involves a barrier crossing, and the motion within the barrier region can be described by means of GLE in which the potential is linearized, that is, it is purely repulsive in accordance with modern hard-core theories in which the repulsive forces govern the process.[17] A well-selected hard-core reference system using nonspherical bodies and some additional thermodynamic perturbation treatment for taking into account the attractive forces allow the properties of simple molecular liquids to be described (Ref. 1, Chapter II). But Kramers' theory breaks down if the barrier height becomes low, for example, $E_0 < k_B T$, and this holds also for recent extensions of Kramers' theory to the non-Markovian case (Section IV). Such a small barrier height can be expected for solutions of aromatic compounds because of strong mode coupling.

For experimental studies, synchrotron radiation is a most helpful tool because of its spectral continuum that allows the full long-wavelength absorption region to be illuminated and this in a high repetition rate that allows for rapid fluorescence photon counting. The Berlin Electron Storage Ring for Synchrotron Radiation (BESSY) in single-bunch operation produced basic results for the analysis of time-resolved fluorescence rise and decay processes.

Direct two-color measurements of the coupling of the excited solute with the solvent are possible on the basis of the quantum statistical theory of dynamical processes presented in Ref. 1, Chapter III. The rate of internal twisting can be altered by system parameters like the temperature and pressure and the nature of the solvent so that its time scale Δt matches the characteristic time τ_c somewhere in the picosecond area, that is, in the (far) infrared (FIR) spectral region; along this time scale, the reacting solute molecule changes the amount and the direction of its electric dipole moment. Now let us select two spectral regions from the pulses of the synchrotron radiation as performed[18] at the BESSY infrared beam line: The excitation pulse in the ultraviolet region experiences well-defined eigenstates since $\lambda / c = 2\pi/\omega \ll \Delta t \simeq \tau_c$, the Franck–Condon state can be achieved in a femtosecond time scale and the relaxation process can be started. The second, FIR (picosecond) pulse starting also at time $t = 0$ experiences no well-defined (rotational) eigenstates since

$2\pi/\omega \simeq \Delta t \simeq \tau_c$, but a continuum in the energy domain. Under such conditions the FIR band shape of the absorption due to dipole–dipole coupling between the time-dependent vector of the FIR electric field and the time-dependent vector of the molecular dipole moment can be analyzed by a generalized fluctuation-dissipation theorem (Ref. 1, Eq. 3.E23), since the molecular dipole operator follows a GLE (Ref. 1, Eq. 3.F8) and off-diagonal elements of the density operator must be taken into account (Ref. 1, Eq. 3.E3).

Another proposed two-color synchrotron radiation experiment is multiphoton ionization photoelectron spectroscopy by which, after excitation, the hypersurface profile should be measured from the energy distance between occupied orbitals and the ionized state. Photoelectron spectroscopy (PES) techniques applied to two-color multiphoton ionization (MPI) have been used to study the dynamic behavior of electronically excited molecules, since one-electron ionization transitions are always allowed from any excited state.[19] The main part of the intended device consists of (1) a vacuum chamber with a nozzle beam for introducing a gas sample into the ionization region, (2) a synchrotron radiation beam line for crossing the nozzle beam for broadband $S_1 \leftarrow S_0$ excitation, (3) an ionizing tunable monochromatic Laser beam crossing the molecular beam (free molecules or clusters) after excitation in a time-of-flight adaptable distance, (4) a photoelectron energy analyzer, (5) some other detection devices for the total ion current and (time-of-flight) mass spectrometry, and (6) a data acquisition system (for further details see Ref. 19).

Both proposed two-color experiments have not yet been performed for problems of internal twisting; therefore, from the experimental viewpoint, this contribution is confined mainly to fluorescence spectroscopic results, that is, quantum yields, spectral quanta distribution, polarization, and rise and decay functions.

B. Electronic and Dynamic Aspects

The understanding of the photoprocess responsible for the dual fluorescence can be developed in two steps. The first step is the static and structural nature and involves estimates of energy surfaces, mainly the location of minima and barriers. It is based on traditional quantum chemistry. The second step deals with dynamics of the process and requires the use of additional theoretical tools of chemical dynamics such as stochastic description.

In the framework of the Born–Oppenheimer approximation, the motion of the nuclei in a molecule being in the ith electronic state characterized by the energy surface $E_{el}^{(i)}(\mathbf{R})$ and the wavefunction $\psi_{el}^{(i)}(\mathbf{x}, \mathbf{R})$ is determined by the shape of the surface. After preparation of the initial state by transfer from one surface to another (absorption), the negative gradient of the surface at the point describing the molecular geometry momentarily determines forces acting on the nuclei. If there is *no* interaction with the environment, the total energy of the molecule remains constant during this process although the kinetic energy of the nuclei is changing. In principle, the classical description of the motion of nuclei can be replaced by more sophisticated quantum-mechanical descriptions. Nevertheless, for the photophysical process in the solution, the main complication concerning nuclear motion arises from the collisions with the neighboring molecules. Consequently, the exchange of the energy of the nuclear motion takes place between the solute molecule and its neighbors, in which case the kinetic energy of the nuclei can be continuously removed by the surroundings. This can be looked at as a part of an infinite thermal bath. The vibrational energy exchange with the surroundings will lead to a rapid thermalization of the vibrational motions.

On the other hand, if isolated molecules have more than one minimum on the energy surface of the excited state owing to the intramolecular geometry relaxtion, how can a fraction of molecules move from one minimum to another? If these minima are deep enough, the vibrational levels will be populated according to the Boltzmann distribution at thermal equilibrium. In the limit of small displacements from the minimum of the potential well, the energy surface can be approximated by the harmonic potential. In general, tunneling from one minimum to the other might occur, but this process is fairly slow and unlikely to be detected at room temperature. Perhaps it can be of some importance at low temperatures, in which case the thermally activated travel over barriers can be avoided. Again, the role of the solvent will be of most importance. Through collisions with neighboring molecules, the nuclei of the molecules can obtain additional kinetic energy so that the fraction of the molecules that has sufficient energy might go over to the other minimum. Such thermally activated mechanisms for moving from one minimum to another are then responsible for the occurrence of a reaction on one energy surface. Local minima are also depopulated by radiative or radiationless transitions. A change of the electronic state has a new effect on the nuclear motion, since after a jump from one hypersurface to another the forces acting on nuclei are determined by the new surface.

The rate for moving from one minimum to another on the surface of the first singlet excited state S_1 is then in competition with radiative rates if the barrier between two minima is sufficiently small. Consequently, for the study of dual fluorescence, two aspects should be investigated: (1) the possibility of existence of two minima on the S_1 energy surface with wavefunctions exhibiting a different electronic nature—locally excited and charge-separated—the characteristics of which will be influenced differently by polar solvents, and (2) the theoretical tool for determining the rate constants depopulating both minima under the influence of the polar solvent.

The knowledge of the two-minima energy surface is sufficient theoretically to determine the microscopic and static rate of reaction of a charge transfer in relation to a geometric variation of the molecule. In practice, the experimental study of the charge-transfer reactions in solution leads to a macroscopic reaction rate that characterizes the dynamics of the intramolecular motion of the solute molecule within the environment of the solvent molecules. Stochastic chemical reaction models restricted to the one-dimensional case are commonly used to establish the dynamical description. Therefore, it is of importance to recall (1) the fundamental properties of the stochastic processes under the Markov assumption that found the analysis of the unimolecular reaction dynamics and the Langevin–Fokker–Planck method, (2) the conditions of validity of the well-known Kramers' results and their extension to the non-Markovian effects, and (3) the situation of a reaction in the absence of a potential barrier.

II. DUAL FLUORESCENCE

A. Some Basic Concepts

Large geometrical intramolecular relaxations in the excited state after absorption, accompanied by electronic redistribution, are of interest to both photochemists and photophysicists. Recent theoretical models allow the classification of some of these cases and the prediction of the conditions when large-amplitude motions like intramolecular twisting are associated with a considerable change of energy and with strong electronic reorganization.[14] Experimentally, the nature of these excited states as well as their dynamics can best be studied for cases where both precursor *and* product states are luminescent at shorter and longer wavelengths, respectively, and the second fluorescence band is strongly solvent dependent. In these cases the experiments clearly demonstrate that the product excited state is of a charge-transfer nature.[2,7] This section presents an overview of these basic experiments and thus is preparation for the detailed quantum chemical outline presented in Section III.

The dynamics will be deduced from the time dependencies of the fluorescence intensities of both precursor and product states using advanced kinetic theories. These are presented in Section IV. One of the main points is whether or not an energy barrier separates precursor from product state. It will be shown that the experimental results of the systems studied till now are more consistent with a precursor-to-product transition with almost no barrier.

1. *The Notion of TICT States in Aromatic Donor–Acceptor Compounds*

Owing to the extensive work of Grabowski and co-workers,[7,11,20–23] the dual fluorescence of *N,N*-dimethylaminobenzonitrile (DMABN,**1**), discovered in 1962 by Lippert et al.,[5] has recieved much attention. Contrary to earlier proposals involving ground- or excited-state complex formation,[24–26] they showed that after excitation a twisted amine configuration has to be reached in order to observe the second ("anomalous") fluorescence band (F_A or A band) (Fig. 2.1). The long-wavelength fluorescence band is missing for planar model compounds related to DMABN like **2**,[20,21,27] **3**,[28] or **4**,[29] and also the corresponding ester **5**,[28,30] for which twisting of the amino group is blocked. Similar model compounds that are twisted in the ground state, like **6**[31] or **7**,[7,12,32] lack or nearly lack the "normal" short-wavelength fluorescence (F_B or B band). In addition to nitriles, dimethylanilines para-substituted with an ester group like **8** (DMABEE),[10,30,33] aldehyde,[34] or keto group[13] also show the two fluorescence bands.

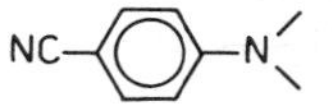

1, DMABN 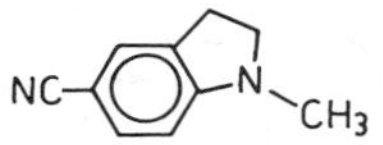 2 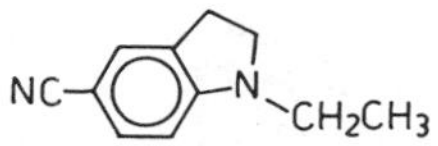 3

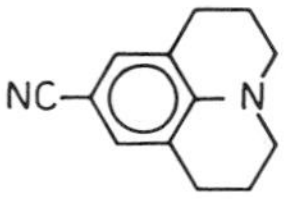

4 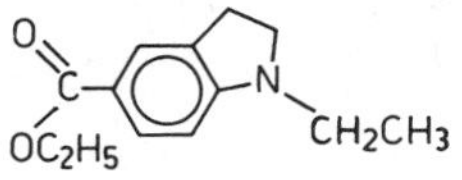 5 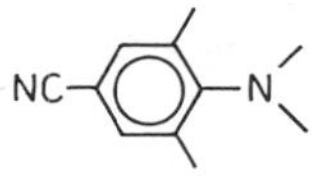 6

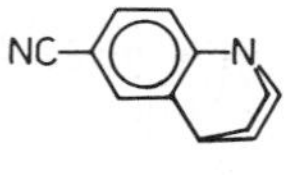

7 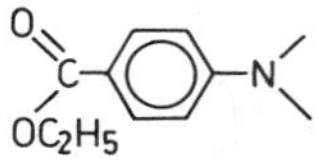 8, DMABEE

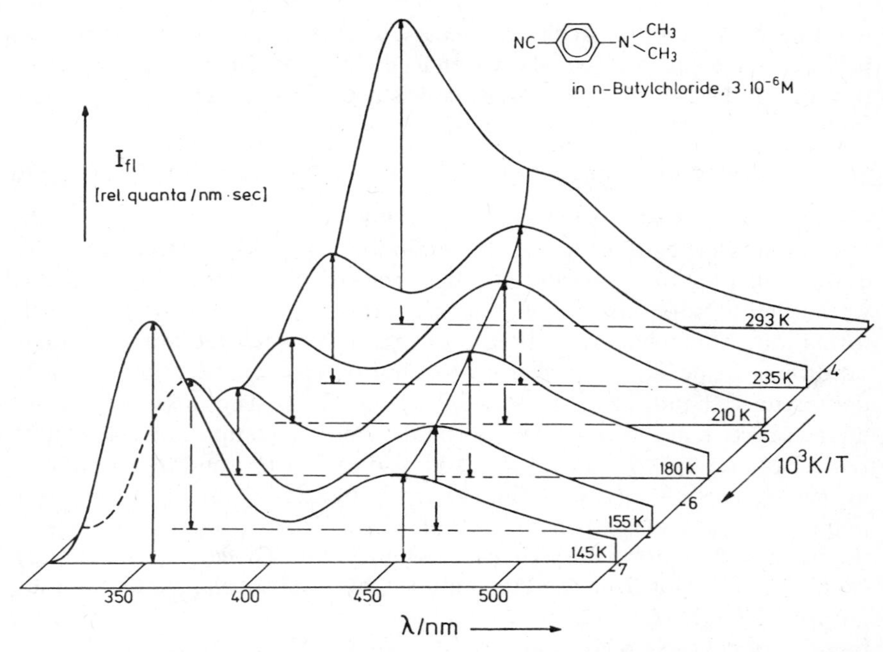

Fig. 2.1. Dual fluorescence of a dilute solution of DMABN in the polar solvent *n*-butyl chloride as a function of temperature.[9] Over a large range, both bands are independent of concentration and excitation wavelength. Upon increasing the solvent polarity, the long wavelength A band shifts to the red and gains in relative intensity.

An anomalously strong sensitivity to a change of solvent polarity has been found for the spectral position of the emission maximum of the long-wavelength A band of all the compounds. From this, a very large dipole moment of the emitting A* state was concluded[5] and later measured independently,[12] corresponding to a large intramolecular charge separation.

Grabowski and co-workers proposed the notion of a TICT excited state.[7] The model implies that the charge transfer is at a maximum for a geometry with two perpendicular π subunits. For this perpendicular geometry, the $\pi-\pi$ interaction is broken, and thus either no charge or a full electronic charge could have been transferred from one to the other subunit in the excited state.

At perpendicular geometries, this excited state of charge-separated nature usually exhibits an energy minimum.[2,7,14,35] Thus, a molecule, which starts off in the planar conformation by Franck–Condon excitation, can spontaneously relax in the excited state by a twisting motion toward a quite different geometry. This is normally called an adiabatic photoreaction[3,36] or horizontal radiationless transition[37] by photochemists and photophysicists.

In contrast to the excited state, the ground state exhibits a maximum at the perpendicular geometry. In fact, it is well known that aromatic amines like DMABN possess a ground-state rotation barrier whose maximum is located at the perpendicular conformation. Thus, the emission from a TICT state occurs, in most cases, from an energy minimum of the excited state to an energy maximum of the ground-state twist potential, which explains in part the structureless feature of the long-wavelength fluorescence A band.

Of course, the surrounding polar solvent plays a major role, because it can stabilize the TICT state, which carries a large dipole moment. It thus strongly influences both energetics (fluorescence spectra) and kinetics (TICT-state formation rate) by affecting the excited-state potential energy surfaces.

2. The Complexity of the Reaction Mechanism

There are three processes conneced with each other in the kinetics of the TICT adiabatic photoreaction mechanism in fluid media:

a. Initial Stage. After absorption, the solute starts to twist. In general this process will be phonon induced, that is, accelerated by collisions within the encounter molecules of the cage. This initial enhancement of the reaction velocity will occur in any solute/cage system independent of whether or not, the reaction mechanism requires activation energy. Since the solvent does not hinder the reorientation in this stage, the system behaves as if its effective viscosity η_{eff} is low.

b. Intramolecular Relaxation. Eventually, the cage hinders further progress of internal twisting. Friction occurs and the nearest-neighbor solvent molecules will release the necessary space for further relaxation only step by step. The system viscosity seems to be higher in this intermediate stage compared with the initial stage. Only where at least one twisting group of the solute is small there might be enough space for almost unhindered twisting, so that there is almost no friction and the solvent viscosity seems to be rather small. Such behavior is known as a "free volume effect" and can be observed for rotors of very small size.[38]

c. Solvent Reorientation. Finally, a substantial change in electronic distribution of the excited state follows internal twisting. In the electric field of the high dipole moment of the final "biradicaloid" excited state, the surrounding polar molecules (or polar groups of solvent molecules) try to reorient and/or to come nearer together by rotational and translational diffusion. In this final stage the system viscosity reaches its maximum value. Steps b and c, of course, do not occur independently of each other.

3. *Potential Energy Surfaces*

Aromatic amines are not the only type of molecule to undergo a spontaneous twist in the excited state, but DMABN is the best-studied example. Other such compounds are stilbene-type molecules, where the double bond twists in the excited state. The electronic structure for both cases, twisting double bonds and twisting single or partly double bonds of π-donor linked to π-acceptor (TICT molecules), can be related to each other and characterized in the framework of quantum-chemical treatment as described in Section III.

For both cases the theory predicts minima in the lowest excited state S_1 at twisted conformations with wavefunctions of zwitterionic and charge-separated nature, respectively. The zwitterionic S_1 state of the twisted double bond linked to two equal subunits can become easily polar due to asymmetrical geometrical changes, one-end substitution, or solvent polarity.[39] The lowest excited states of orthogonally twisted stilbene are very similar to the excited states of orthogonal ethylene, which are of zwitterionic nature. Similarly, the analogy can be drawn between the states of orthogonally twisted TICT molecules and those of perpendicular aminoborane $H_2N–BH_2$.[40] The S_1 state of the 90° twisted aminoborane is of a charge-separated nature. Since the initial excitation at planar geometries of stilbene and TICT molecules is expected to occur into one of the locally excited states ^1L_a or ^1L_b (in Platt's notation for aromatic hydrocarbons) and, since the zwitterionic or charge-separated state decreases in energy upon twisting from planarity, crossings or avoided crossings result. Consequently, the S_1 state can exhibit a barrier separating two minima, one close to the planar geometry and the other at the twisted conformation, as a memory of these avoided crossings. The wavefunctions corresponding to the two minima are of completely different nature. The height of the energy barriers in the exited state along the reaction path from the originally reached Franck–Condon conformation toward the twisted conformation is essential for the nature of the process considered.

In 1969, Suppan reviewed experimental data about dipole moment changes in excited states of substituted aromatic molecules and suggested a theoretical approach according to which charge transfer occurs if the lowest vacant orbitals are very close in energy.[41] In 1978, Birks introduced the term "horizontal radiationless transition," which was applied to intramolecular rotation in stilbene and polyene derivatives.[37] In this simple model, the relative position of two excited states $^1A^*$ and $^1B^*$ of different symmetry and their energy dependence as a function of the twist angle have been studied in order to propose a qualitative shape of the S_1 energy surface (Fig. 2.2). Supported by experimental and theoretical data, it was assumed that the ground state ^1A and excited state $^1B^*$ have minima at planar and maxima at perpendicular geometries, while the $^1A^*$ state descends along the twisting coordinate exhibiting a minimum at the orthogonal geometry. Consequently, dependent on the shape of S_1, which is closely connected with the energy ordering of $^1B^*$ and $^1A^*$ at planar geometries, fluorescence, horizontal radiationless, and other transitions have been discussed for some specific examples. This scheme is instructive although oversimplified. Note that there is a large number of low-energy locally excited states in *trans*-stilbene. Recently, Hohlneicher

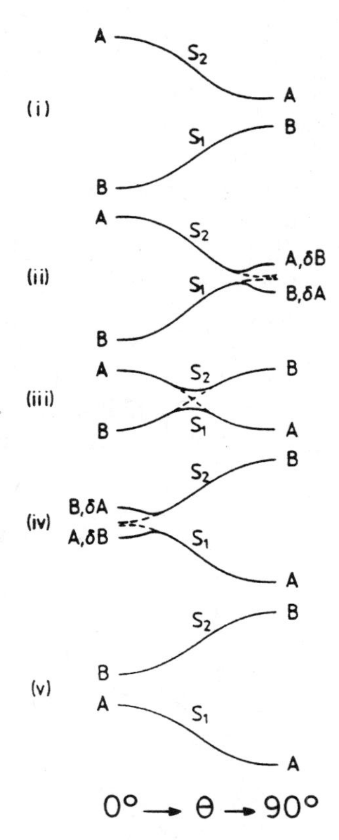

Fig. 2.2. Schematic diagrams of the angular dependence of the S_1 and S_2 potentials along θ from the initial to the final state of adiabatic photoreactions.[37]

and Dick[42] correlated the S_4 excited state of *trans*-stilbene with the orthogonal zwitterionic state of A symmetry called "phantom" state by Saltiel et al.[43] Since the ^1L_a state (S_1) is below the ^1L_b state (S_2) at planar geometries and since the ^1L_a state will strongly interact with the "phantom" state because it possesses the same symmetry, a barrier appears in S_1 due to the avoided crossing. For a symmetrical path like a twofold rotational symmetry, ^1L_b has the wrong symmetry to interact with the "phantom" state and increases in energy considerably. Only for the asymmetrical path will this crossing be weakly avoided. Indeed, two consecutive activation barriers are involved in the *trans*→*cis* photoisomerization of 4-nitro-4′-dimethylaminostilbene in toluene solutions.[44] In some dyes like pentamethine–cyanine and 4,4′-dialkylamino-4′-oxytriphenylmethane twisted conformers in both S_0 and S_1 states are reported.[45]

Depending on the relative energy of the charge-separated and locally excited states as well as on the influence of the polar solvent, the S_1 state of TICT molecules might exhibit a barrier (Fig. 2.3).[14] For DMABN, the vertical absorption reaches the Franck–Condon state at θ_{FC} close to planarity. The Franck–Condon state is S_2 with *A* symmetry (^1L_a-type), but according to Kasha's rule, the initial state for the photoreaction is S_1 with *B* symmetry (^1L_b-type state) as can easily be verified from the polarization spectra in Fig. 2.4. (The F_B band of DMABN has a small or negative polarization *P* when excited into the main absorption ^1L_a-type band.) The symmetry labels are with respect to a twofold rotational axis, which is present throughout the twist. The long-wavelength fluorescence F_A, however, is strongly polarized in the long axis of the molecule (*P* approaching 0.5) and therefore emitted from an excited state of *A* symmetry. Hence, a certain activation energy is needed for the forward

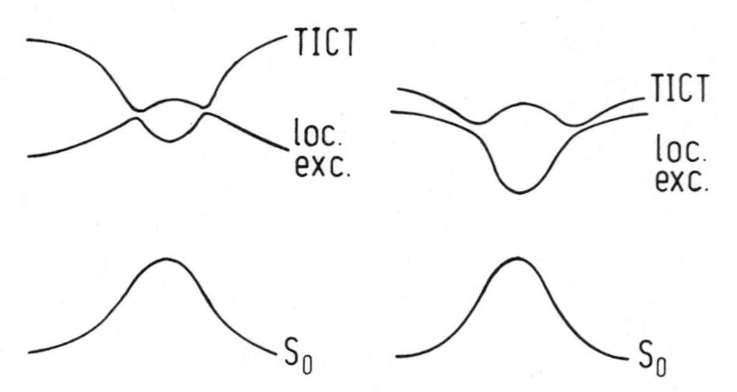

Fig. 2.3. Schematic diagrams of two typical cases of relative energies of a locally excited state and a TICT state as a function of twist angle.

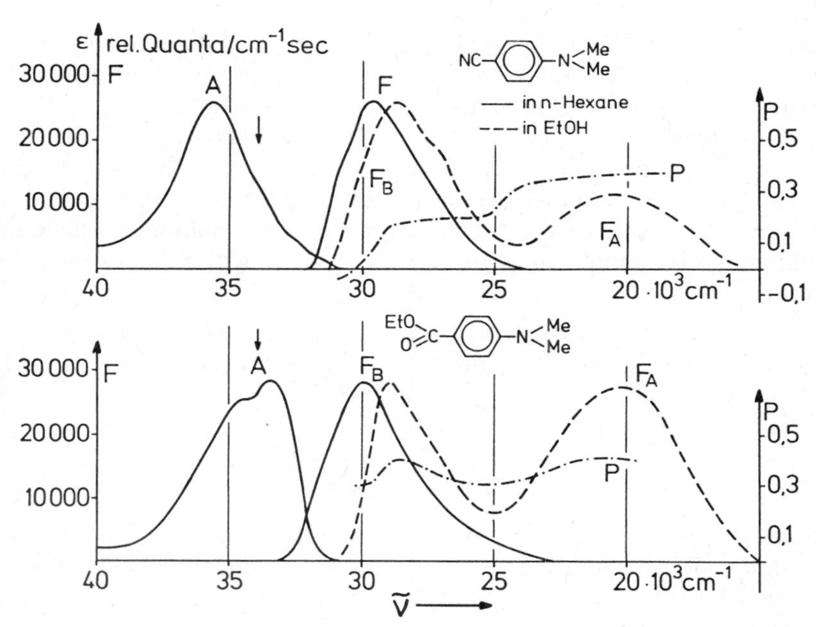

Fig. 2.4. Absorption A, fluorescence F, and degree of polarization of fluorescence P of DMABN and DMABEE (p-dimethylaminobenzoic acid ethylester) in n-hexane at room temperature, $c = 5 \times 10^{-6} M$, and ethyl alcohol at 140 K, $c = 1 \times 10^{-4} M$, respectively. Arrows indicate excitation wavenumber for measurements of P.[6] Subscripts A and B refer to final and initial emission, respectively.

reaction on S_1 (left-hand side of Fig. 2.3), which will be reduced if the polarity of the solvent increases (right-hand side of Fig. 2.3), owing to the stabilization of the charge-separated TICT state, which should lead to a solvent-polarity-dependent TICT formation rate.[46] For DMABN in glycerol (in contrast to Fig. 2.4) P is large and positive even in the short-wavelength emission range F_B[11] like in DMABEE (lower part of Fig. 2.4). It can be stated, therefore, that under the influence of steric hindrance, substituents, and/or solvent polarity the vertical absorption could lead directly into an area of the S_1-hypersurface, where $\partial V/\partial\theta < 0$ (barrierless case), and the horizontal radiationless process could start even in the gas phase[12,32]; we can distinguish, therefore, between *phonon-induced* and *spontaneous* adiabatic photoreactions in solutions.

4. *Kinetic Distinctions*

The presence or absence of an energy barrier along the pathway of relaxation is a decisive feature determining the type of kinetics encountered. As will be shown in detail in Section IV, a large barrier $E_0 \gg k_B T$ implies a stationary reaction rate linked to time-independent probability distribution functions. One of the possible stochastic approaches is that of Kramers, as will be outlined shortly in the next section.

A small or nonexisting barrier $E_0 \ll k_B T$, on the other hand, implies that the initial conditions (Franck–Condon geometry, and so on) can influence the reaction rate. A consequence of the resulting nonstationary probability distribution functions are time-dependent reaction rates (see Section IV).

a. The Large-Barrier Case. 1. THE KRAMERS EQUATION. The Kramers equation[16,47–49] for the thermally activated, one-dimensional barrier crossing rate constant k reads (Fig. 2.5)

$$k = \nu_0\{[1 + (2\tau_v \hat{\omega})^{-2}]^{1/2} - (2\tau_v \hat{\omega})^{-1}\}e^{-\beta E_0}\,, \tag{2.1}$$

where $\beta = (k_B T)^{-1}$ and E_0 is the activation energy of the forward reaction k. Expression (2.1) follows from the Langevin (Fokker–Planck) equation for the time evolution of the twisting angle θ of the two parts of the molecule,

$$I\ddot{\theta} + I\Gamma\dot{\theta} + \frac{dV_1}{d\theta} = F(t)\,, \tag{2.2}$$

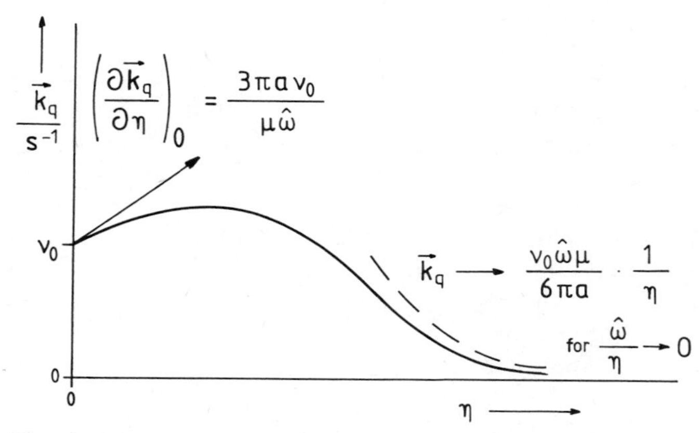

Fig. 2.5. The viscosity dependence of the forward rate constant following the Kramers equation with $k = k_q(\eta)\exp(-\Delta E_0/k_B T)$. In small viscosity domains $\eta \cdots \eta + \Delta\eta$ one often approximates $k \propto \eta^{\alpha}$, where $-1 < \alpha = \text{const} < +1$.

in which the potential V_1 of S_1 along the one reaction coordinate θ is composed of three parabolas with

$$\omega^2 = (2\pi\nu_0)^2 = I^{-1}\left(\frac{d^2V_1}{d\theta^2}\right)_{\theta_0}, \tag{2.3}$$

where ν_0 is the valley frequency, and the second (inverse) parabola with a maximum at the twist angle $\hat{\theta}_c$ and the frequency $\hat{\omega}$

$$\hat{\omega}^2 = -I^{-1}\left(\frac{d^2V_1}{d\theta^2}\right)_{\hat{\theta}_c} \tag{2.4}$$

determines the barrier; $I = \mu r^2$ is the reduced moment of inertia, $I\Gamma$ is the angular drag coefficient, and $F(t)$ is a randomly fluctuating torque with

$$\langle F(t)F(t+\tau)\rangle = 2I\Gamma k_B T\delta(\tau)\,. \tag{2.5}$$

The momentum correlation time

$$\Gamma^{-1} = \tau_v = \frac{\mu}{6\pi a\eta} \tag{2.6}$$

depends on the reduced mass μ, the viscosity $\eta = \eta_0 \exp(+\beta E_\eta)$, and the effective cylindrical molecular volume $4\pi r^2 a$.

The Kramers equation (2.1) in the high-viscosity limit yields *the Smoluchowski limit* (see also Eqs. 4.171 and 4.178)

$$k_\infty = \frac{\nu_0\mu\hat{\omega}}{6\pi a\eta_0}\exp\left[-\beta(E_0 + E_\eta)\right] \propto \eta^{-1}, \tag{2.7}$$

and in the *low-viscosity limit*[50–53] it takes the following form:

$$k_0 = \nu_0[1 - (2\tau_v\hat{\omega})^{-1}]e^{-\beta E_0}\,. \tag{2.8}$$

It has been[50] shown that

$$\left(\frac{dk}{d\eta}\right)_{\eta\to 0} = +\frac{3\pi a\nu_0}{\mu\hat{\omega}}\exp(-\beta E_0) > 0\,. \tag{2.9}$$

Unfortunately, even for isomerization reactions it has been shown that the Smoluchowski limit fits neither for alkane[54] nor for alcohol solutions,[47] and that a frequency-dependent friction should be used in the vicinity of θ_c.[55–57]

2. COUPLED MODES. For simplicity in most publications concerning adiabatic photoreactions, the pathway is assumed to be one-dimensional. Under this assumption the intramolecular downhill gradient of the excited-state potential energy hypersurface might be achieved via an energy barrier or it might be barrierless. If a barrier is present, consideration of more dimensions might allow the barrier to be avoided. For aromatic compounds in solution, a one-dimensional well-defined pathway is an oversimplified approximation since the number of atoms involved in the reaction and, therefore, the number of strongly coupled modes in most cases might be rather high,[58] allowing the system to escape from the initial to the final fluorescing minimum by avoiding the barrier, giving rise to a so-called frequency-dependent tunneling effect, which additionally alters Kramers' rate constant.

Figure 2.6 summarizes the results in simple terms: The Kramers equation describes the situation where the shape of the probability density distribution $P(x, t)$ is fully equilibrated inside the starting potential well, thus it is time independent, and the reaction across the barrier changes only the weight of $P(x, t)$. In the high viscosity regime observed activation energy of the reaction E_{obs} is expected to be given by the sum of E_0 and the activation energy for solvent viscosity E_η [see Eq. (2.7)].

b. The Barrierless Case. If $E_0 \ll k_B T$, the assumptions made for deriving Eq. (2.7) no longer hold, and a new stochastic description has to be sought. As will be detailed in Section IV, the system is far from equilibrium, and the shape of the probability density distribution $P(x, t)$ evolves with time and in space. This is shown in Fig. 2.7, where an initial δ-shaped distribution moves toward the origin of a parabola and is gradually transformed into a broader distribution.

For this case, the stochastic description (Section IV) predicts that the rate constants to be observed can be time dependent $[k(t)]$. Moreover,

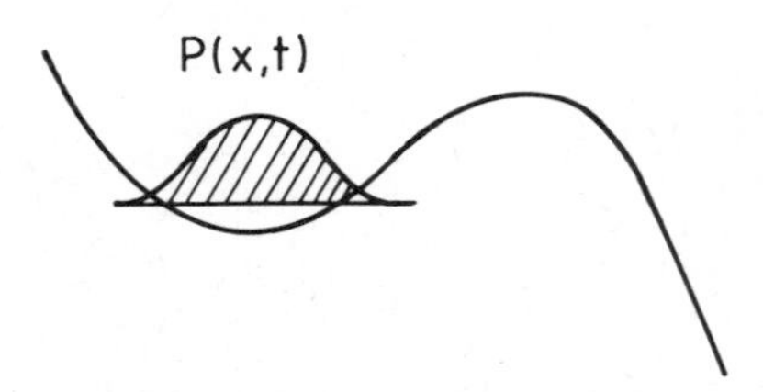

Fig. 2.6. Schematic representation of a reaction treated in the Kramers approximation. The shape of the probability density distribution is assumed to have reached equilibrium (i.e., time independence) at the bottom of the reactant valley. Only the weight of $P(x, t)$ (total number of reactants) diminishes by activated diffusion across the barrier.

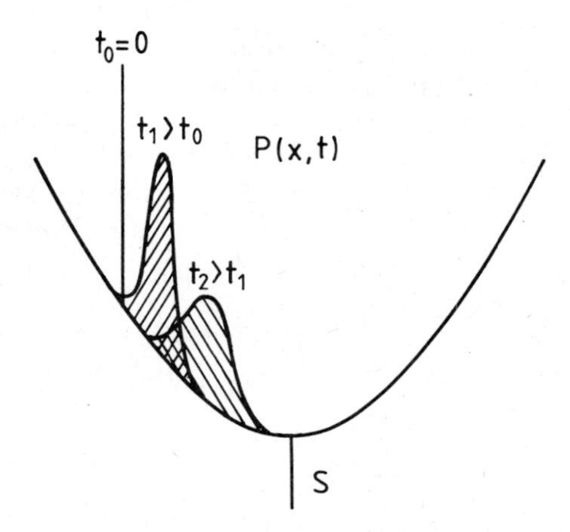

Fig. 2.7. Schematic representation of the Bagchi–Fleming–Oxtoby model used for barrierless reactions. As the probability density distribution $P(x, t)$ (shown δ shaped in this example for $t = 0$) moves toward the origin with a nonradiative sink S, it broadens due to the Brownian motion.

the activation energy E_{obs} of the process may be nonzero, but is always smaller than or equal to E_η.

B. Steady-State Fluorescence

1. *Quantum Yields and Solvatochromism of DMABN*

The temperature dependence of the dual fluorescence of DMABN as shown in Fig. 2.1 displays several characteristic features. The short wavelength B band at around 350 nm shows an intensity minimum at a certain temperature T_m around 200 K, whereas the quantum yield of the long wavelength A band slightly but monotonically increases with temperature. Thus, above T_m, the total fluorescence quantum yield increases with temperature, a feature which is rather uncommon in photophysics.

This behavior can be rationalized in terms of the kinetic scheme I proposed by Grabowski et al.[7,21] It involves the equilibrium of two excited states A* and B* and three temperature-independent

$$B^* \underset{k_{AB}}{\overset{k_{BA}}{\rightleftharpoons}} A^*$$

$$k_f^B \downarrow \; \downarrow k_0^B \qquad\qquad k_f^A \downarrow \; \downarrow k_0^A$$

Kinetic Scheme I

(k_f^B, k_0^B, k_0^A) and three temperature-dependent rate constants (k_f^A, k_{BA}, k_{AB}). This scheme is similar to the kinetic scheme generally used for describing excimer kinetics[59] (with the exception of a temperature dependent k_f^A).

Solving the corresponding differential equations for steady-state conditions leads to the following expressions for the fluorescence quantum yields ϕ_A and ϕ_B of A and B bands:

$$\phi_B = \frac{k_f^B(k_f^A + k_0^A + k_{AB})}{(k_f^A + k_0^A)(k_f^B + k_0^B + k_{BA}) + k_{AB}(k_f^B + k_0^B)}, \tag{2.10}$$

$$\phi_A = \frac{k_f^A k_{BA}}{(k_f^A + k_0^A)(k_f^B + k_0^B + k_{BA}) + k_{AB}(k_f^B + k_0^B)}, \tag{2.11}$$

$$\frac{\phi_A}{\phi_B} = \frac{k_f^A k_{BA}}{k_f^B(k_f^A + k_0^A + k_{AB})}. \tag{2.12}$$

The temperature dependence of these quantum yields can be derived for limiting cases where certain members of the sums can be neglected. Most relevant is the behavior above and below T_m, the high-temperature (HT) and low-temperature (LT) region. For the LT region, $k_{BA}(T) \gg k_f^B + k_0^B$, but $k_{AB}(T) \ll k_f^A + k_0^A$, that is, the reaction proceeds efficiently from B* to A*, but the back reaction from A* to B* does not take place within the excited-state lifetime. Then, the temperature dependence of ϕ_A and ϕ_B reduces to that of $k_f^A(T)$ and $1/k_{BA}(T)$, respectively, if $k_0^A \gg k_f^A$ and $(k_f^A + k_0^A)k_{BA} \gg (k_f^B + k_0^B)k_{AB}$ as in the case of DMABN:

$$\phi_B|_{LT} = \frac{k_f^B}{k_{BA}}, \tag{2.13a}$$

$$\left.\frac{d(\ln \phi_B)}{d(1/T)}\right|_{LT} = -\frac{d[\ln k_{BA}(T)]}{d(1/T)} = +\frac{E_{BA}}{k_B}, \tag{2.13b}$$

$$\phi_A|_{LT} = \frac{k_f^A}{k_0^A}, \tag{2.14a}$$

$$\left.\frac{d(\ln \phi_A)}{d(1/T)}\right|_{LT} = \frac{d[\ln k_f^A(T)]}{d(1/T)} = -\frac{E_{fA}}{k_B}, \tag{2.14b}$$

where the temperature-dependent rate constants are given by

$$k_{BA}(T) = k_{BA}^{\infty} \exp(-E_{BA}/k_B T) \tag{2.15}$$

$$k_{AB}(T) = k_{AB}^{\infty} \exp(-E_{AB}/k_B T) \qquad (2.16)$$

$$k_f^A(T) = k_{fA}^0 + k_{fA}^1 \exp(\mp E_{fA}/k_B T) \qquad (2.17)$$

E_{BA} and E_{AB} are the activation energies for forward and reverse reaction, respectively, and E_{fA} accounts for the observed temperature dependence of the radiative TICT transition. The significance of the experimentally observed nonzero E_{fA} is still controversial. It may be sufficient in the present context to mention that it is most likely connected with an intensity gain of the radiative transition k_f^A via vibronic coupling.[7]

In the HT region where $k_{BA}(T) \gg k_f^B + k_0^B$ and $k_{AB}(T) \gg k_f^A + k_0^A$, the excited-state equilibrium can be fully established within the excited-state lifetime. The slopes are then given by

$$\phi_B|_{HT} = \frac{k_f^B k_{AB}}{k_0^A k_{BA}}, \qquad (2.18a)$$

$$\left.\frac{d(\ln \phi_B)}{d(1/T)}\right|_{HT} = \frac{d \ln[k_{AB}(T)/k_{BA}(T)]}{d(1/T)} = \frac{(E_{BA} - E_{AB})}{k_B}, \qquad (2.18b)$$

$$\phi_A|_{HT} = \frac{k_f^A}{k_0^A}, \qquad (2.19a)$$

$$\left.\frac{d(\ln \phi_A)}{d(1/T)}\right|_{HT} = \frac{d \ln[k_f^A(T)]}{d(1/T)} = -\frac{E_{fA}}{k_B}. \qquad (2.19b)$$

This is represented in Fig. 2.8 using the decomposed fluorescence spectra of Fig. 2.1. From the measured slopes, experimental values $E_1 = E_{BA}$ and $E_2 = E_{AB}$ for the activation energies of both forward k_{BA} and backward processes k_{AB} as well as for k_f^A can be derived.

The measured quantity E_{BA} represents the energy needed for the reaction to occur in a reasonable time. It has a different physical meaning for large- and low-barrier case. Roughly speaking, in the large-barrier case, $E_1 = E_{BA}$ would correspond to $E_0 + E_\eta$, whereas in the barrierless case E_{BA} is only related to E_η and does not correspond to a real potential energy maximum along the reaction path. In this case, one could call E_{BA} a "dynamical" activation energy.

For DMABN in polar solvents, the observed activation energy E_{AB} for the backward reaction is larger than E_{BA}, that is, the adiabatic photoreaction is exothermic. For example, from the spectra and activation energies shown in Figs. 2.1 and 2.8, the potential energy diagram displayed in Fig. 2.9 can be constructed. Only part of the strong redshift of the A band is

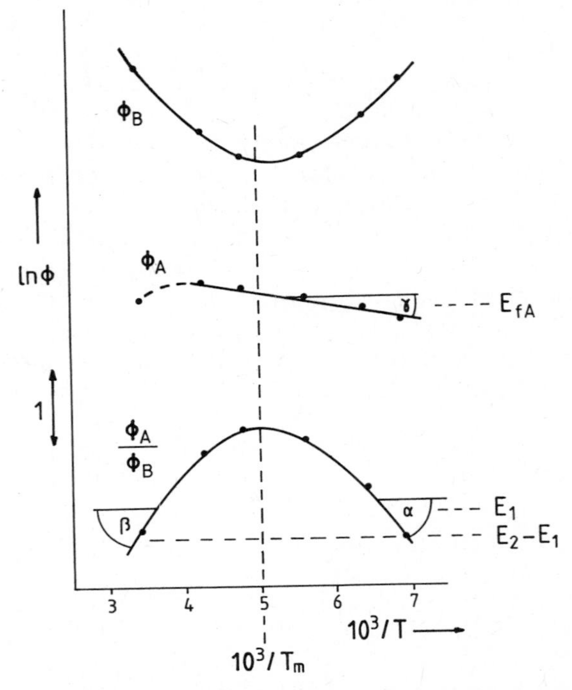

Fig. 2.8. Temperature dependence of the quantum yields of B fluorescence (ϕ_B) and A fluorescence (ϕ_A) of DMABN in *n*-butyl chloride (data from Fig. 2.1). From the slopes, the formal activation energies E_1 and E_2 for $B^* \rightarrow A^*$ and reverse reaction, respectively, can be derived.

thus due to the energetic stabilization of the A* state, the larger part results from the energy destabilization of the twisted ground state owing to combined effects of ground-state intramolecular rotational barrier and intermolecular solute–solvent interactions directly after photon emission.

The polarity of the solvent strongly affects both the shapes of the excited state and of the ground-state surfaces: For increasing solvent polarity, the A* state is more strongly stabilized than the B* state, resulting in a deformation of the excited state potential ($E_2 - E_1$ increases). The effect on the spectra is shown in Fig. 2.10, where the very strong redshift of the A band with increasing solvent polarity is readily apparent.

A simple way to quantify these redshifts and relate them to dipole moments of fluorescing species is the Lippert–Mataga equation[60–62]:

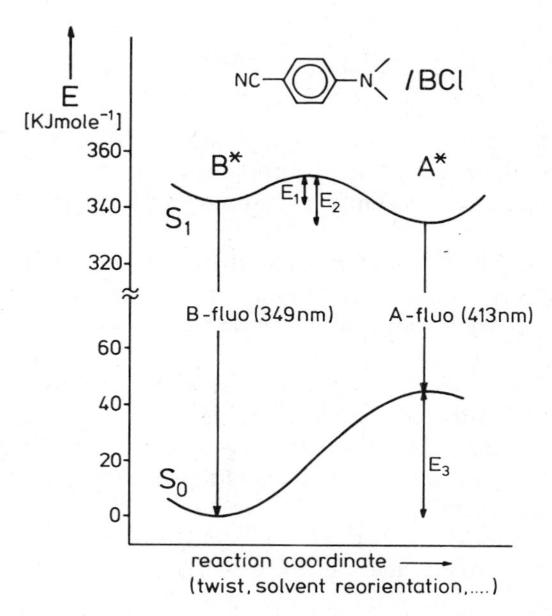

Fig. 2.9. Empirical energy diagram for DMABN in *n*-butyl chloride (energetics based on room-temperature fluorescence band maxima and on activation energies). In the "small-barrier case," E_1 is to be viewed as a "dynamical" activation energy resulting from solvent viscosity. The Franck–Condon ground state (after emission from A*) is anomalously destabilized (large E_3).

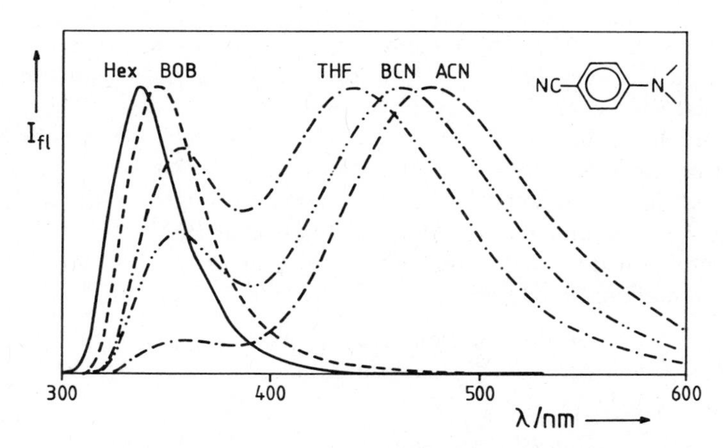

Fig. 2.10. Room-temperature fluorescence spectra of DMABN in aprotic solvents of different polarity: *n*-hexane, ——; di-*n*-butyl ether, – – –; tetrahydrofurane, – · – ·; *n*-butyronitrile, – · · · –; and acetonitrile, – – ·.

$$\tilde{\nu}_A - \tilde{\nu}_F = 2\,\frac{(\mu_e - \mu_g)^2}{hca^3}\left\{\frac{\epsilon - 1}{2\epsilon + 1} - \frac{n^2 - 1}{2n^2 + 1}\right\}, \tag{2.20}$$

where a is the Onsager distance, μ_e and μ_g are the electric dipole moments of the fluorescing and absorbing species, respectively, and ϵ and n^2 are the dielectric constants at infinite and transition wavelength, respectively.

This equation has been reformulated by Rettig[63] using a microstructural-solvent-interaction (MSI) model recently proposed by Nolte and Dähne.[64] The charges in a solute molecule polarize the surrounding dielectric. The potential energy of the system solute/solvent is lowered by the resulting electrostatic interaction energy. In the simplified MSI model, the polarization of the surrounding solvent cage is approximated as the sum of incremental interactions of the actual charge distribution of the molecule with the solvent. The charge distribution is calculated using a quantum-chemical method, which is independent of the Onsager factor $1/a^3$ appearing in Eq. (2.20). The advantage of the application of this "charges polarize the cage" (CPC) model is that the point dipole approximation is lifted as well as the restriction to spherical or ellipsoidal Onsager cavities.[61,63] Solutes of different size and shape can be compared directly, as will be described in the next section.

2. *The TICT Excited-State Dipole Moment*

Consider a cage of solvent molecules with a thickness Δr_S around a solute molecule with the total surface S (Fig. 2.11). The thickness Δr_S is of minor importance since it will not appear explicitly in our final results. The surface S will be divided rather arbitrarily in subsurfaces S_i of subgroups i for each atom except hydrogen, that is, for C, O, and N in our examples, which carry the fractional charge Q_i of the subgroup i. Only five sizes $S_i = k_s(4 - l_i)$ are allowed for each subgroup surface according to the number l_i of bonds to another subgroup center nuclei j, so that $4 - l_i$ counts the sum of the number of bonds to H atoms, of $2p_2$-orbitals for double bonds, and of lone-pair orbitals of the ith central atom. Some common subgroups are described, for example, by

$l_i = 4$ for tetrahedral, centered atoms
(e.g., central C in neopentane) $\qquad S_{i,4} = 0$

$l_i = 3$ for trigonal atoms
(e.g., central C in iso-butane) $\qquad S_{i,3} = 1k_S$

$l_i = 2$ for atoms in unsaturated chains
(all C atoms in benzene) $\qquad S_{i,2} = 2k_S$

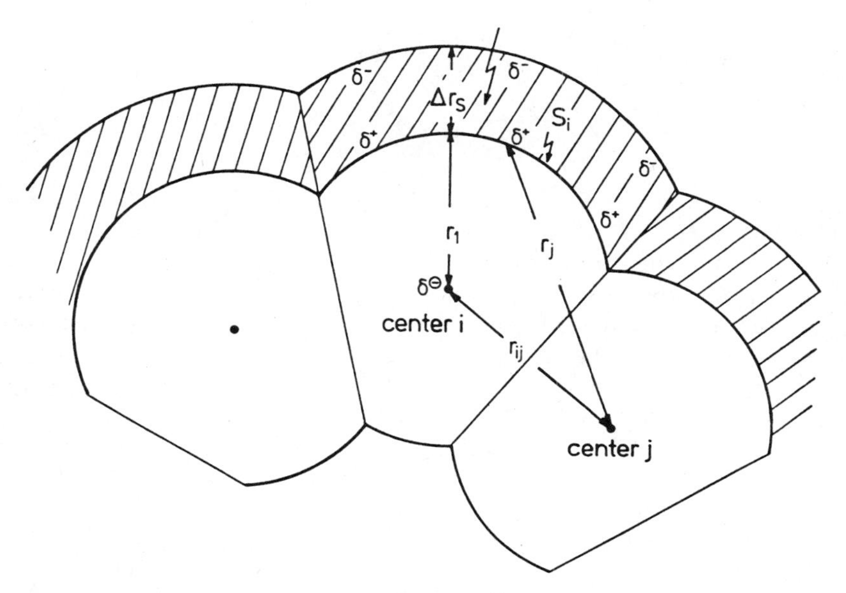

Fig. 2.11. Geometry of the adapted MSI model.[63] The molecule is constructed from subgroups i with fractional charges. It is surrounded by a solvent layer S with thickness Δr_S, inside which the electrostatic potential ϕ_M^S is active.

$l_i = 1$ for end-standing atoms (C of methyl groups) $\qquad S_{i,1} = 3k_S$

$l_i = 0$ for CH_4, NH_3, H_2O $\qquad S_{i,0} = 4k_S$

The total surface is then constructed from the subsurfaces by adjustment of k_S

$$S = \sum_i S_i = k_S \sum_i (4 - l_i) \tag{2.21}$$

Each fractional charge Q_i contributes to a dielectric potential $\Delta\Phi_i = Q_i\,\Delta V_i/r_i$ in a volume element $\Delta V_i = S_i\,\Delta r_S$ of the solvent cage in the distance $r_i = r_1$ from center i. The mean potential $\langle\Phi_i\rangle$ inside the cage increment ΔV_i around the ith subgroup is given by summation over all centers j (distance r_j from S_i),

$$\langle\Phi_i\rangle = \frac{Q_i}{\langle r_i\rangle} + \sum_{j\neq i} \frac{Q_i}{r_j} \tag{2.22}$$

and polarizes the cage by a constant factor k_P, so that the total solute–solvent interaction energy becomes

$$W = -k_P k_S \, \Delta r_S \sum_i \langle \Phi_i \rangle (4 - l_i) \tag{2.23}$$

The nature of the solvent is introduced by the factor $k_P = k'L$, where, according to Liptay's theory,[70,71] the solvent polarity function L depends on ϵ and n^2 of the solvent as well as on the polarizability α of the solute. Following a proposal[66,67] by Bakhshiev, α can be approximated by setting $\alpha_i = 0.5r_i^3$, eventually yielding[70,71] for the total solute–solvent interaction energy W after solvent relaxation (with $k'k_S \, \Delta r_S = k = \text{const}$):

$$W = -k \frac{\epsilon - 1}{\epsilon + 2} \sum_i \langle \Phi_i \rangle^2 (4 - l_i) \tag{2.24}$$

Thus, k is the only parameter to be adjusted, for example, by comparison with a known dipole moment standard.[63]

For the fluorescence solvatochromism emitted from a TICT state with the dipole moment μ_e to a FC ground state with dipole moment $\mu_g^{FC} \sim 0$ the CPC model results, in improving Eq. (2.20),

$$hc\tilde{\nu}_F = \text{const} - kA\mu_e^2 \left\{ \frac{\epsilon - 1}{\epsilon + 2} - \frac{1}{2} \frac{n^2 - 1}{n^2 + 2} \right\} \tag{2.25}$$

$$A = \mu_{th}^{-2} \sum_i \langle \Phi_i \rangle^2 (4 - l_i) \tag{2.26}$$

$$\mu_{th} = \sum_i Q_i R_i \tag{2.27}$$

The approximate dipole moment values μ_{th} can be readily obtained by semiempirical methods (for example Hückel–MO or CNDO/s-CI[63,69]).

The advantage in applying the CPC model is that by combination of the Onsager–Ooshika–Lippert–Mataga–Liptay method,[60–62,70–73] with an initial guess of the excited-state charge distribution, experimental excited-state dipole moments can be determined from the anomalous Stokes red shift in polar solvents rather accurately, since all relative errors cancel each other to a higher percentage because they occur in both the numerator and denominator. No simplifying assumptions like the choice of an Onsager radius a are necessary, only the factor k of proportionality has to be determined. Moreover, it can be shown that for a comparison of different elongated systems the CPC model leads to a more realistic description of the asymptotic behavior of interaction energies.[63]

By this method, very high dipole moments have been obtained for the A* state.[63] The dipole moment of the B* state, on the other hand, is much smaller, as derived from the diminished solvent polarity red shift of the B band (Fig. 2.10). On the other hand, there are also other intra- and intermolecular interactions than dipole–dipole forces influencing the shape of the excited-state potential surface.[225,226] Electrooptical absorption and emission measurements[225] on DMABN and related compounds have shown that the amount of the dipole moment μ_a of the TICT state A* is about three times larger than the dipole moment μ_b of the almost planar fluorescing B* state (S_1 of type ^1L_b). But the dipole moment μ_a^{FC} of the absorbing state S_2 (of ^1L_a character) is of similar magnitude as that of μ_a. In contradiction to the expected increase of $\mu(t)$ along the reaction coordinate from B* to A* some transient dielectric loss measurements[226] have been interpreted by $\mu_b \geq \mu_a$. The results from the electrooptical measurements ($\mu_a \gg \mu_b$) are consistent with a simple zero-order states crossing scheme.[225c] The question if and why $\mu_a^{FC} \approx \mu_a$ may be answered by comparing DMABN and other nitriles with the corresponding esters[225d] where no level crossing occurs.[2,6,33]

3. *Other Solute Examples*

DMABN and related dialkylanilines are not the only compounds that show this type of dual fluorescence with its strong solvent polarity dependence indicative of considerable charge transfer. Figure 2.12 shows

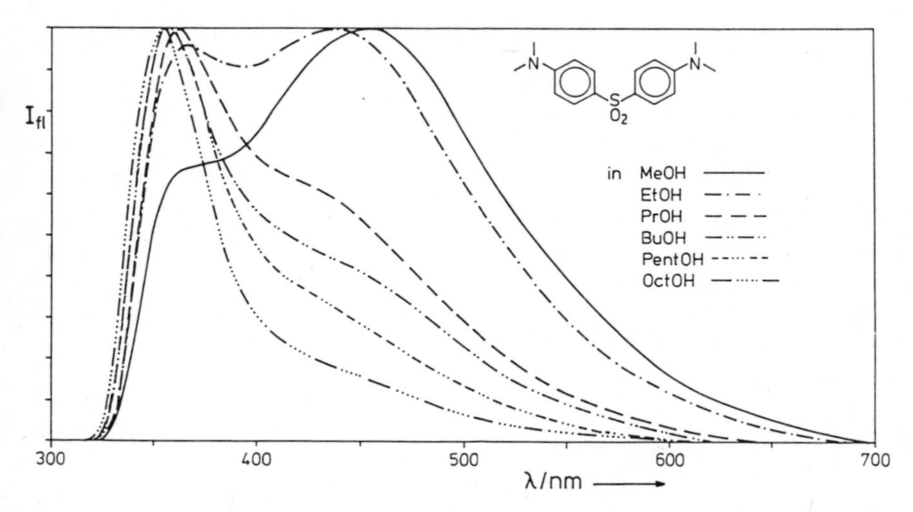

Fig. 2.12. Corrected fluorescence spectra of DMAPS in homologous alcohol solvents.[74]

an aromatic sulfone where the second fluorescence is strongly enhanced in polar solvents. Solvatochromic plots yield a very high slope for the long-wavelength band and a weaker slope for the short-wavelength band.[74] Comparison with the similar dual fluorescence of a sulfone, where the $N(CH_3)_2$ groups are exchanged for NH_2 groups, leads to the conclusion that the rotating donor moiety is the entire anilino group and not solely the dimethylamino substituent as in DMABN.[8,74] Michler's Ketone, which has a structure similar to the sulfone shown in Fig. 2.12, but with the SO_2 group exchanged for C=O, also shows TICT fluorescence identified by its solvatochromy which will be discussed in more detail in Section V.

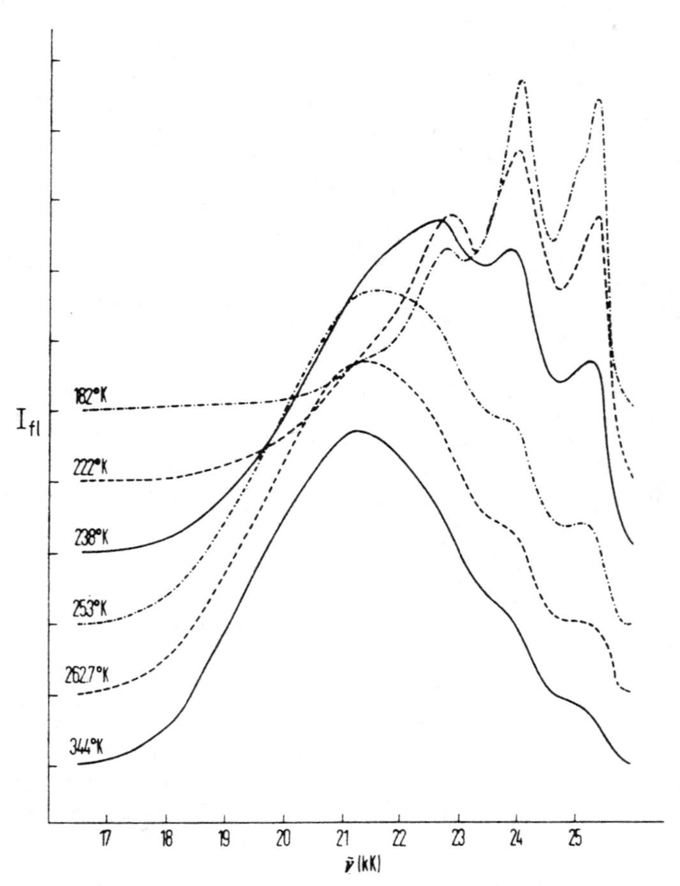

Fig. 2.13. Fluorescence spectra of BA in glycerol at different temperatures.[75]

Dual fluorescences have also been discovered in biaromatic compounds without any substituent. The first example was 9,9-bianthryl, where the dual fluorescence is strongly temperature/viscosity dependent as shown in Fig. 2.13.[75] At low temperatures, only the structured F_B band remains, which is quite similar to the emission of monomeric anthracene. The position of the second band (the main fluorescence in polar solvents at higher temperatures) is strongly polarity dependent, testifying of its charge-transfer character. Figure 2.14 shows the solvatochromic plot.[75] The horizontal portion for weak polarities reflects the solvent insensitivity of the F_B band dominating in this range. We have thus two fluorescing states, one with zero and the other with a very high dipole moment. It has been argued that to proceed from the nonpolar to the polar state a symmetry-breaking process induced by the solvent has to occur.[76,77] Comparison with slightly unsymmetric derivatives of 9,9′-bianthryl led to the conclusion that the polar state is reached after the transfer of a maximal amount of electronic charge, and its minimum energy conformation is expected at 90° twist.[77] Recent laser induced fluorescence spectra of 9,9′-bianthryl in a supersonic jet established that the nonpolar state has a nonorthogonal equilibrium geometry (torsional angle 66°) as opposed to the orthogonal ground state geometry.[68] In the jet experiments,

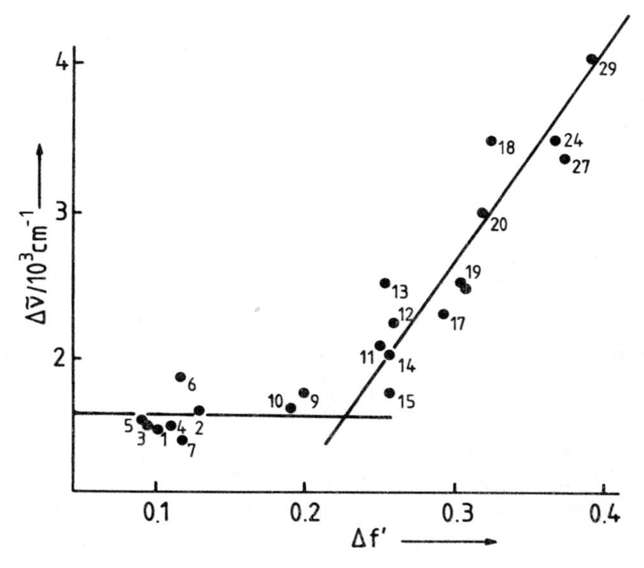

Fig. 2.14. Anomalous Stokes red shift $\Delta\tilde{\nu} = \tilde{\nu}_A - \tilde{\nu}_F$ versus solvent polarity $\Delta f' = (\epsilon - 1)/(2\epsilon + 1) - (n^2 - 1)/2(2n^2 + 1)$ of different 22 neat solvents as numbered in Ref. 75.

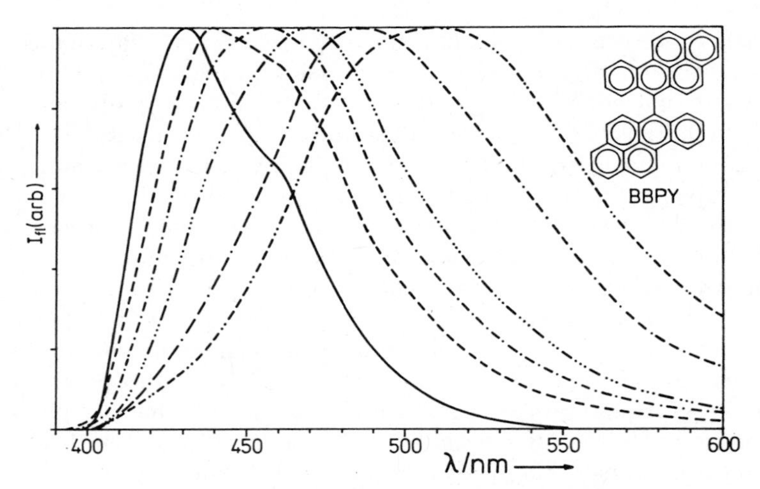

Fig. 2.15. Room-temperature fluorescence spectra of BBPY in *n*-hexane, ——; diethyl ether, – – –; *n*-butyl chloride, – · – ·; tetrahydrofuran, – · · ·; *n*-butyronitrile, – – · – –; and acetonitrile; – – – · · – – –.[69]

population of the highly polar state has only been observed for solute-solvent clusters, that is, if a small number of polar solvent molecules surround the solute 9,9′-bianthryl. Some other biaryls can exhibit a similar dual fluorescence.[69] Figure 2.15 shows the example of a bibenz-pyrenyl. It shows this type of dual fluorescence even in a nonpolar environment.[69]

C. Kinetic Investigations

1. *The Conventional Kinetic Model (Time-Independent Rate Constants)*

According to the kinetic scheme I, the excited B* and A* states can reach an equilibrium, at least above the characteristic temperature T_m introduced in Fig. 2.8. Since only the B* state is excited directly, the equilibration involves, first, the depopulation of B* toward A*, and then the repopulation of B* from A* to reach equilibrium.

At short times, before equilibration, the population of A* rises with a rate approximately given by the TICT formation rate k_{BA}, and the decay of B* is governed by the same rate. It is this time range that is most interesting for direct studies of the intramolecular rotational relaxation.

By solving the differential equations relating to the kinetic scheme I, equations (2.28)–(2.32) can be derived describing the temporal profiles $i_f^B(t)$ and $i_f^A(t)$ for B and A fluorescence bands[59]:

$$i_f^B(t) = \frac{k_f^B}{\lambda_1 - \lambda_2}\{(x - \lambda_2)e^{-\lambda_1 t} + (\lambda_1 - x)e^{-\lambda_2 t}\}, \tag{2.28}$$

$$i_f^A(t) = \frac{k_f^A k_{BA}}{\lambda_1 - \lambda_2}\{-e^{-\lambda_1 t} + e^{-\lambda_2 t}\}, \tag{2.29}$$

$$\lambda_{1,2} = \tfrac{1}{2}\{x + y \pm ((x - y)^2 + 4k_{AB}k_{BA})^{1/2}\}, \tag{2.30}$$

$$x = k_{BA} + k_f^B + k_0^B, \tag{2.31}$$

$$y = k_{AB} + k_f^A + k_0^A. \tag{2.32}$$

Around and above T_m, the decay of the B band becomes biexponential, because the prefactors $(x - \lambda_2)$ and $(\lambda_1 - x)$ in Eq. (2.28) become comparable in magnitude. The fast initial decay corresponds to the equilibration part, and the slower second component corresponds to the decay of the equilibrated system. Of course, for this later time period, the equilibrated system behaves as a single kinetic unit (high-temperature region: $k_{BA} \gg k_f^B + k_0^B$ and $k_{AB} \gg k_f^A + k_0^A$), and B and A bands have identical decay times given by Eq. (2.33):

$$\tau_e^{-1} = \frac{(k_f^B + k_0^B) + K_e(k_f^A + k_0^A)}{1 + K_e}. \tag{2.33}$$

The weighting factor K_e is the equilibrium constant between A* and B* states and is given by k_{BA}/k_{AB}.

Figure 2.16 shows an example for such a biexponential decay measured with time correlated single-photon counting.[78,79] Several picosecond laser experiments have explored this early time behavior of the equilibration process.[46,79–84]

The nature of the solvent, like alcohols (hydrogen-bonded chains and networks) or other polar but aprotic solvents (simpler dielectric relaxation properties) or mixed solvents (where translational diffusion processes play an important role, too), can lead to different degrees of complexity of the observed decay and rise times, the interpretation of which demands a careful choice of experiments to discriminate between reaction mechanisms including only nonspecific or specific solute–solvent interactions. For example, Meech and Phillips[84] tried to distinguish between a diffusion transient model and a multiple-exciplex model in alcohols. A different approach (stochastic kinetics) was used by Heisel and Miehé[78,79] (see Section IV). Other authors[46] plotted the rate constant for arriving at the final state A* of DMABN in mixtures of octane/butyronitrile versus the

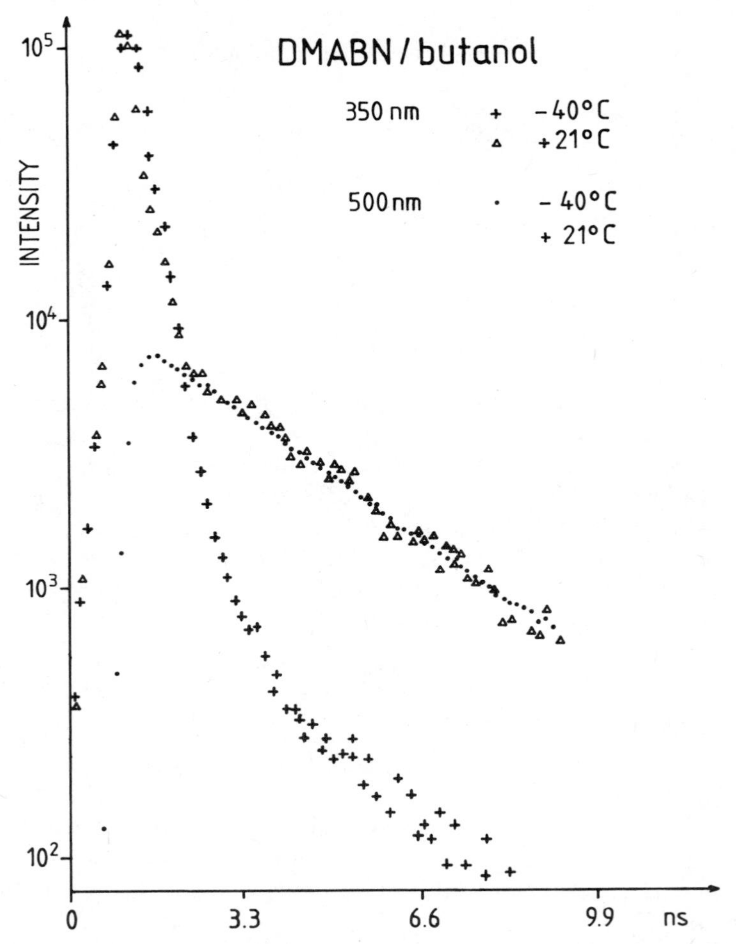

Fig. 2.16. Observed decay of short-wavelength and long-wavelength bands of DMABN in *n*-butanol at 21°C and −40°C.[78] At room temperature, after about 1 ns, the excited-state equilibrium is established, and both bands show similar long-time decays. At −40°C, however, the equilibration is nearly stopped.

apparent polarity of those mixtures as measured by $E_T(30)$ values. They concluded that, for this system, a potential barrier separates the initial and the final state; the height of the barrier decreases with increasing solvent polarity. The $E_T(30)$ values, however, are derived from absorption (femtosecond timescale) of a betaine dye and may not be linearly related to the parameters determining diffusion-controlled fluorescence processes (see, for example, Ref. 85 for additional references).

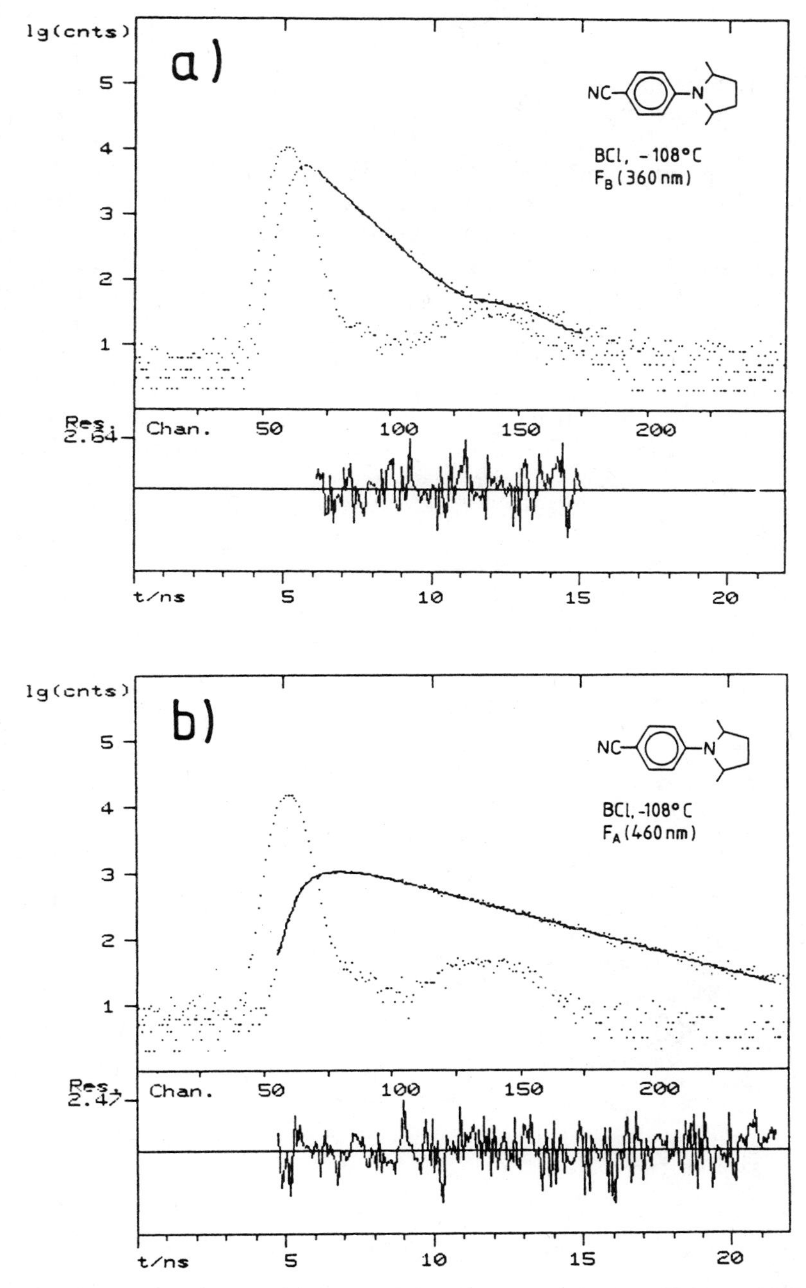

Fig. 2.17. Fluorescence decay curves of 2,5-dimethylpyrrolidinobenzonitrile at −108°C in *n*-butyl chloride, excited by synchrotron radiation from BESSY and observed at (*a*) 360 nm and (*b*) 460 nm (A band).[50] The computed lines are mono- to biexponential fits [decay time in (*a*) 1.04 ns, rise time in (*b*) 0.91 ns].

At temperatures sufficiently below T_m, the back reaction k_{AB} is not fast enough to compete with k_0^A and k_f^A, thus the entire decay curves can be thought of as representing equilibration without reaching it within the time given by the lifetime of the excited A* state. Figure 2.17 shows an example measured with time-correlated single-photon counting. The excitation source used was synchrotron radiation, which, in comparison with conventional nanosecond flashlamps, has the advantage of much stronger intensity, higher repetition rate, and narrower pulse shape, and which, compared to high repetition-rate laser sources, allows for an unprecedented flexibility in the excitation wavelength.[86] Figure 2.17*a* shows that for compounds with a larger rotating amino moiety than in DMABN, this TICT formation rate in cooled *n*-butyl chloride is slow enough for the delayed appearance of the A* emission to be quite evident (Fig. 2.17*b*).

The analysis of the temporal fluorescence profiles in Figs. 2.17*a* and 2.17*b* shows that the kinetic scheme used is a fairly good description at least for the case shown: The ratio of the preexponential factors for rise and decay parts of the A fluorescence is 0.96, thus very close to the theoretical value of 1.0 predicted by Eq. (2.29), and the decay time of the B fluorescence is similar to the rise time of the A fluorescence.

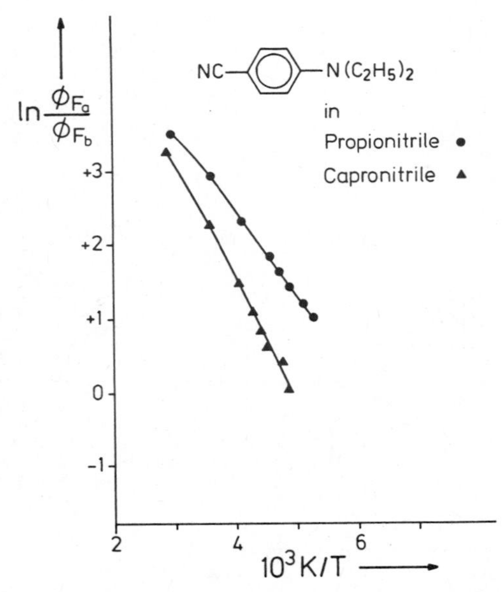

Fig. 2.18. Low-temperaure slopes of the ratio of the quantum yields of dual fluorescence for diethylaminobenzonitrile in two polar solvents, propionitrile and capronitrile, of different E_η.[87]

The interesting result of both kinetic and static measurements is that in all cases studied the experimental energy E_{BA} as the apparent value taken from an Arrhenius plot (for the applicability of this method see Ref. 229) is smaller or similar to E_η. Moreover, E_{BA} is very strongly influenced by changes of E_η. An example of a static measurement on a compound closely related to DMABN is shown in Fig. 2.18. The low-temperature slopes of ϕ_B yield for E_{BA} 9.6 kJ/mol and 13.8 kJ/mol in propionitrile and capronitrile, respectively.[87] The corresponding E_η values measured near room temperature are 7.5 and 11.0 kJ/mol. They are expected to increase for lower temperature and to become comparable to the E_{BA} values measured in the low-temperature region. The ratio of the two E_{BA} values and the ratio of the two E_η values (1.44 and 1.47) are indeed very close to each other, indicating that an increase of E_η leads to a similar increase of E_{BA}. This is further confirmed by values of E_{BA} measured using the method of integrated time-resolved data as summarized in Table I.[88]

In addition to Eq. (2.10), the fluorescence quantum yield ϕ_B is also given by

$$\phi_B = \int_0^\infty i_f^B(t)\, dt\,. \tag{2.34}$$

Substituting Eq. (2.28) into Eq. (2.34) (with $(\lambda_1 - x) \ll (x - \lambda_2)$ for the low-temperature range) the values of the quantum yield obtained from Eqs. (2.10) (steady-state) and (2.34) (time-resolved) can be directly compared. Their temperature dependence, when plotted in Arrhenius

TABLE I
Activation Energies E_{BA} for DMABN in Different Polar Solvents, as Determined from Integrated Time-Resolved Measurements,[88] and Comparison to E_η, the Activation Energy for Solvent Viscosity

Solvent	E_{BA} (kcal M^{-1})	E_η (kcal M^{-1})
Methanol	2.2	2.5
Ethanol	2.9	3.2
Butanol	3.5	4.7
Pentanol	4	5.4
Octanol	4.6	6.4
Decanol	5	6.4
n-Butyl chloride	1.7	1.7
Propionitrile	1.5	1.8
Butyronitrile	2.5	2.2
Capronitrile	2.4	2.6

form, should yield a straight line. In the neighbourhood and above the temperature T_m, where the steady-state quantum yield ϕ_B bends away from the low-temperature straight line (see Fig. 2.8), time-resolved measurements allow to remove that part of ϕ_B that is due to fluorescence before equilibration (by integrating only the fraction due to the initial fast component of the biexponential decay; see Fig. 2.16). The temperature dependence of the resulting ϕ_B values deliver a direct continuation of the straight Arrhenius line in the low-temperature range, into the region of T_m, and can yield more precise E_{BA} values.[88]

As can be seen in Table I, the E_{BA} values thus derived correlate with E_η for various series of homologous solvents. This and the fact that $E_{BA} \lesssim E_\eta$ directly lead to the inequality $E_0 \ll E_\eta$, and we conclude that TICT formation, in the cases considered here, proceeds as a barrierless adiabatic photoreaction.

2. *Time-Dependent "Rate Constants"*

As will be shown in detail in Section IV, the large-barrier case leads to "well-behaved" time-independent reaction rates $k = \text{const}$, but for the no-barrier case, time-dependent reaction rates $k = k(t)$ are predicted. These can also appear in other processes, for example, in Förster energy transfer, where $k(t)$ is proportional to $t^{1/2}$.[59]

If time-dependent rate constants are involved, the solutions (2.28)–(2.32) of the rate equations are no longer valid, and we have to go back to the rate equations reformulated in terms of $k(t)$. For the irreversible case below T_m (negligible back reaction), they read

$$\frac{dn_B}{dt} = -(k_f^B + k_0^B)n_B - k_{BA}(t)n_B \tag{2.35}$$

$$\frac{dn_A}{dt} = -(k_f^A + k_0^A)n_A + k_{BA}(t)n_B \,, \tag{2.36}$$

where n_B and n_A are the number of fluorescing B and A molecules.

The rise time of the A fluorescence differs from a convolution (C_1) of the exponential A decay with the experimental decay function of the precursor B fluorescence manifesting a time-dependent rate constant. This is nicely shown in Fig. 2.19 on a picosecond time scale, where these effects are especially strong. The rise of the A fluorescence is significantly faster than the rise of C_1. In contrast, similar rise times would be expected if a n_B population proportional to $i_f^B(t)$ were feeding the A state with a time-independent rate constant.[78,79] In Section IV, a more-detailed analysis of the results will derive the explicit time dependence of $k(t)$. It turns out that, for early times, $k(t)$ possesses a maximum. In Section V, the experimental implications will be discussed.

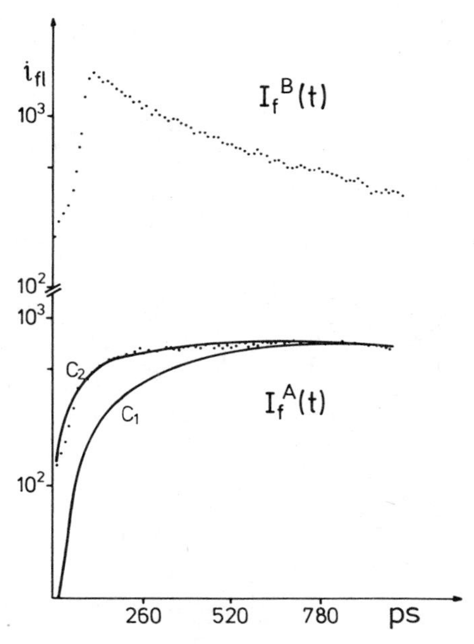

Fig. 2.19. Decay of B band and rise and decay of A band of DMABN observed in *n*-pentanol at −50°C. The intensity of the A band rises much faster (fitted by curve C_2) than would be expected on the basis of the observed decay of the B band [curve C_1 calculated by convolution with $i_f^B(t)$].[78]

3. *Time-Resolved Solvatochromism*

As previously noted (see Section II.B.1), the A emission exhibits an important solvatochromic effect due to the appearance of a large dipole moment in the TICT state. The polar interactions between the solute molecule and the polar environment lead to the reorientation of the solvent molecules and to a relaxation of the electronic energy of the TICT state whose manifestation is a spectral shift during the lifetime of the excited state. The competition between the energy relaxation, whose dynamics is strongly viscosity dependent, and the deactivation of the TICT state has been made evident for DMABN[78,89]:

1. From the temperature dependence of the position of the stationary emission: Whereas for butyl chloride solutions, where even at low temperatures the viscosity is small, only the red shift attributed to an increase of the medium polarity with decreasing temperature is observed, a blue shift is manifest for alcoholic solutions at low temperatures (Fig. 2.20), indicating that the relaxation is not achieved before emission of photons.

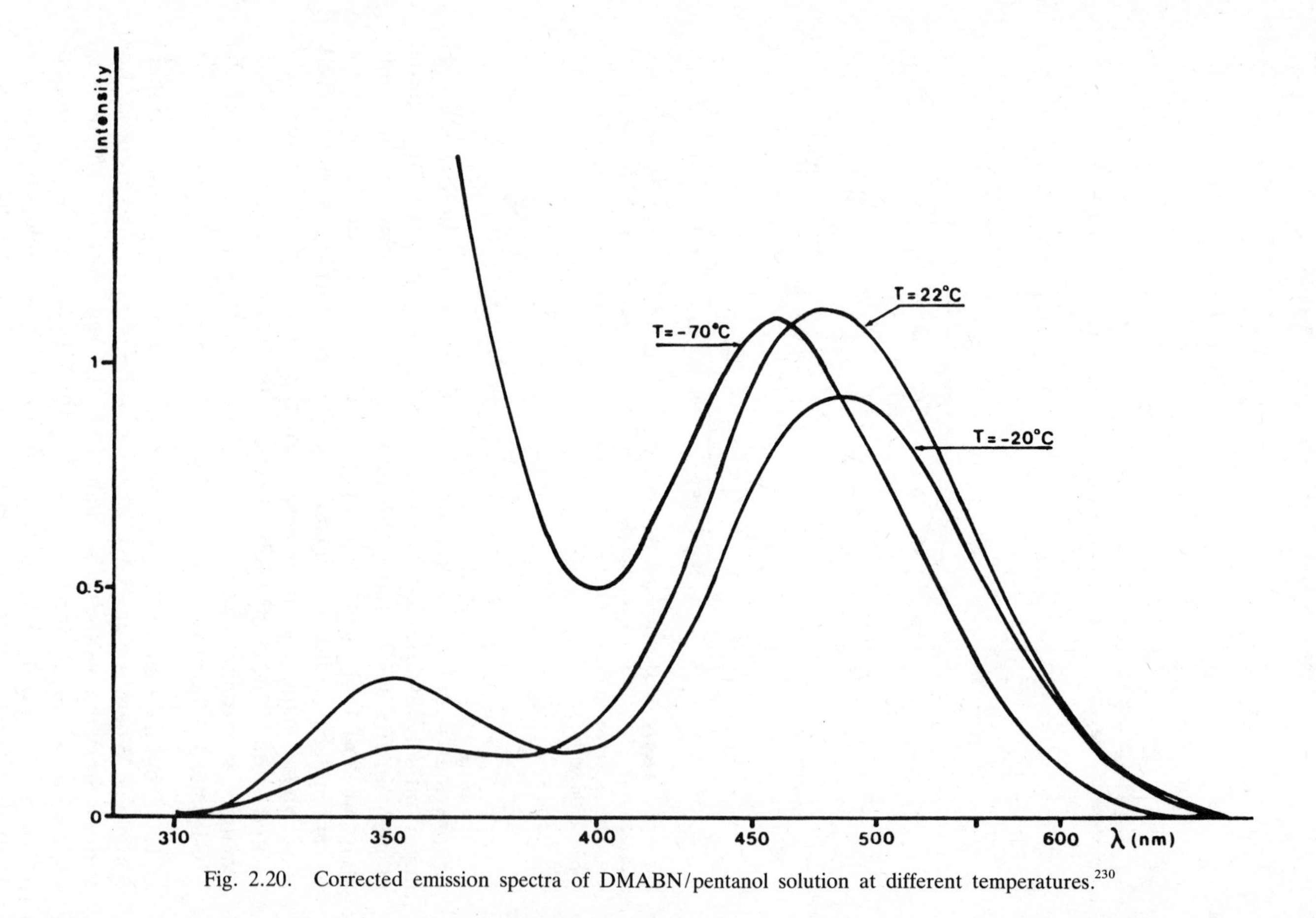

Fig. 2.20. Corrected emission spectra of DMABN/pentanol solution at different temperatures.[230]

2. From the fluorescence decay curves $i(\tilde{\nu}, t)$ recorded at different wavelengths: Figure 2.21 where decay curves for DMABN/propanol solution at −80°C are reported, clearly illustrates that for wavelengths isolated in the A spectrum there is an initial deviation from the exponential behavior that is observed at long times and that characterizes the decay time of the total A* emission. For short wavelengths the decay is considerably shortened at the very beginning of the fluorescence because of the electronic energy relaxation to a final state of lower energy, whereas on the long-wavelength side an increasing rise time is apparent because a portion of the light emitted in the red tail originates from states resulting from the damping process and hence delayed with respect to the initially excited state.

A relevant result here is that the emissions of the relaxed and unrelaxed DMABN charge-transfer states have the same spectral shape: the instantaneous spectral distributions deduced from the decays $i(\tilde{\nu}, t)$ show no significant time dependence of their contours, which differ only slightly from that of the stationary distribution (Fig. 2.22).

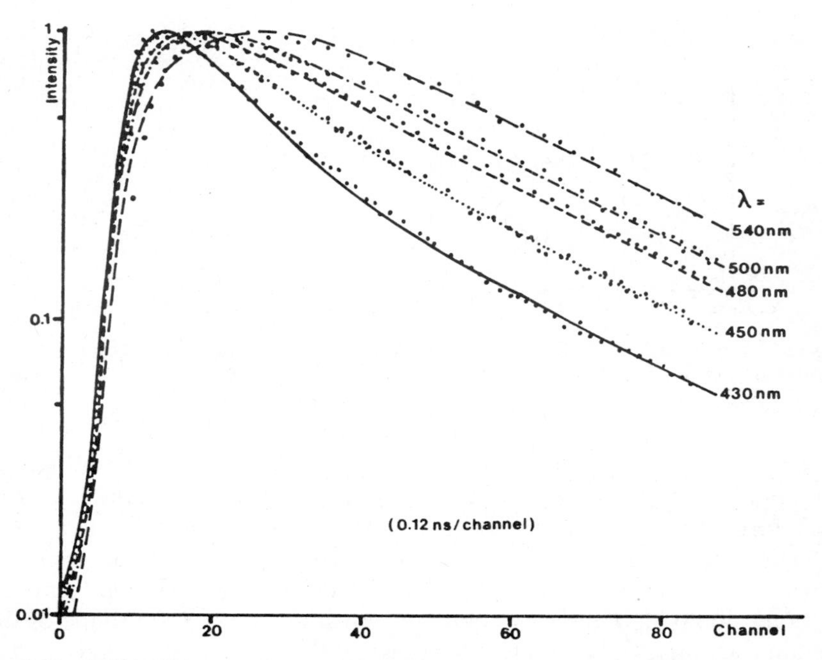

Fig. 2.21. DMABN/propanol at −80°C: experimental (· · ·) and calculated (——) decay curves at different wavelengths. The fitted parameters are $\tilde{\nu}_0 = 22{,}700\ \text{cm}^{-1}$, $\tilde{\nu}_\infty = 20{,}200\ \text{cm}^{-1}$, and $\tau_R = 1.5$ ns.[89]

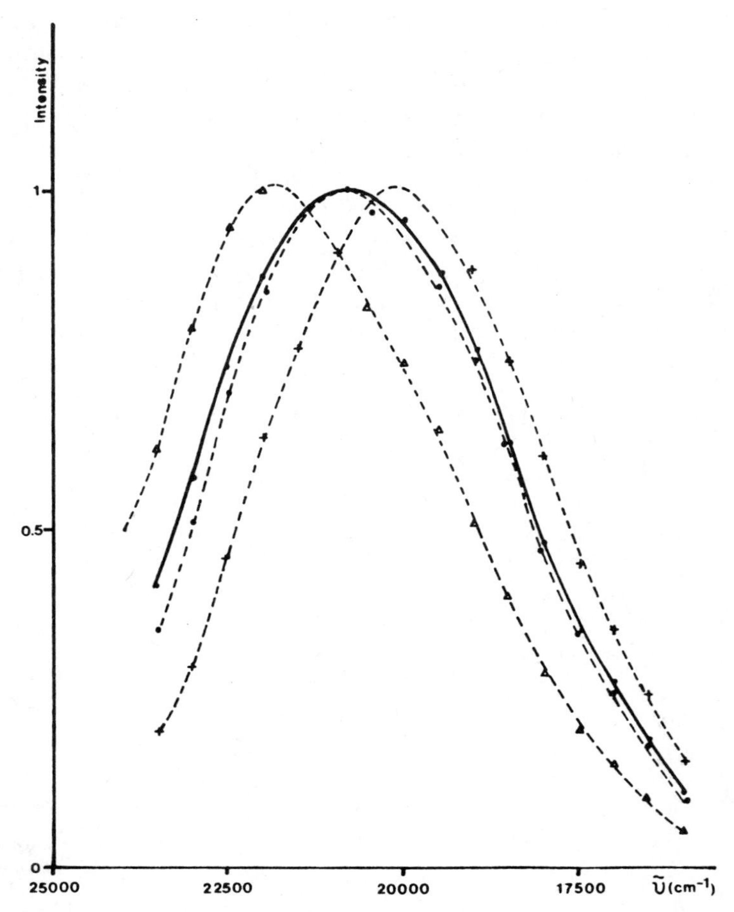

Fig. 2.22. Normalized stationary (solid line) and instant spectra at various times t (gate width = 0.1 ns) for DMABN/propanol at −80°C: △, $t = 0.5$ ns; ●, $t = 1.5$ ns; +, $t = 5$ ns.[230]

Two different approaches have been proposed for describing the temporal evolution of the relaxation process, one[90] based on the reaction-field formalism, the other[91] using a kinetic model assuming that every solvent reorientation step implies a decrease of the solute energy by a constant amount and that the solvent relaxation occurs exponentially without modifying either the solute deactivation rates or the shape of the instant emission spectra (see Section IV.D.2.b.3). They lead to the conclusion that the wavenumber $\tilde{\nu}_m(t)$ characterizing the position of the emission at time t is given by

$$\tilde{\nu}_m(t) = \tilde{\nu}_0 - (\tilde{\nu}_0 - \tilde{\nu}_\infty)(1 - e^{-t/\tau_R}) \tag{2.37}$$

where $\tilde{\nu}_0$ and $\tilde{\nu}_\infty$ define, respectively, the position at times $t = 0$ (unrelaxed state) and $t \rightarrow \infty$ (relaxed state) and τ_R is the spectral relaxation time constant related to the solvent relaxation time $\tau_r \approx (\epsilon_\infty/\epsilon_0)\tau_D$, where ϵ_∞ and ϵ_0 are the optical and low-frequency dielectric constants and τ_D is the dielectric relaxation time constant.

The decay curve $i(\tilde{\nu}, t)$ at a given wavenumber $\tilde{\nu}$ is then

$$i(\tilde{\nu}, t) = k_F i(t) S(\tilde{\nu} - \tilde{\nu}_m(t)) \tag{2.38}$$

with the deexcitation law $i(t)$ and the radiation rate k_F independent of the spectral shift process. $S(\tilde{\nu} - \tilde{\nu}_m(t))$ represents the normalized instant spectral distribution whose shape is invariant with time and which has to be defined for each solution and each temperature.

Taking into account that $S(\tilde{\nu} - \tilde{\nu}_m(t))$ can also be written

$$S(\tilde{\nu} - \tilde{\nu}_m(t)) = \delta(\tilde{\nu} - \tilde{\nu}_m(t)) \circledast S(\tilde{\nu}) \tag{2.39}$$

and with $i(t) = e^{-t/\tau_F}$, the stationary emission is given by

$$S^{st}(\tilde{\nu}) = \int i(\tilde{\nu}, t)\, dt = k_F \frac{\tau_R}{\Delta\tilde{\nu}_s} \left(\frac{\tilde{\nu} - \tilde{\nu}_\infty}{\Delta\tilde{\nu}_s}\right)^{(\tau_R - \tau_F)/\tau_F} \circledast S(\tilde{\nu})\,, \tag{2.40}$$

where the term in parentheses is equal to zero outside the range defined by $\tilde{\nu}_\infty < \tilde{\nu} < \tilde{\nu}_0$ and where $\Delta\tilde{\nu}_s = \tilde{\nu}_0 - \tilde{\nu}_\infty$ represents the maximum spectral shift. From Eq. (2.40) it is clear that the position of the stationary spectrum is a function of $\tilde{\nu}_0$, $\Delta\tilde{\nu}_s$, and the ratio (τ_R/τ_F) of the deactivation and relaxation rates. The broadening of the stationary emission with respect to the instant spectrum is maximum for $\tau_R = \tau_F$, increases with $\Delta\tilde{\nu}_s$, and gains in importance for small initial widths. This is illustrated in Fig. 2.23, representing the stationary emission calculated for two values of the width $\Delta\tilde{\nu}_0$ at half-maximum of an instant distribution assumed to be of the form

$$S(\tilde{\nu}) = \frac{1}{\Delta\tilde{\nu}_0} \cos^2 \frac{\pi}{2} \frac{\tilde{\nu}}{\Delta\tilde{\nu}_0} \qquad \text{for } -\Delta\tilde{\nu}_0 < \tilde{\nu} < \Delta\tilde{\nu}_0\,. \tag{2.41}$$

In this figure, one can see that with $\Delta\tilde{\nu}_s = 5000\ \text{cm}^{-1}$, the maximum broadening (for $\tau_R = \tau_F$) is only of the order of 20% for $\Delta\tilde{\nu}_0 = 5000\ \text{cm}^{-1}$, whereas it becomes near to 60% for $\Delta\tilde{\nu}_0 = 3000\ \text{cm}^{-1}$.

For DMABN in *n*-butyl chloride and in various alcohols, the greatest

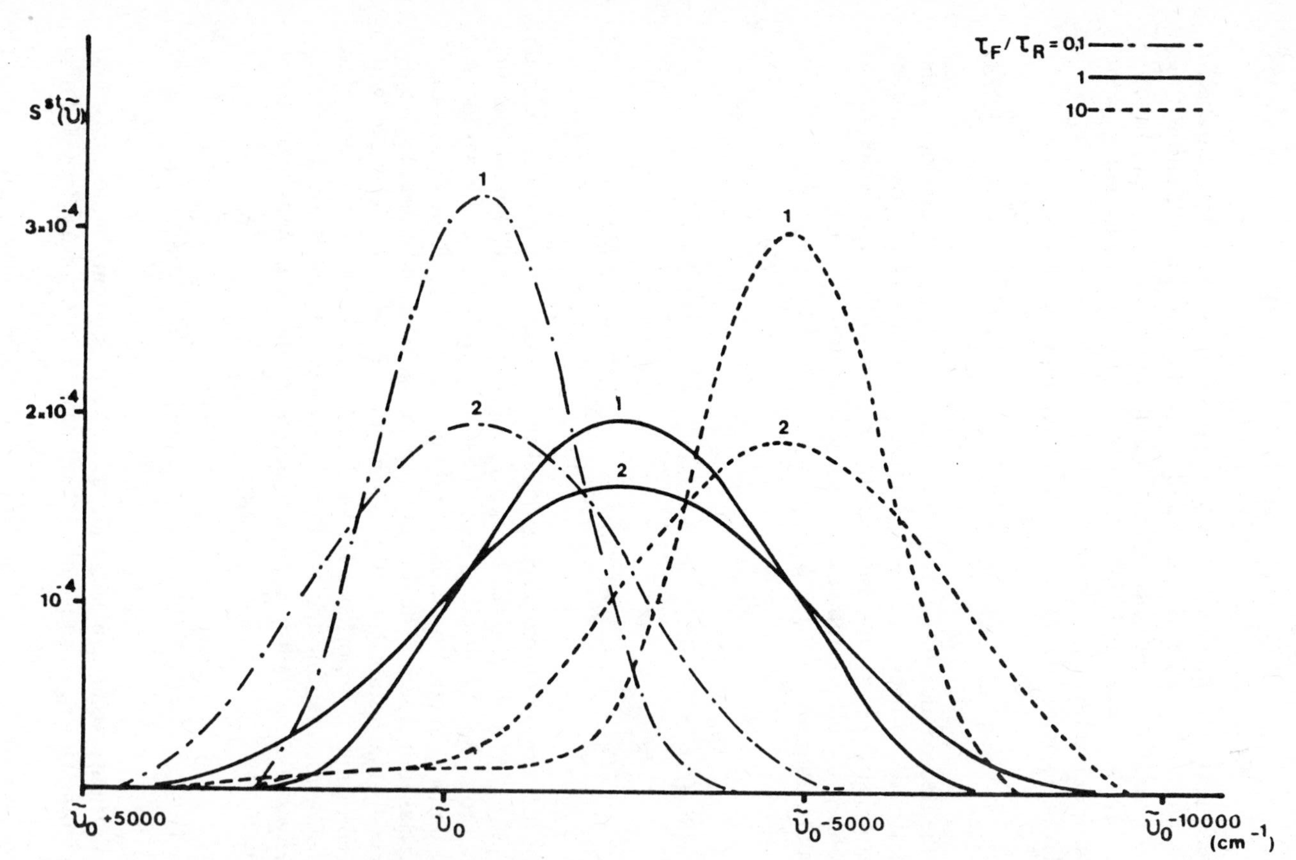

Fig. 2.23. Stationary spectra $S^{ST}(\tilde{\nu})$ calculated [Eq. (2.40)] with $\Delta\tilde{\nu}_s = 5000\ \mathrm{cm}^{-1}$ for different values of the ratio τ_F/τ_R and assuming $k_F = 1/\tau_F$: $S(\tilde{\nu})$ given by Eq. (2.41) with $\Delta\tilde{\nu}_0 = 3000\ \mathrm{cm}^{-1}$ (curves 1) and $5000\ \mathrm{cm}^{-1}$ (curves 2).[230]

difference observed between the widths of the instantaneous and stationary emissions of the TICT state was of the order of 10%: the spectra have generally an equivalent $\Delta\tilde{\nu}_0$ value at about 5000 cm^{-1} except at very low temperature in alcohols where $\Delta\tilde{\nu}_0$ is smaller but where the relaxation is very slow ($\tau_F/\tau_R \ll 1$).

In Table II are reported the values of $\tilde{\nu}_0$, $\tilde{\nu}_\infty$, and τ_R obtained for different temperatures as well as the experimental and calculated wavenumber $\tilde{\nu}_m^{st}$ of the peak of the stationary spectrum. Figure 2.21, where the solid lines represent calculated decays, shows that the experimental results can well be accounted for by the expressions (2.37) and (2.38). These results indicate that the relaxation of the electronic energy of the TICT state of DMABN due to interaction with the polar medium can well be described by a single exponential law not only for the *n*-butyl chloride solution but also for the solutions in alcohols. This relaxation process, leading to final states having an electronic energy markedly lower than that of the unrelaxed charge-transfer states, is responsible for the presence of an intramolecular potential barrier for the reverse reaction to the locally excited B* state; the barrier is made evident by the

TABLE II
Best Fit Values[89] of the Peak Position of the Unrelaxed ($\tilde{\nu}_0$) and Relaxed ($\tilde{\nu}_\infty$) TICT State Emission and of the Spectral Relaxation Time τ_R for Different DMABN Solutions.[a]

	$\tilde{\nu}_0$ (cm^{-1})	$\tilde{\nu}_\infty$ (cm^{-1})	τ_R (ns)	$\tilde{\nu}_m^{st}$ (cm^{-1}) Experimental	Calculated
n-Butyl chloride					
−125°C	23,500	21,800	0.75	21,900	21,850
Propanol					
−40°C	23,500	19,700	0.4	20,000	20,000
−60°C	23,200	19,700	0.75	20,300	20,370
−80°C	22,700	20,200	1.5	20,800	20,850
−100°C	22,800	20,300	8.	21,700	21,800
−115°C	22,900	21,000	≥15	22,400	22,300
Butanol					
−40°C	23,200	19,700	0.7	20,400	20,200
−55°C	23,000	19,700	1.4	20,800	20,600
−70°C	23,000	20,500	3.5	21,500	21,300
−80°C	23,000	20,500	5.5	22,100	21,900
Pentanol					
−20°C	22,800	20,000	0.6	20,400	20,300
−40°C	22,500	20,000	1.5	20,800	20,700

[a]Comparison between the experimental and calculated values of the maximum ($\tilde{\nu}_m^{st}$) of the stationary emission.

measurement of an activation energy for the A* → B* reaction higher than that of the solvent mobility (see Section II.B.1 and Ref. 88).

The final remarks concern the B emission: (1) compared to that of the TICT state it is only slightly dependent on the solvent polarity (the peak position varies from 29,300 cm^{-1} in cyclohexane to 28,200 cm^{-1} in methanol at a temperature near to the freezing point); (2) it does not present any wavelength-dependent decay curves. This essentially proceeds from the fact that the dipole moments in the ground and in the excited B* states are in the same direction and are not very different.

D. Cage Effects

1. *Rearrangements in Mixed Solvents*

We have already noted that, in general, the electric dipole moment of a polar aromatic compound increases by excitation (that is in contrast to merocyanines[65] for which $\mu_g \gg \mu_e$).[60,61] This behavior has been demonstrated impressively by picosecond time-resolved absorption and gain

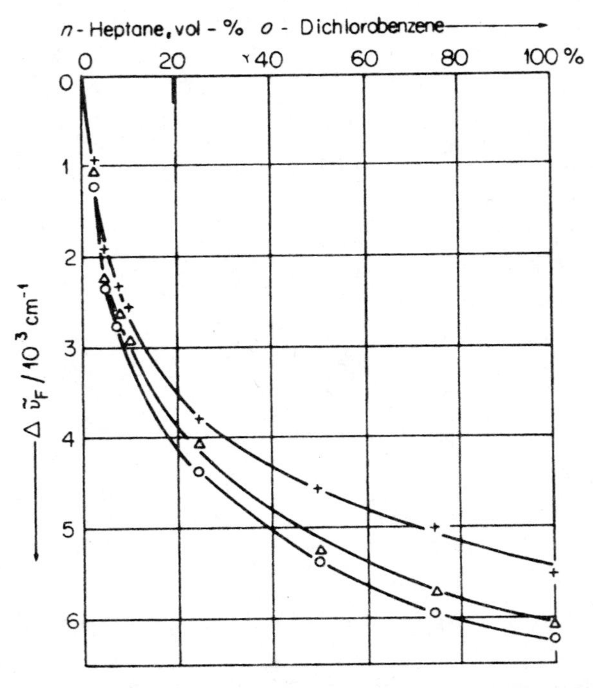

Fig. 2.24. The red shift $\Delta\tilde{\nu}_F$ of the fluorescence maximum of 4-dimethylamino-4′-nitrostilbene in mixtures of *n*-heptane and *ortho*-dichlorobenzene at 80°C, +; 20°C, △, and −5°C, ○.[94]

spectroscopy in solutions and polymer films.[92] Hence, reorientation occurs in polar solvents after absorption and after emission. In mixtures of inert and polar solvents diffusion-controlled reactions[93] will start after absorption resulting in an enrichment of polar solvent molecules in the cages around the excited solute molecules,[94] giving rise to an additional anomalous Stokes red shift (Fig. 2.24). The red shift increases linearly with the first drops of a polar solvent added to a solution in an inert solvent, and eventually reaches its final value asymptotically with increasing polar solvent concentration. The shape of this red shift concentration dependence is that of a *Langmuir adsorption isotherm*, since the number of sites on the surface of the solute is limited as has been outlined in detail in Section 1.1.3 of Ref. 3 and in Ref. 95.

The enrichment of the concentration of the polar solvent component in the cage and, therefore, the relative amount of the red shift of the fluorescence band is a function of viscosity, since the diffusion-controlled reaction time must be smaller than the excited-state lifetime. This lifetime limitation of the red shift is even more severe if the higher value of the excited-state dipole moment is not a property of the initial Franck–Condon state but of the final state of an adiabatic reaction. Nevertheless, the additional red shift has been observed for the fluorescence of TICT biradical excited states due to their nanosecond lifetime together with a quenching effect of the total fluorescence since the A* to S_0 transition is weak (symmetry forbidden) (Fig. 2.25).

2. *Side Reactions*: *Complex Formation*

Solute/solvent 1:1 complexes can be formed, of course, in the ground as well as in excited states. Since solutes that show photochemical internal twisting normally possess polar substituent groups, the probability of complex formation is rather high. Thus, studies of the photophysics of internal twisting must be carried out carefully enough to avoid errors due to complex formation.

If 1:1 complexes are formed (1) in the ground state, then two types of photoreactions will occur parallel to each other, that of the bare and that of the complexed solute, each in its own type of cage. Mixed alcohol/alkane systems, for example, show an indication of preformed solute-solvent complexes as evidenced by the picosecond experiments of Wang and Eisenthal[80,81] on DMABN. Planar model systems like the indolines *3* and *5* (Sec. II.A.1) indicate that an additional channel opens for the B* state in alcoholic solvents which increases the nonradiative decay path. This can explain the observed reduction of the fluorescence quantum yield in protic solvents: about 0.1 in the aprotic polar solvent n-butyl chloride,[6] about 0.01 in 1,2-propanediol, and about 0.001 in water.[228]

In the excited state, the initially uncomplexed solute monomer can

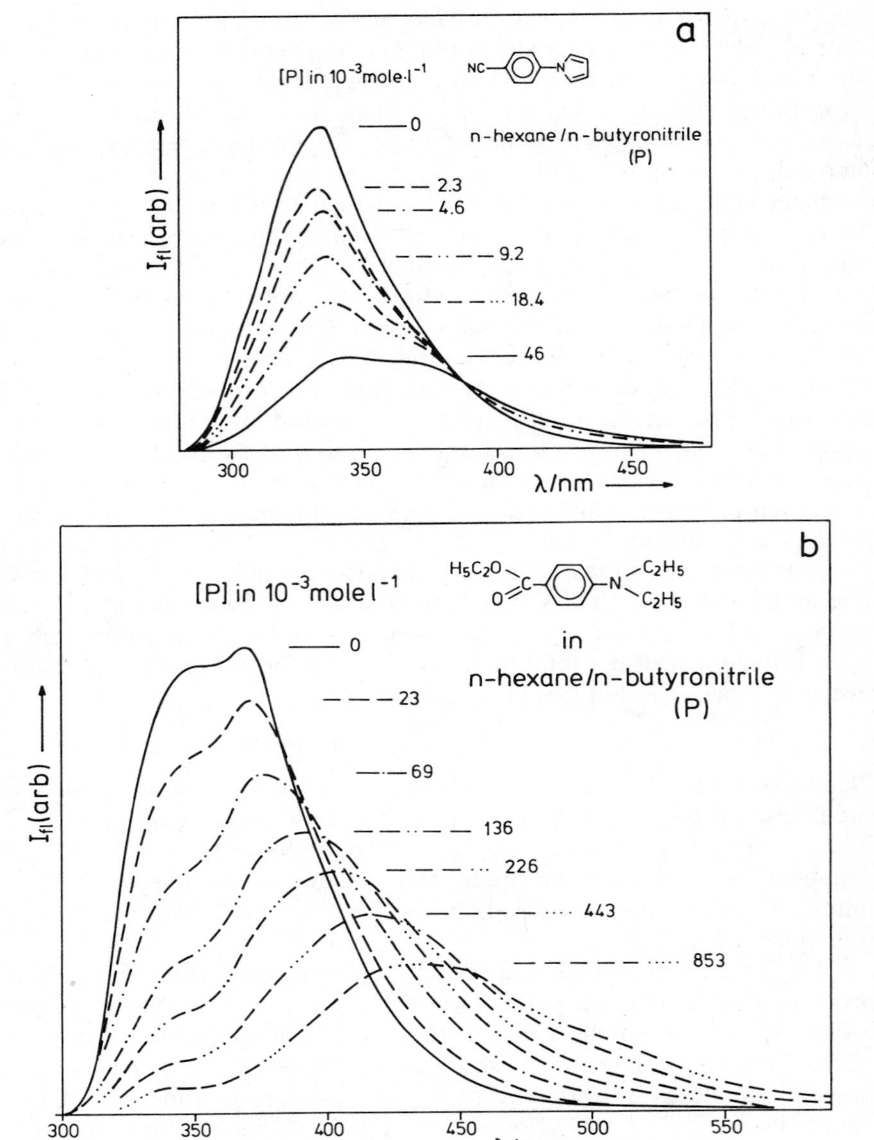

Fig. 2.25. Quenching effect on the short-wavelength F_B fluorescence band of two different TICT state-forming compounds by adding traces of a polar solvent P to a nonpolar solution. Successive spectra correspond to repeated addition of small amounts of *n*-butyronitrile to initially 50 mL *n*-hexane solution, $\sim 5 \times 10^{-6}$ M/L. Note the strong difference in response observed for the two solutes.

form (2) a solute–solvent exciplex or (3) a solute–solute excimer[227] after absorption. Excimer formation can be avoided using sufficiently small solute concentrations. But exciplex formation between an excited solute molecule and a molecule of a solvent or impurity component is a dangerous source of error. Varma and co-workers[96,97] repeatedly pointed out that 1:1 exciplexes could be formed by specific intermolecular interactions, and this especially due to overlap of electronic wavefunctions of the constituents. In mixed solvents, the exciplex partner for an excited solute molecule could be either a member of the initial cage or could belong to a new cage formed by diffusion during the excited-state lifetime. On the other hand, since photochemical twisting occurs not only in well-checked solutions but also in the gaseous phase, and this for various families of solute compounds, there remains no doubt that TICT formation is a most general and important effect, and occurs independently of the necessity for complex formation (see Section IV).

In the particular cases of DMABN and DMABEE, specific interactions most probably can occur with certain solvent partners after the adiabatic photoreaction has achieved its final state of orthogonality and

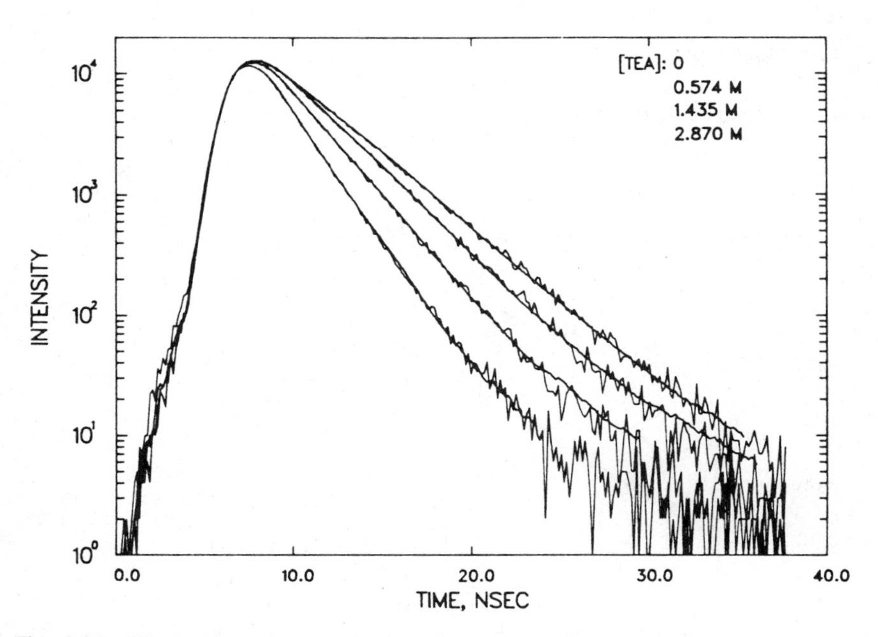

Fig. 2.26. Fluorescence decays of DMABN in iso-octane in the presence of varying amounts of triethylamine (TEA), at 340 nm, that is, B fluorescence (no A fluorescence at all).[98]

the dialkylamino group possesses only one electron in its lone-pair orbital.[224] But the observed linear relationship of the quantum yield ratio ϕ_A/ϕ_B with added small concentrations of a polar solvent component[96,97] is not a proof for exciplex formation, as we just have seen for the Langmuir-adsorption-isotherm-type example. Instead, correlations with steric effects of quencher molecules (with shielded or unshielded lone pairs)[98] show that saturated tertiary amines (and possibly even ethers) can form specific interactions with the molecule in its TICT state. In these cases, fluorescence quenching by three electron $(>N\!:\!N<)^+$ bonding (Figs. 2.26 and 2.27) can occur and is a well-known effect.[98–100]

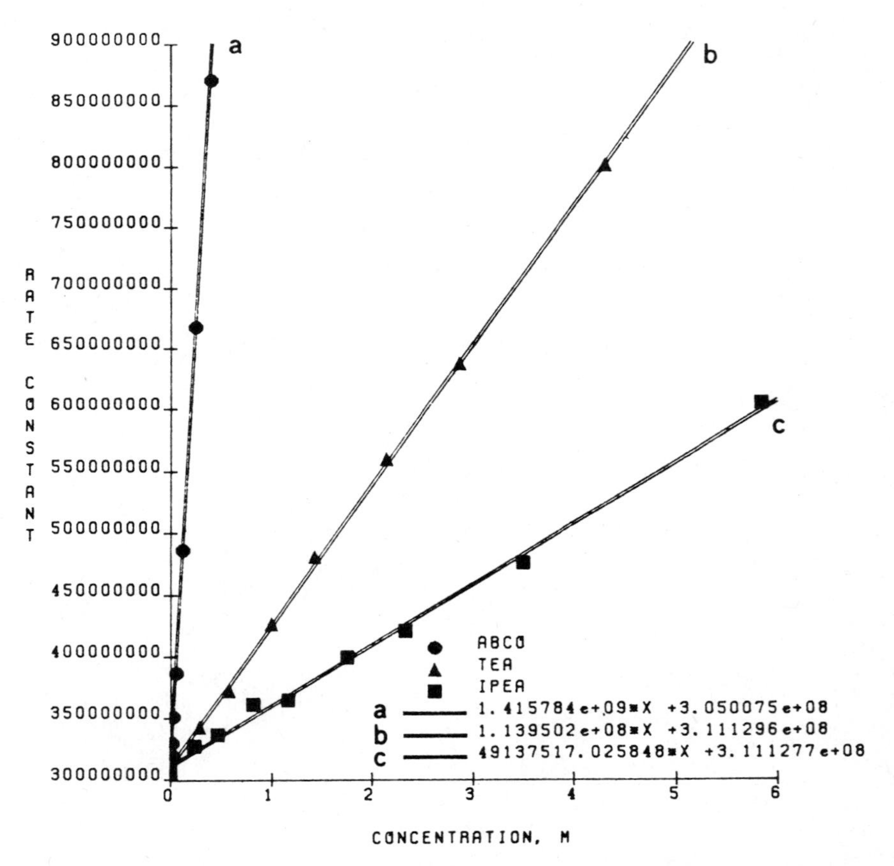

Fig. 2.27. Stern–Volmer plots of DMABN fluorescence decay rates as a function of amine concentrations. TEA–triethylamine; ABCO—quinuclidine; IPEA—ethyldiisopropylamine.[98]

In contrast to Varma and co-workers[96,97] we wish to point out that exciplex formation *can*, but is not forced to, occur with internal twisting. In appropriate solutions, of course, it may be connected with the TICT reaction mechanism, but the formation of the twisted biradical excited state plays the governing role.

In the following sections, therefore, our goal will be first to look for molecules that possess electronically excited states the potential energy of which decreases sufficiently steeply with twisting accompanied by a charge separation. After that we will concentrate on solvent effects using stochastic models and spectroscopic observations.

III. ELECTRONIC STRUCTURE OF THE BONDED π-DONOR–π-ACCEPTOR PAIRS

A. Outline of the Theoretical Treatment

The course of the studied photo process in this work is primarily determined by the shapes of the lowest singlet excited-state surface S_1 along with that of the ground state S_0. Therefore, the locations of minima and barriers on these surfaces are required. The shapes of the potential surfaces are directly connected with the changes in the electronic wavefunctions as a function of molecular geometry. A knowledge of the surface shapes and energies does not give sufficient information. An understanding of the nature of the electronic wavefunctions associated with the particular molecular geometries in various electronic states is needed. This will, first, help to choose which particular pathways should be explored in order to find minima and other interesting regions of the energy surfaces. It is unrealistic to expect that all regions of the nuclear configuration space could be explored. Second, the knowledge of the leading features of the electronic wavefunctions corresponding to the minima on S_1 is essential for an understanding of the environmental influence on the electronic states of the free molecules (polar solvent effects).

There are two classes of geometries at which S_1 (or the lowest triplet T_1) surfaces of molecules studied here tend to have minima: (1) S_1 has a minimum at a geometry similar to that for which the ground state has a minimum (spectroscopical minimum); (2) S_1 has a minimum at "biradicaloid" geometries,[101–104] for which the ground state S_0 almost always has a maximum. The simplest definition of the "biradicaloid" geometries is that the ground states S_0 of molecules have two electrons in a pair of nearly degenerate nonbonding orbitals: one bond is missing. This can be achieved by stretching a σ bond or by twisting a π bond. If the geometry is changed toward a "normal" situation so that two nonbonding orbitals overlap giving rise to bonding and antibonding linear combinations, two originally unpaired electrons can occupy the bonding orbital, stabilizing the energy of the ground state. In contrast, the energy of S_1 (or T_1) increases going from "biradicaloid" to "normal" geometries, since one electron occupies a bonding and the other an antibonding orbital, which is usually more destabilized than the former at "normal" geometries. Therefore, the local biradical minima in S_1 (or T_1) result.

The return to the ground state S_0 will occur either from spectroscopical or from "biradicaloid" or under certain conditions from both minima. Therefore, the theory should provide a proper energy ordering of low-lying singlet—S_0, S_1, S_2—and triplet states at these geometries as well as the information about the nature of the corresponding wavefunctions.

Since in this work the most studied molecular systems involve the π-donor and π-acceptor subunits, we will consider the relation between a "nonbonding" donor–acceptor combination, on the one hand, and the dative bond, on the other hand.

The interaction of a π-donor with a π-acceptor, bonded to each other so that the π electrons from both subunits are allowed to interact, gives rise to a π-dative bond in the ground state S_0. Such systems usually prefer the planar geometries unless steric effects produce a deviation from planarity. Twisting to orthogonality causes breaking of the bond because the π-electron donation from one subunit to another is prohibited. The dative bonds are relatively weak (partially double bonds) because the bonding is mainly due to charge separation, and, therefore, thermal *cis–trans* isomerization by twisting occurs easily and the isolation of isomers is difficult. By analogy to the twisted double bonds, it is to be expected that the S_1 and T_1 excited states will also favor twisted geometries, although it is evident that the twist of the partial double bonds will have their own characteristics in the excited states.[14,40] It is easy to guess that, owing to a large electronegativity difference between two "radical" centers, different energy ordering of states will occur than for biradicals with two equivalent or slightly different "radical" centers. We will show that instead of two pairs of close-lying states S_0, T_1 of covalent and S_1, S_2 of zwitterionic nature typical for biradicals[101–104] the noninteracting π-donor–π-acceptor combination will have a large energy gap between the S_0 and the close-lying S_1, T_1 charge-separated states.[40]

On the other hand, since biradical excited states are in principle easy polarizable,[39,105] they will have something in common with the charge-separated states of a broken dative bond. In order to find out the essentials about the excited states of the dative bonds, especially about the geometries corresponding to minima from which emission might take place, we will proceed in four steps.

1. A simple two-orbital two-electron model will be reviewed for molecules at "biradicaloid" geometries.[14,101–104] Since we are primarly interested in one bond breaking, it is useful to consider the states of only two electrons that become unpaired during this process and all other electrons in the molecule are assumed to be inactive (frozen-core approximation). There will be in principle an infinite number of stationary states for these two electrons, but, again, in order to simplify matters, we consider only those states in which two electrons are allowed to move in two orbitals. This will make it possible to formulate the conditions under which the charge-separated TICT states can occur. Also, comparison between the nature of TICT wavefunctions and the wavefunctions of other biradicaloids produced by breaking of a double bond or a charged bond will be made (Ref. 14).

2. The results of ab initio large-scale calculations for two series of prototype systems will be discussed, illustrating that the basic ideas obtained from the crude two-orbital, two-electron model remain valid, although the presence of other than two electrons and of many more than two orbitals will change the details, whose importance should be carefully considered. A twisting of ethylene in the presence of a point charge simulating the external field serves as a good example for comparison between the TICT states and the states of biradicaloids with large charge separation.[14] The same principles will be found by comparison of singlet and triplet states of twisted propylene,[107] formaldiminium cation,[108] and aminoborane.[40] The analysis of the correlated wavefunction of propylene will be shown in some detail.[107,109]

3. Since the molecules for which anomalous fluorescence has been observed contain subunits with many more than one center orbital (they are usually aromatic or other π systems), complications due to presence of locally excited states within a donor and an acceptor subunit will be qualitatively introduced. A breaking of such complex dative π bonds gives rise to wavefunctions that are complicated mixtures of locally excited and of charge-separated zero-order states. Relative weights of individual contributions of these two different types of zero-order states to the final S_1 state of the molecule differ substantially for planar and twisted geometries. It is known that at planar geometries the locally excited states are lower in energy than the charge-separated state, and it will be shown that at perpendicular geometry the opposite might be easily true if the donor–acceptor subunits are chosen carefully.

4. Finally, an "effective" influence of the environment (solvent polarity) on the wavefunctions and energies of the low-lying excited states of the complex dative bond will be qualitatively considered. An influence of a polar solvent stabilizing the charge-separated state might play a key role for the process considered. An example of an extreme strong

influence of the polar solvent on the excited states of 9,9′-bianthryl and its consequence for the luminescence process will also be presented.

B. The Two-Electron, Two-Orbital Model

1. 3 × 3 Interaction

This model corresponds to the minimal basis description of bonds, and it has been in use in organic chemistry for a long time.[101–104] Recently, it has been elaborated in more detail[14] and used as qualitative guidance for characterization of minima on the low-lying excited energy surface of biradicaloids. In the case that two orbitals $\mathscr{A}$ and $\mathscr{B}$ interact weakly, the two-electron states are not adequately described by the single configuration because all three singlet configurations ($\mathscr{A}^2$, $\mathscr{B}^2$, $\mathscr{A}\mathscr{B}$) must be treated equally (3 × 3 CI). The triplet is of course represented by a single configuration. If two orbitals are degenerated the "broken" system is denoted as perfect biradical; if they differ in energy, the system is referred to as a biradicaloid. Since we are primarily interested in dative bonds, $\mathscr{A}^2$ and $\mathscr{B}^2$ differ in energy and we will use the convention that $\mathscr{B}^2$ is lower in energy than $\mathscr{A}^2$.

Detailed consideration of this model is very instructive because we deal with the exact solution of the problem and we will be primarily interested in the nature of the exact two-electron wavefunctions.

There are four ways to place two electrons in two orbitals without a spin consideration: $|\mathscr{A}^2\rangle = \mathscr{A}(1)\mathscr{A}(2)$, $|\mathscr{B}^2\rangle = \mathscr{B}(1)\mathscr{B}(2)$, $|\mathscr{A}\mathscr{B}\rangle = \mathscr{A}(1)\mathscr{B}(2)$ and $|\mathscr{B}\mathscr{A}\rangle = \mathscr{B}(1)\mathscr{A}(2)$. The orbitals $\mathscr{A}$ and $\mathscr{B}$ can have any form; we will assume only that they are orthogonal:

$$S_{\mathscr{A}\mathscr{B}} = \langle \mathscr{A} | \mathscr{B} \rangle = 0 . \tag{3.1}$$

The Hamiltonian operator for the two electrons can be written as

$$\hat{H} = \hat{h}(1) + \hat{h}(2) + e^2/r_{12} , \tag{3.2}$$

where $\hat{h}(1)$ and $\hat{h}(2)$ are one-electron Hamiltonian operators. In order to find the eigenstates of the system, the Hamiltonian matrix in the basis of four configurations $|\mathscr{A}^2\rangle$, $|\mathscr{B}^2\rangle$, $|\mathscr{A}\mathscr{B}\rangle$, and $|\mathscr{B}\mathscr{A}\rangle$ has to be considered. The diagonal elements are

$$\langle \mathscr{A}^2 | \hat{H} | \mathscr{A}^2 \rangle = 2h_{\mathscr{A}\mathscr{A}} + J_{\mathscr{A}\mathscr{A}} , \tag{3.3}$$

$$\langle \mathscr{B}^2 | \hat{H} | \mathscr{B}^2 \rangle = 2h_{\mathscr{B}\mathscr{B}} + J_{\mathscr{B}\mathscr{B}} , \tag{3.4}$$

$$\langle \mathscr{A}\mathscr{B} | \hat{H} | \mathscr{A}\mathscr{B} \rangle = \langle \mathscr{B}\mathscr{A} | \hat{H} | \mathscr{B}\mathscr{A} \rangle = h_{\mathscr{A}\mathscr{A}} + h_{\mathscr{B}\mathscr{B}} + J_{\mathscr{A}\mathscr{B}} , \tag{3.5}$$

where diagonal elements of one-electron Hamiltonian are defined as

$$\langle \mathscr{A}|\hat{h}|\mathscr{A}\rangle = h_{\mathscr{A}\mathscr{A}} , \tag{3.6}$$

$$\langle \mathscr{B}|\hat{h}|\mathscr{B}\rangle = h_{\mathscr{B}\mathscr{B}} , \tag{3.7}$$

and $J_{\mathscr{A}\mathscr{A}}$, $J_{\mathscr{B}\mathscr{B}}$, and $J_{\mathscr{A}\mathscr{B}}$ are Coulomb repulsion integrals;

$$J_{\mathscr{A}\mathscr{B}} = \langle \mathscr{A}(1)\mathscr{B}(2)| \frac{e^2}{r_{12}} |\mathscr{A}(1)\mathscr{B}(2)\rangle \tag{3.8}$$

represents the repulsion between two charge densities due to one electron in the orbital $\mathscr{A}$ and the other in orbital $\mathscr{B}$. $J_{\mathscr{A}\mathscr{A}}$ ($J_{\mathscr{B}\mathscr{B}}$) is the time-averaged repulsion of two electrons in orbital $\mathscr{A}$ ($\mathscr{B}$). The energies of configurations $|\mathscr{A}^2\rangle$, $|\mathscr{B}^2\rangle$, and $|\mathscr{A}\mathscr{B}\rangle$, given by Eqs. (3.3)–(3.5) are self-explanatory: The energy of $|\mathscr{A}^2\rangle$ or $|\mathscr{B}^2\rangle$ is twice the one-electron energy plus the time-averaged repulsion of two electrons in one orbital. The energy of $|\mathscr{A}\mathscr{B}\rangle$ is the sum of one-electron energies for orbitals $\mathscr{A}$ and $\mathscr{B}$ plus the electron repulsion integral due to one electron in $\mathscr{A}$ and one in $\mathscr{B}$. For evaluation of the off-diagonal elements, one needs three additional quantities:

$$h_{\mathscr{A}\mathscr{B}} = \langle \mathscr{A}|\hat{H}|\mathscr{B}\rangle , \tag{3.9}$$

the hybrid integral

$$(\mathscr{A}\mathscr{A}|\mathscr{A}\mathscr{B}) = \langle \mathscr{A}(1)\mathscr{A}(2)| \frac{e^2}{r_{12}} |\mathscr{A}(1)\mathscr{B}(2)\rangle , \tag{3.10}$$

and the exchange integral

$$K_{\mathscr{A}\mathscr{B}} = \langle \mathscr{A}(1)\mathscr{A}(2)| \frac{e^2}{r_{12}} |\mathscr{B}(1)\mathscr{B}(2)\rangle \tag{3.11}$$

representing the repulsion between the overlap of charge densities $e\mathscr{A}(1)\mathscr{B}(1)$ with $e\mathscr{A}(2)\mathscr{B}(2)$ due to first and second electron. Then, the matrix for four configurations (without spin consideration) takes the following form:

$$\begin{array}{c} |\mathscr{A}^2\rangle \\ |\mathscr{B}^2\rangle \\ |\mathscr{A}\mathscr{B}\rangle \\ |\mathscr{B}\mathscr{A}\rangle \end{array} \begin{pmatrix} 2h_{\mathscr{A}\mathscr{A}} + J_{\mathscr{A}\mathscr{A}} & K_{\mathscr{A}\mathscr{B}} & h_{\mathscr{A}\mathscr{B}} + (\mathscr{A}\mathscr{A}|\mathscr{A}\mathscr{B}) & h_{\mathscr{A}\mathscr{B}} + (\mathscr{A}\mathscr{A}|\mathscr{A}\mathscr{B}) \\ K_{\mathscr{A}\mathscr{B}} & 2h_{\mathscr{B}\mathscr{B}} + J_{\mathscr{B}\mathscr{B}} & h_{\mathscr{A}\mathscr{B}} + (\mathscr{B}\mathscr{B}|\mathscr{B}\mathscr{A}) & h_{\mathscr{A}\mathscr{B}} + (\mathscr{B}\mathscr{B}|\mathscr{B}\mathscr{A}) \\ h_{\mathscr{A}\mathscr{B}} + (\mathscr{A}\mathscr{A}|\mathscr{A}\mathscr{B}) & h_{\mathscr{A}\mathscr{B}} + (\mathscr{B}\mathscr{B}|\mathscr{B}\mathscr{A}) & h_{\mathscr{A}\mathscr{A}} + h_{\mathscr{B}\mathscr{B}} + J_{\mathscr{A}\mathscr{B}} & K_{\mathscr{A}\mathscr{B}} \\ h_{\mathscr{A}\mathscr{B}} + (\mathscr{A}\mathscr{A}|\mathscr{A}\mathscr{B}) & h_{\mathscr{A}\mathscr{B}} + (\mathscr{B}\mathscr{B}|\mathscr{B}\mathscr{A}) & K_{\mathscr{A}\mathscr{B}} & h_{\mathscr{A}\mathscr{A}} + h_{\mathscr{B}\mathscr{B}} + J_{\mathscr{A}\mathscr{B}} \end{pmatrix} . \tag{3.12}$$

the first two configurations are already spin adapted, and the linear combinations of the latter two $(\mathscr{AB} \pm \mathscr{BA})/\sqrt{2}$ leads to singlet and triplet wavefunctions $({}^{1,3}|\mathscr{AB}\rangle)$ for plus and minus sign, respectively. Since we treat $\mathscr{A}^2$ and $\mathscr{B}^2$ equally, it is reasonable to consider their linear combinations, as well $({}^1|\mathscr{A}^2 \pm \mathscr{B}^2\rangle)$.

For an arbitrary choice of orbitals $\mathscr{A}$ and $\mathscr{B}$ that we assume to be orthogonal $(S_{\mathscr{AB}} = 0)$, the Hamiltonian matrix in terms of this spin-adapted basis is

$$\begin{matrix} {}^1|\mathscr{A}^2 - \mathscr{B}^2\rangle \\ {}^1|\mathscr{A}^2 + \mathscr{B}^2\rangle \\ {}^1|\mathscr{AB}\rangle \\ {}^3|\mathscr{AB}\rangle \end{matrix} \begin{pmatrix} E(T) + 2K'_{\mathscr{AB}} & \delta_{\mathscr{AB}} & \gamma^-_{\mathscr{AB}} & 0 \\ \delta_{\mathscr{AB}} & E(T) + 2(K'_{\mathscr{AB}} + K_{\mathscr{AB}}) & \gamma_{\mathscr{AB}} & 0 \\ \gamma^-_{\mathscr{AB}} & \gamma_{\mathscr{AB}} & E(T) + 2K_{\mathscr{AB}} & 0 \\ 0 & 0 & 0 & E(T) \end{pmatrix} . \tag{3.13}$$

In the matrix [Eq. (3.13)] we use the following abbreviations:

$$\gamma_{\mathscr{AB}} = 2h_{\mathscr{AB}} + (\mathscr{AA}|\mathscr{AB}) + (\mathscr{BB}|\mathscr{BA}) , \tag{3.14}$$

$$\gamma^-_{\mathscr{AB}} = (\mathscr{AA}|\mathscr{AB}) - (\mathscr{BB}|\mathscr{BA}) , \tag{3.15}$$

$$\delta_{\mathscr{AB}} = h_{\mathscr{AA}} - h_{\mathscr{BB}} + (J_{\mathscr{AA}} - J_{\mathscr{BB}})/2 , \tag{3.16}$$

$$K'_{\mathscr{AB}} = [(J_{\mathscr{AA}} + J_{\mathscr{BB}})/2 - J_{\mathscr{AB}}]/2 , \tag{3.17}$$

$$E_0 = h_{\mathscr{AA}} + h_{\mathscr{BB}} + (J_{\mathscr{AA}} + J_{\mathscr{BB}})/4 + J_{\mathscr{AB}}/2 , \tag{3.18}$$

$$E(T) = E_0 - K'_{\mathscr{AB}} - K_{\mathscr{AB}} , \tag{3.19}$$

where $h_{\mathscr{AA}}$, $h_{\mathscr{BB}}$, $J_{\mathscr{AB}}$, $h_{\mathscr{AB}}$, $(\mathscr{AA}|\mathscr{AB})$, and $K_{\mathscr{AB}}$ have been defined by Eqs. (3.6)–(3.11). $J_{\mathscr{AB}}$ and $K_{\mathscr{AB}}$ are always positive for the real orbitals. E_0 is the average energy of four configurations: ${}^1|\mathscr{A}^2 - \mathscr{B}^2\rangle$, ${}^1|\mathscr{A}^2 + \mathscr{B}^2\rangle$, ${}^1|\mathscr{AB}\rangle$, and ${}^3|\mathscr{AB}\rangle$.

General formulation of the matrix for nonorthogonal orbitals $S_{\mathrm{AB}} \neq 0$ and the analytical solution will be discussed subsequently.

The quantities $\gamma_{\mathscr{AB}}$, $\gamma^-_{\mathscr{AB}}$, and $\delta_{\mathscr{AB}}$ represent deviations from an ideal biradical. $\gamma_{\mathscr{AB}}$ is a measure of the interaction between the orbitals $\mathscr{A}$ and $\mathscr{B}$ and will be important in consideration of gradual twisting of a bond. The quantity $\delta_{\mathscr{AB}}$ is a measure of the electronegativity difference between orbitals $\mathscr{A}$ and $\mathscr{B}$. It is equal to half of the energy difference of the configurations ${}^1|\mathscr{A}^2\rangle$ and ${}^1|\mathscr{B}^2\rangle$. Since we have chosen $\mathscr{A}$ to be always less electronegative than $\mathscr{B}$, $\delta_{\mathscr{A}\mathscr{B}} \geq 0$. This quantity is large for the dative bonds under consideration. $\gamma^{-}_{\mathscr{A}\mathscr{B}}$ has less obvious physical significance. It represents the deviation from the most localized choice of orbitals $\mathscr{A} = \mathrm{A}$ and $\mathscr{B} = \mathrm{B}$ and from the most delocalized choice of orbitals $\mathscr{A} = \mathrm{a}$ and $\mathscr{B} = \mathrm{b}$ (cf. Chart. I):

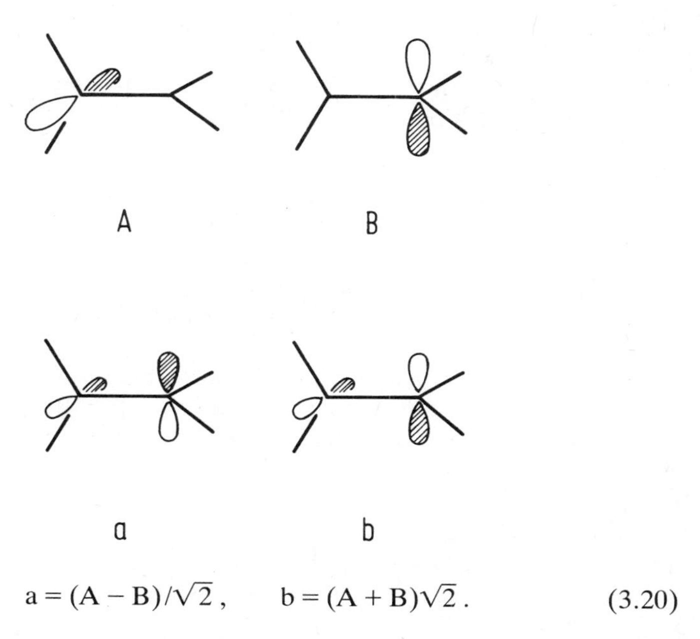

$$\mathrm{a} = (\mathrm{A} - \mathrm{B})/\sqrt{2}\,, \qquad \mathrm{b} = (\mathrm{A} + \mathrm{B})\sqrt{2}\,. \tag{3.20}$$

It is easy to show that for both limiting choices (the most localized and the most delocalized choice) $\gamma^{-}_{\mathscr{A}\mathscr{B}}$ vanishes.[14] The choice of orbitals has no significance, since we work with the exact solution of the 3×3 model. Nevertheless, it is sometimes more convenient to work with localized and sometimes with delocalized orbitals, as we will see.

It is worth mentioning that exchange and Coulomb integrals $K_{\mathscr{A}\mathscr{B}}$ and $J_{\mathscr{A}\mathscr{B}}$ are both minimized and maximized for the most localized and the most delocalized choice of orbitals, respectively. This is understandable since localized orbitals A and B try to avoid each other in the space in contrast to the delocalized orbitals a and b.

From any arbitrary choice of orbitals $\mathscr{A}$ and $\mathscr{B}$ it is possible to construct the most localized and the most delocalized orbitals through an unitary transformation. It is possible to show that the following quantities remain invariant to this transformation[14]:

$$\mathrm{E}_0,\; K_{\mathscr{A}\mathscr{B}} - J_{\mathscr{A}\mathscr{B}},\; K'_{\mathscr{A}\mathscr{B}} + K_{\mathscr{A}\mathscr{B}},\; (K'_{\mathscr{A}\mathscr{B}} - K_{\mathscr{A}\mathscr{B}})^2 + (\gamma^{-}_{\mathscr{A}\mathscr{B}})^2,\; \delta^2_{\mathscr{A}\mathscr{B}} + \gamma^2_{\mathscr{A}\mathscr{B}}\,.$$

It is clear that, for example, the interpretation of the quantity $K'_{\mathcal{AB}}$ depends on the choice of orbitals.

2. *Energies and Wavefunctions of Biradicaloids*

If the real orthogonal orbitals $\mathcal{A}$ and $\mathcal{B}$ interact, $\gamma_{\mathcal{AB}} \neq 0$, or have different energies, $\delta_{\mathcal{AB}} \neq 0$, or are both independent on the choice of orbitals, we refer to the system as a biradicaloid.

Since triplet and singlets do not interact, the wavefunctions for singlets $|S_0\rangle$, $|S_1\rangle$, and $|S_2\rangle$ obtained by the diagonalization of (3.13) are

$$|S_i\rangle = C_{i,-}{}^1|\mathcal{A}^2 - \mathcal{B}^2\rangle + C_{i,+}{}^1|\mathcal{A}^2 + \mathcal{B}^2\rangle + C_{i,0}{}^1|\mathcal{AB}\rangle\,, \qquad i = 0, 1, 2\,. \tag{3.21}$$

The off-diagonal elements of the matrix (3.13) contain quantities $\delta_{\mathcal{AB}}$, $\gamma_{\mathcal{AB}}$, and $\bar{\gamma}_{\mathcal{AB}}$. Since $\delta^2_{\mathcal{AB}} + \gamma^2_{\mathcal{AB}}$ is invariant to the unitary transformation, it is clear that for any perturbation an appropriate choice of $\mathcal{A}$ and $\mathcal{B}$ can make either $\delta_{\mathcal{AB}}$ or $\gamma_{\mathcal{AB}}$ vanish. This will be, of course, neither the most localized nor the most delocalized orbitals. On the other hand, since $\bar{\gamma}_{\mathcal{AB}}$ has no clear physical significance, the most convenient working choice of orbitals is the one for which $\bar{\gamma}_{\mathcal{AB}}$ vanishes. As was already mentioned, this is the case of the most localized $\mathcal{A} = \mathrm{A}$ and $\mathcal{B} = \mathrm{B}$ and the most delocalized orbitals $\mathcal{A} = \mathrm{a}$ and $\mathcal{B} = \mathrm{b}$. For these choices, consequently, we have to deal with two independent perturbations that are related to each other in two different basis as

$$\delta_{\mathrm{AB}} = \gamma_{\mathrm{ab}} \quad \text{and} \quad \gamma_{\mathrm{AB}} = \delta_{\mathrm{ab}}\,. \tag{3.22}$$

The simplest way to illustrate physical meaning of these quantities is to consider the perturbations of orthogonally twisted ethylene for which $\delta_{\mathrm{AB}} = \gamma_{\mathrm{AB}} = \delta_{\mathrm{ab}} = \gamma_{\mathrm{ab}} = 0$ holds via (1) return to planarity or (2) substitution at one end of the C=C bond. For (1), localized orbitals interact, $\gamma_{\mathrm{AB}} \neq 0$, but their energies are the same, $\delta_{\mathrm{AB}} = 0$. Since delocalized orbitals become eventually HOMO and LUMO of planar ethylene, they do not have the same energy, $\delta_{\mathrm{ab}} \neq 0$, but they do not interact, $\gamma_{\mathrm{ab}} = 0$. For (2), orthogonal-substituted ethylene, the situation is different. In the localized basis $\delta_{\mathrm{AB}} \neq 0$, but the interaction is not present due to the symmetry $\gamma_{\mathrm{AB}} = 0$. (A and $\mathcal{BB}$ belong to different irreducible representations.) For the delocalized description the energies of these orbitals are the same $\delta_{\mathrm{ab}} = 0$ since the orbitals are equally distributed over both carbon atoms. But $\gamma_{\mathrm{ab}} \neq 0$, since a and b are not canonical orbitals.

In the case of return to planarity for substituted twisted ethylene none of the perturbations vanish:

$$\delta_{AB} \neq 0\,, \quad \delta_{ab} \neq 0\,, \quad \gamma_{AB} \neq 0\,, \quad \text{and} \quad \gamma_{ab} \neq 0\,. \tag{3.23}$$

Since our primary interest concerns breaking of the dative bond, we will deal with *heterosymmetric biradicaloids* for orthogonally twisted geometries ($\delta_{AB} = \gamma_{ab} \neq 0$; $\gamma_{AB} = \delta_{ab} = 0$) and with *nonsymmetric biradicaloids* for the nonorthogonal geometries ($\delta_{AB} = \gamma_{ab} \neq 0$ and $\gamma_{AB} = \delta_{ab} \neq 0$). Before discussion of nonsymmetrical biradicaloids, we shall briefly address *homosymmetric biradicaloids* ($\gamma_{AB} = \delta_{ab} \neq 0$ and $\delta_{AB} = \gamma_{ab} = 0$). A twisting of the ethylenic double bond for angles different than 90° can serve as a prototype.

We will now use the localized basis A and B, since this choice of orbitals makes it possible to use the VB language, which, from the chemical point of view, is more instructive. In this sense we will name $^{1,3}|AB\rangle$ wavefunctions "dot–dot" structures, indicating the presence of unpaired electrons separated in space, and we will refer to $^{1}|A^2\rangle$ and $^{1}|B^2\rangle$ as to "hole–pair" structures.

3. *Heterosymmetric Biradicaloids*

The Hamiltonian matrix has the following form:

$$\begin{array}{l} ^{1}|A^2 - B^2\rangle \\ ^{1}|A^2 + B^2\rangle \\ ^{1}|AB\rangle \end{array} \begin{pmatrix} E(T) + 2K'_{AB} & \delta_{AB} & 0 \\ \delta_{AB} & E(T) + 2(K'_{AB} + K_{AB}) & 0 \\ 0 & 0 & E(T) + 2K_{AB} \end{pmatrix}. \tag{3.24}$$

Since $\gamma_{AB} = 0$, $^{1}|AB\rangle$ is an eigenstate that is characterized by the "dot–dot" structure with one electron in each of the two localized orbitals A and B. The other two functions mix, giving rise to states

$$|2\rangle = \cos\beta\, ^{1}|A^2 + B^2\rangle + \sin\beta\, ^{1}|A^2 - B^2\rangle\,, \tag{3.25}$$

$$|1\rangle = -\sin\beta\, ^{1}|A^2 + B^2\rangle + \cos\beta\, ^{1}|A^2 - B^2\rangle\,, \tag{3.26}$$

$$\beta = \tfrac{1}{2}\tan^{-1}(\delta_{AB}/K_{AB})\,. \tag{3.27}$$

Mixing of $^{1}|A^2 - B^2\rangle$ and $^{1}|A^2 + B^2\rangle$ is given by the parameter β. The degree of mixing, which corresponds to the degree to which the wavefunction becomes polar, is given by the perturbation δ_{AB}/K_{AB}. For

$\delta_{\mathrm{AB}} = 0$, $\beta = 0$ and

$$|2\rangle = {}^1|\mathrm{A}^2 + \mathrm{B}^2\rangle \,, \tag{3.28}$$

$$|1\rangle = {}^1|\mathrm{A}^2 - \mathrm{B}^2\rangle \,, \tag{3.29}$$

both states contain equal contributions of "hole–pair" structures and they are neutral. For very large δ_{AB}, $\beta = \pi/4$ and

$$|2\rangle = {}^1|\mathrm{A}^2\rangle \,, \tag{3.30}$$

$$|1\rangle = {}^1|\mathrm{B}^2\rangle \,, \tag{3.31}$$

and both these states are represented by the individual "hole–pair" structures.

Let us now address the energy ordering of these states:

$$E(\mathrm{S}_2) = E(\mathrm{T}) + 2K'_{\mathrm{AB}} + K_{\mathrm{AB}} + \sqrt{K^2_{\mathrm{AB}} + \delta^2_{\mathrm{AB}}} = E(|2\rangle) \,, \tag{3.32}$$

$$E(\mathrm{S}_0),\, E(\mathrm{S}_1) = \begin{cases} E(\mathrm{T}) + 2K'_{\mathrm{AB}} + K_{\mathrm{AB}} - \sqrt{K^2_{\mathrm{AB}} + \delta^2_{\mathrm{AB}}} = E(|1\rangle) \,, & (3.33) \\ E(\mathrm{T}) + 2K_{\mathrm{AB}} = E({}^1|\mathrm{AB}\rangle) \,. & (3.34) \end{cases}$$

Evidently, the states $|1\rangle$ and ${}^1|\mathrm{AB}\rangle$ can be either S_0 or S_1. Moreover, the energy gap $\mathrm{S}_0 - \mathrm{S}_1$ and the nature of the corresponding wavefunctions depend essentially on the perturbation $\delta_{\mathrm{AB}}/K_{\mathrm{AB}}$. For an appropriate choice of the perturbation, a touching of two states S_0, S_1 occurs if the following condition is satisfied[14]:

$$\delta_c = 2\sqrt{K'_{\mathrm{AB}}(K'_{\mathrm{AB}} - K_{\mathrm{AB}})} \,. \tag{3.35}$$

It is interesting to distinguish three cases: (1) "strong" heterosymmetric biradicaloid $\delta_{\mathrm{AB}} > \delta_c$, (2) "weak" heterosymmetric biradicaloid $\delta_{\mathrm{AB}} < \delta_c$, and (3) "critical" heterosymmetric biradicaloid $\delta_{\mathrm{AB}} = \delta_c$.

Since we are interested in the large electronegativity difference between orbitals A and B, let us consider first the case of $\delta_{\mathrm{AB}} > \delta_c$ in some detail: The singlet excited state S_1 is of "dot–dot" nature and the ground-state wavefunction is dominated by the "hole–pair" structure ${}^1|\mathrm{B}^2\rangle$ [cf. Eqs. (3.33) and (3.34)]. If

$$\delta_c < \delta_{\mathrm{AB}} > 2\sqrt{K'_{\mathrm{AB}}(K'_{\mathrm{AB}} + K_{\mathrm{AB}})} \,, \tag{3.36}$$

the energy of the triplet state $E(\mathrm{T})$ is considerably higher than the energy of the ground state $E(\mathrm{S}_0)$.

The question of whether those species can be considered as biradicaloids at all can be raised, especially since the ground-state wavefunction is characterized by the hole-pair structure $^1|B^2\rangle$ with the negative charge on B (usually called an ion-pair or zwitterionic state) and the energy of S_0 is considerably lower than $E(T)$. However, the condition (3.36) can be fullfilled only if the electronegativity difference δ_{AB} between orbitals A and B is sufficiently large, and this is usually the case when interaction between π-donor and π-acceptor is interrupted (twisted dative bonds). In such cases "dot–dot" structure $^1|AB\rangle$ involves formal charge separation $\cdot A^- B^+ \cdot$, so that the S_1 *state* has charge separation as well as separated unpaired electrons. Therefore, we refer to such states as to *charge-transfer biradicaloids*.[40] The wavefunction of the ground state S_0 dominantly represented by the "hole–pair" AB: structure $^1|B^2\rangle$ does not contain any separated formal charges.

In summary, the simple model yields a large S_0–S_1 energy gap if the condition (3.36) is satisfied. Moreover, the triplet state T does not lie close to S_0 but close to S_1, and this pair of close-lying states is characterized by "dot–dot" structures $^{1,3}|AB\rangle$ involving separated formal charges between donor and acceptor groups (TICT states). Such a situation is expected to be present for orthogonally twisted aminoborane[40] or for *p*-*N*,*N*-dimethylaminobenzonitrile (Chart II).

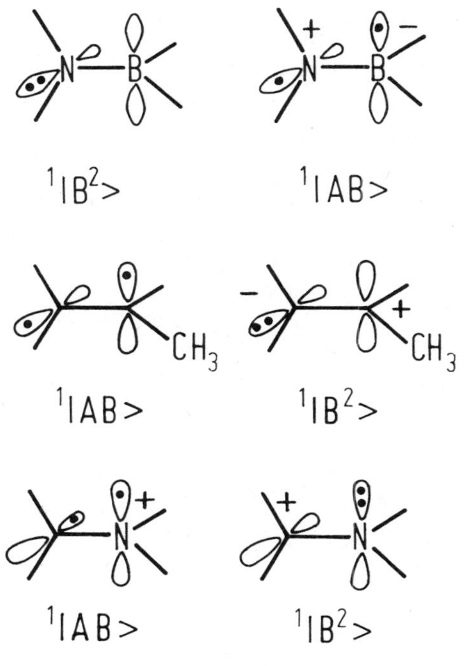

$^1|B^2\rangle$ $^1|AB\rangle$

$^1|AB\rangle$ $^1|B^2\rangle$

$^1|AB\rangle$ $^1|B^2\rangle$

A comparison with the case of "weak" heterosymmetric biradicaloids—case 2, $\delta_{AB} < \delta_c$—is straightforward. Appropriate examples are twisted propylene or twisted one-end pyramidalized ethylene. The electronegativity difference between A and B is relatively small. Consequently, from a comparison of equations (3.33) and (3.34) it follows that the ground state S_0 is of $^1|AB\rangle$ nature, $E(^1|AB\rangle) < E(|1\rangle)$. The "dot–dot" structure $^1|AB\rangle$ is of covalent nature since the electronegativity difference between A and B is relatively small and there is no formal separation of charges. The triplet state represented by $^3|AB\rangle$ lies close to the ground state $[E(S_0) - E(T) = 2K_{AB}]$. The singlet excited state S_1 is a mixture of the "hole–pair" structures $^1|A^2\rangle$ and $^1|B^2\rangle$ with the dominant contribution from $^1|B^2\rangle$. Since this structure contains strong charge separation (an extra electron on B), the S_1 state is polar. This situation is known in literature as "sudden polarization" effect.[39,106,110,111] The second excited state is polarized in opposite direction since the wavefunction $|2\rangle$ is dominated by the $^1|A^2\rangle$ "hole–pair" structure, with the lack of an electron (cf. Chart II). The energy difference between two excited states is

$$E(S_2) - E(S_1) = 2\sqrt{K_{AB}^2 + \delta_{AB}^2}\,. \tag{3.37}$$

The energy difference as well as degree of mixing between $^1|A^2 + B^2\rangle$ and $^1|A^2 - B^2\rangle$ are given by the perturbation parameter $\beta = \delta_{AB}/K_{AB}$ [Eq. (3.27)]. For small values of K_{AB} (twisted ethylene) a relatively small electronegativity difference δ_{AB} between orbitals A and B causes strong "polarization" of the S_1 and S_2 wavefunctions. For larger values of K_{AB}, in order to reach a given degree of polarization, δ_{AB} must increase correspondingly. In all cases of weak heterosymmetric biradicaloids the $S_0 - S_1$ energy gap is relatively large.

Finally, let us briefly consider case 3 for which the energy gap $S_0 - S_1$ vanishes ($\delta_{AB} = \delta_c$), which is not that interesting in the context of the TICT states but is of general interest for radiationless transitions, especially for photochemistry involving translocation of charges (Ref. 14 and references therein). In order to satisfy the condition (3.35) the electronegativity difference between orbitals A and B must be chosen so that the energy of the "dot–dot" structure $^1|AB\rangle$ is equal to the energy of the "hole–pair" structure $^1|B^2\rangle$. This is likely to occur when neither of these structures contains charge separation but rather charge translocation (cf. Ref. 14). The appropriate example is the orthogonally twisted protonated Schiff base H_2C–NH_2^+ (Chart II).

4. *Homosymmetric Biradicaloids*

The Hamiltonian matrix for the singlet state is

$$\begin{matrix} {}^1|A^2 - B^2\rangle \\ {}^1|A^2 - B^2\rangle \\ {}^1|AB\rangle \end{matrix} \begin{pmatrix} E(T) + 2K'_{AB} & 0 & 0 \\ 0 & E(T) + 2(K'_{AB} + K_{AB}) & \gamma_{AB} \\ 0 & \gamma_{AB} & E(T) + 2K_{AB} \end{pmatrix} . \tag{3.38}$$

Two orthogonal orbitals that interact, $\gamma_{AB} \neq 0$, are not completely localized; they correspond to Löwdin orthogonal orbitals. Two symmetrical wavefunctions mix (${}^1|A^2 + B^2\rangle$ and ${}^1|AB\rangle$) giving rise to S_0 and S_2 while antisymmetric wavefunction ${}^1|A^2 - B^2\rangle$ remains an eigenstate:

$$|S_2\rangle = \cos\alpha \, {}^1|A^2 + B^2\rangle + \sin\alpha \, {}^1|AB\rangle , \tag{3.39}$$

$$|S_1\rangle = {}^1|A^2 - B^2\rangle , \tag{3.40}$$

$$|S_0\rangle = -\sin\alpha \, {}^1|A^2 + B^2\rangle + \cos\alpha \, {}^1|AB\rangle , \tag{3.41}$$

$$\alpha = \tfrac{1}{2} \tan^{-1}(\gamma_{AB}/K'_{AB}) , \tag{3.42}$$

The mixing parameter α is determined by the perturbation given by γ_{AB}/K'_{AB}. For $\gamma_{AB} = 0$ (90° twisted ethylene), $S_0 = {}^1|AB\rangle$ and $S_2 = {}^1|A^2 + B^2\rangle$. For very large negative $\alpha = -\pi/4$, $S_0 = ({}^1|AB\rangle + {}^1|A^2 + B^2\rangle)/\sqrt{2}$ corresponding to the "closed-shell" ground state.

The corresponding energies are

$$E(S_2) = E(T) + K'_{AB} + 2K_{AB} + \sqrt{K'^2_{AB} + \gamma^2_{AB}} , \tag{3.43}$$

$$E(S_1) = E(T) + 2K'_{AB} , \tag{3.44}$$

$$E(S_0) = E(T) + K'_{AB} + 2K_{AB} - \sqrt{K'^2_{AB} + \gamma^2_{AB}} . \tag{3.45}$$

Note that S_0 is degenerated with T if $|\gamma_{AB}| = 2\sqrt{K_{AB}(K_{AB} + K'_{AB})}$. For larger values of $|\gamma_{AB}|$, $E(S_0) < E(T)$, which means deviation from biradicaloid geometries.

5. *Nonsymmetric Biradicaloids*

In this case both perturbations γ_{AB} and δ_{AB} are present, and we have mixing of all three singlet states, since the corresponding Hamiltonian matrix takes the following form:

$$\begin{matrix} {}^1|A^2 - B^2\rangle \\ {}^1|A^2 + B^2\rangle \\ {}^1|AB\rangle \end{matrix} \begin{pmatrix} E(T) + 2K'_{AB} & \delta_{AB} & 0 \\ \delta_{AB} & E(T) + 2(K'_{AB} + K_{AB}) & \gamma_{AB} \\ 0 & \gamma_{AB} & E(T) + 2K_{AB} \end{pmatrix}. \tag{3.46}$$

The energies can be obtained explicitly. Nevertheless, the inspection of graphical solutions for the equation

$$\begin{aligned} &[E - E(T) - 2K'_{AB} - K_{AB}]^2 - \delta^2_{AB} - K^2_{AB} \\ &\qquad = 2(K_{AB} - K'_{AB})\gamma^2/[E - E(T) - 2K_{AB}] + \gamma^2_{AB} \end{aligned} \tag{3.47}$$

is more instructive.

The solutions $E(S_0)$, $E(S_1)$, and $E(S_2)$ are the cuts of the parabola with the two branches of the hyperbola [left- and right-hand sides of Eq. (3.47), respectively]:

$$x^2 - (\delta^2_{AB} + K^2_{AB}) = 2(K_{AB} - K'_{AB})\gamma^2_{AB}/(x - K_{AB} + 2K'_{AB}) + \gamma^2_{AB}\,, \tag{3.48}$$

where $x = E - E(T) - 2K'_{AB} - K_{AB}$. For $K_{AB} - 2K'_{AB} < 0$, $E(S_0)$ lies on the left of the asymptote $x = K_{AB} - 2K'_{AB}$ and $E(S_1)$ and $E(S_2)$ lie on the right. From Fig. 3.1 it is easy to see how the energy gap $S_0 - S_1$ changes with changes in γ_{AB} and δ_{AB}. The perturbations that introduce mixing of the ${}^1|A^2 - B^2\rangle$, ${}^1|A^2 + B^2\rangle$, ${}^1|AB\rangle$ into the wavefunctions of $|S_0\rangle$, $|S_1\rangle$, and $|S_2\rangle$ states are now given by δ_{AB}, γ_{AB}, K_{AB}, and K'_{AB}.

So far, the preceding considerations have been very useful for characterization of states and for ordering of singlet-state energies relative to $E(T)$ with fixed values of K and K'. Nevertheless, we could not receive any qualitative ideas about the shapes of energy surfaces as the biradicaloids develop from the "normal" molecules. For this purpose the introduction of overlap, which we have neglected so far, is necessary. The Hamiltonian matrix for nonorthogonal orbitals **A** and **B** for nonsymmetric biradicaloid is

$$\begin{matrix} {}^1|\mathbf{A}^2 - \mathbf{B}^2\rangle \\ {}^1|\mathbf{A}^2 + \mathbf{B}^2\rangle \\ {}^1|\mathbf{AB}\rangle \end{matrix} \begin{pmatrix} (1 - S^2_{AB})^{-1}E^P_T + 2K'_{AB} & (1 - S^4_{AB})^{-1/2}\delta_{AB} & (1 - S^4_{AB})^{-1/2}\gamma^-_{AB} \\ (1 - S^4_{AB})^{-1/2}\delta_{AB} & (1 + S^2_{AB})^{-1}E^P_T + 2(K'_{AB} + K_{AB}) & (1 + S^2_{AB})^{-1}\gamma'_{AB} \\ (1 - S^4_{AB})^{-1/2}\gamma^-_{AB} & (1 + S^2_{AB})^{-1}\gamma'_{AB} & (1 + S^2_{AB})^{-1}E^P_T + 2\kappa_{AB} \end{pmatrix}, \tag{3.49}$$

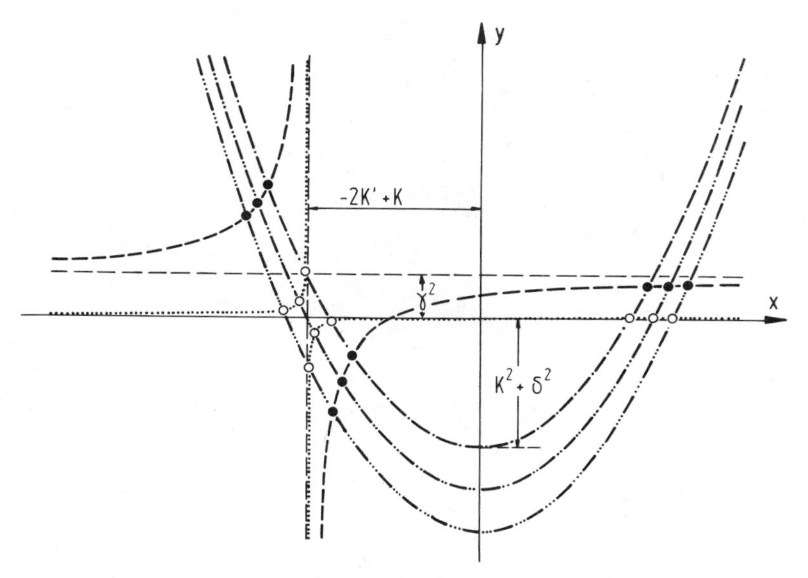

Fig. 3.1. Graphical solutions of Eq. (3.47) for two choices of γ_{AB}^2 and three choices of $K_{AB}^2 + \delta_{AB}^2$. Three solutions S_0, S_1, and S_2 for $\gamma_{AB}^2 \neq 0$ are indicated with solid circles for different choices of $K_{AB}^2 + \delta_{AB}^2$. In the case of $\gamma_{AB} = 0$, a hyperbola degenerates in a pair of lines and the $S_0 - S_1$ energy gap can become very small or zero. The solutions are indicated by open circles.

where

$$\kappa_{AB} = K_{AB} + 2h_{AB}S_{AB} \tag{3.50}$$

$$\gamma'_{AB} = \gamma_{AB} + S_{AB}(h_{AA} + h_{BB})/2 \tag{3.51}$$

$$E_T^P = h_{AA}^P + h_{BB}^P + J_{AB}^P - K_{AB}^P \tag{3.52}$$

E_T^P is the triplet energy at the orthogonal geometry $\gamma_{AB} = 0$. Unfortunately, the analytical, and even graphical, solutions of Eq. (3.49) are cumbersome and a discussion of them would not be very instructive. For qualitative purposes, serving as a guide for more complicated considerations, we can introduce the overlap into the expression for the triplet energy $E(\mathrm{T})$, which we used as the reference energy. Morevoer, it is known that the triplet energy is low at biradicaloid geometries and increases for the geometrical changes leading toward planarity. It is possible to show[14] that $E(\mathrm{T})$ can be approximated by

$$E(\mathrm{T}) \sim E_T^P + CS_{AB}^2/(1 - S_{AB}^2)\,, \tag{3.53}$$

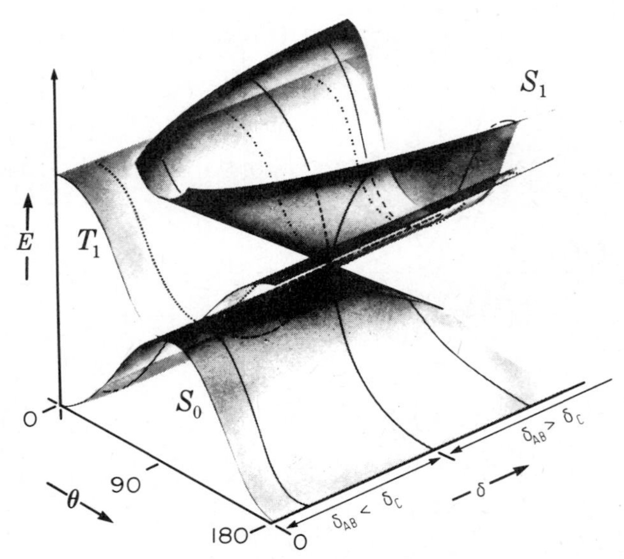

Fig. 3.2. Energies of the singlet and triplet states as a function of the twist angle θ and the electronegativity difference δ_{AB} (schematic).

where constant C contains the average electronegativity difference between two orbitals and has a positive value.

Since $S_{\mathbf{AB}}$ is proportional to the interaction $\gamma_{\mathbf{AB}}$ of two Löwdin orthogonalized orbitals **A** and **B**, we can conclude that since for $\gamma_{\mathbf{AB}} = 0$, $E(T) = E_T^P$, the triplet energy increases with increasing interaction $|\gamma_{\mathbf{AB}}| \neq 0$. Shapes of S_0, S_1, S_2, and T energy surfaces for different electronegativity differences δ are schematically drawn in Fig. 3.2.

C. Ab Initio CI Models of Simple Double, Charged, and Dative π-Bonds

Let us consider the influence of the point charge on the twist of the C=C bond in ethylene. The point charge is placed on the C=C axis at the distance $R = 1.8$ Å from the middle of the C=C bond and its magnitude is varied in order to simulate a variation of the electronegativity difference between two carbon atoms.

The results[14] of the large-scale CI[230] calculations employing an AO basis set of double-zeta quality are shown in Fig. 3.3. The quantities and concepts that we developed from the two-orbital two-electron model can be immediately recognized as the leading factors in largely correlated wavefunctions as well as in the shapes of the corresponding energy surfaces. The details are changed as expected. The analogies can be

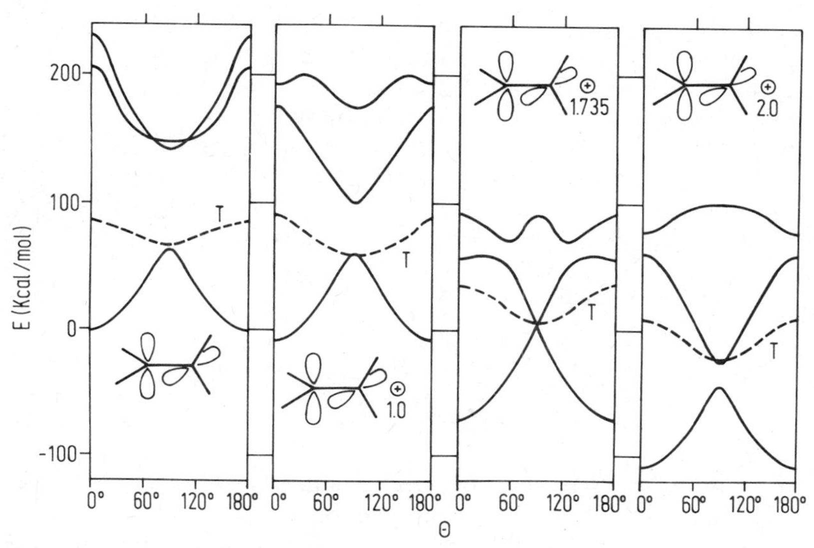

Fig. 3.3. Singlet and triplet state energies of ethylene in the presence of a positive charge (located on the C–C axis 1.85 Å from the midpoint of the bond) as a function of the twist angle obtained from the ab initio multireference double CI calculations. From left to right, no charge, $+1.0|e|$, $+1.735|e|$, and $+2.0|e|$. Huzinaga double-zeta-quality basis set, nine reference configurations, except for $q = 1.735|e|$ where 14 were used (6000 configurations total). Bond lengths are C–C = 1.416 Å and C–H = 1.09 Å, and bond angles = 120° were used corresponding to the S_1 minimum at $\theta = 90°$. No optimization of geometry has been carried out.

drawn between different values of δ and different values of the point charge $q = 1.0|e|$; $1.735|e|$; $2|e|$ as well as between the twist angle θ and the interaction γ (for $\theta = 90°$, $\gamma = 0$). At the 90° twist the analysis of correlated wavefunctions for S_0, S_1, and S_2 is straightforward. The canonical H-F HOMO and LUMO are localized in two parts of the molecule separated by the twisted bond. (A and B are now group orbitals.) The total contributions from three configurations that differ in occupation numbers of HOMO and LUMO vary from 80 to 90% in the wavefunctions of S_0, S_1, and S_2. Although the MOs are obtained from the H-F treatment of the whole molecule and in the CI treatment all electrons except four core electrons of carbon atoms are included, we can still use qualitatively the language of "dot–dot" and "hole–pair" structures developed from the simple model in order to characterize the important features of the wavefunction. Instead of discussing wavefunctions in terms of many thousands of configurations, the coefficients next to the leading configurations in the CI expansion can be transformed into the coefficients next to the VB-like structures [cf. Eqs. (3.57)–(3.59)]. Note that for a qualitative discussion of the wavefunction we do not carry out the full transformation into the VB wavefunctions. This would involve work with nonorthogonal basis and there would be overlap between VB structures. Therefore, from now on we will call "dot–dot" ${}^{1,3}|\mathrm{AB}\rangle$ and "hole–pair" ${}^{1}|\mathrm{A}^2\rangle$, ${}^{1}|\mathrm{B}^2\rangle$ VB-like structures. For the 90° twisted ethylene $\mathrm{S}_0 \approx {}^{1}|\mathrm{AB}\rangle$, $\mathrm{S}_1 \approx ({}^{1}|\mathrm{A}^2\rangle + {}^{1}|\mathrm{B}^2\rangle)/\sqrt{2}$ and $\mathrm{S}_2 \approx ({}^{1}|\mathrm{A}^2\rangle - {}^{1}|\mathrm{B}^2\rangle)/\sqrt{2}$, where

$$ {}^{1}|\mathrm{A}^2\rangle = |\cdots\rangle|\mathrm{A}\alpha\rangle|\mathrm{A}\beta\rangle\,, \tag{3.54} $$

$$ {}^{1}|\mathrm{B}^2\rangle = |\cdots\rangle|\mathrm{B}\alpha\rangle|\mathrm{B}\beta\rangle\,, \tag{3.55} $$

$$ {}^{1}|\mathrm{AB}\rangle = |\cdots\rangle[|\mathrm{A}\alpha\rangle|\mathrm{B}\beta\rangle - |\mathrm{A}\beta\rangle|\mathrm{B}\alpha\rangle]/\sqrt{2}\,. \tag{3.56} $$

The inclusion of other than two electrons in the CI inverses the energies of in- and out-of-phase linear combinations of the "hole–pair" structures[106,112] (compare first part of Fig. 3.3) so that at 90° twist Z state lies below V state in Mulliken notation in contrast to the results of the 3×3 CI model. For the 90° twisted ethylene in presence of the point charge $q = 1.0|e|$, $\mathrm{S}_0 \approx {}^{1}|\mathrm{AB}\rangle$, but $\mathrm{S}_1 \approx {}^{1}|\mathrm{B}^2\rangle$ with the large charge separation with an extra electron in the part of the molecule close to the charge. Both states characterized by ${}^{1}|\mathrm{AB}\rangle$ and ${}^{1}|\mathrm{B}^2\rangle$ have the same energies for $\theta = 90°$ and $q = 1.735|e|$. With increasing value of charge, $\mathrm{S}_0 \approx {}^{1}|\mathrm{B}^2\rangle$ and $\mathrm{S}_1 \approx {}^{1}|\mathrm{AB}\rangle$. For sufficiently large values of q, the external field has polarized ethylene so strongly that the two parts of the molecule can be viewed as a donor linked to an acceptor. At the 90° twist S_1 is of "charge-transfer biradicaloid" nature and can be considered as a TICT state.

Moreover, a comparison of the four cases given in Fig. 3.3 is also instructive as a model for examining the role of a polar solvent on the twisted species or along the twisting coordinate. Since ethylenic excited states S_1, S_2 are highly polarizable,[39,105] they can become polar under the influence of geometrical changes, substitution, or solvent effects. Two cases of weak and strong perturbation should nevertheless be distinguished in connection with breaking of a double and a "partially double," or a single bond, respectively. In the former case S_1 exhibits "sudden polarization" and is characterized by the "hole–pair" VB-like structure ${}^{1}|\mathrm{B}^2\rangle$ with an extra electron in the B part of molecule. In the latter case S_1 TICT state is characterized by the "dot–dot" VB-like structure ${}^{1}|\mathrm{AB}\rangle$, which has one unpaired electron in A and another unpaired electron in B, but is also polar since the charges are formally separated on A and B.[14,40]

An additional illustration of the preceding principles is given in Fig. 3.4, which contains the results of ab initio large-scale CI calculations for three singlets and one triplet state of propylene,[107] protonated Schiff base (formaldiminium cation),[108] and aminoborane.[40] As expected, the triplet state of the 90° twisted aminoborane lies close to, even slightly above, the first excited state S_1. Since this systems serves as a prototype for the TICT states, good quality calculations have been carried out (including the Rydberg orbitals in the AO basis).[40] Twisting motion is energetically unfavorable for the Rydberg states, which are low-lying excited states for the planar geometries of small molecules. Although the simple two-orbital two-electron model guides us in the discussion of the nature of the wavefunction, the necessity of large-scale CI calculations with adequate choice of the AO basis set should not be underestimated for determining proper energy ordering of excited states or barriers on their potential surfaces. Note that the good quality CI calculations often yield close-lying singlet–triplet pair of biradical states, the singlet state lower than the triplet[113–115] in contrast to the Hund's rule and to the simple 3×3 model. Both of the mentioned shortcomings of the 3×3 CI are connected with the wrong ordering of two states split by $2K_{AB}$ when K_{AB} is a small quantity (S_0, T and S_1, S_2). Since the sensitive relationship between

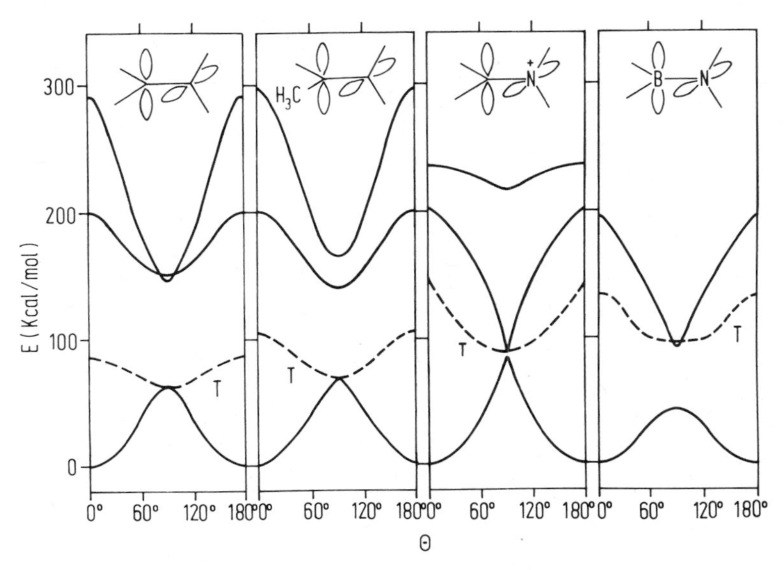

Fig. 3.4. Singlet and triplet energies of ethylene,[108] propene,[107] formaldiminium cation,[108] and aminoborane[40] (from left to right) as functions of the twist angle obtained from ab initio multireference double CI calculations.

repulsion and exchange integrals plays a very important role for the twisted species, caution is necessary for using semiempirical methods even when they consider correlation effects to a reasonable extent.

In addition to energy surfaces, the analysis of the CI wavefunctions in the twisting interval 0–90° for the propylene is given in Fig. 3.5. The coefficients η_{i,A^2}, η_{i,B^2}, and $\eta_{i,AB}$ label contributions of two "hole–pair" and one "dot–dot" VB-like structures in the singlet states S_i $(i = 0, 1, 2)$ of propylene. Three coefficients for three singlet states are obtained from the following transformation[107,109]:

$$\eta_{i,A^2} = C_{i,a^2}\lambda_1^2 + C_{i,b^2}\lambda_3^2 + \sqrt{2}C_{i,ab}\lambda_1\lambda_3 , \tag{3.57}$$

$$\eta_{i,B^2} = C_{i,a^2}\lambda_2^2 + C_{i,b^2}\lambda_4^2 - \sqrt{2}C_{i,ab}\lambda_2\lambda_4 , \tag{3.58}$$

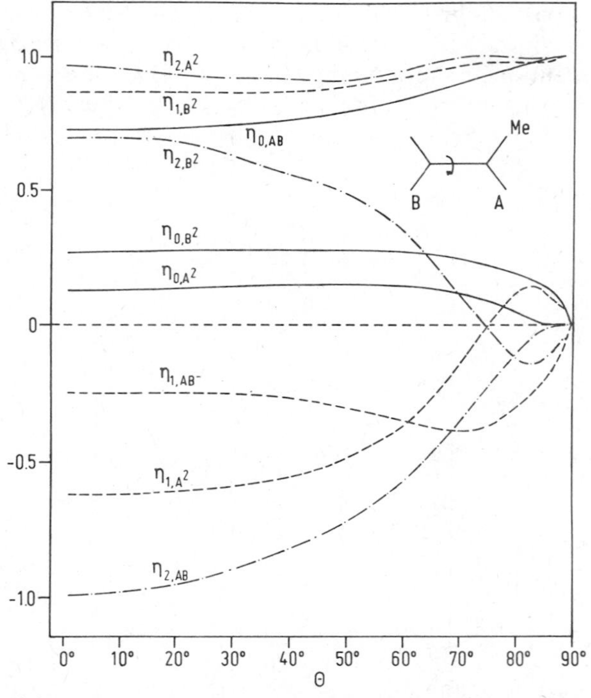

Fig. 3.5. Coefficients η_{k,A^2}, η_{k,B^2}, and $\eta_{k,AB}$ with $k = 0, 1, 2$ for S_0(–), S_1(– – –), and S_2(– · – · –) states of propene from the expansion of wavefunctions ψ_k in terms of group orbital configurations ("VB-like structures") $|A^2\rangle$, $|B^2\rangle$, and $|AB\rangle$ [see Eqs. 3.57–3.59] as a function of twist angle θ. Wavefunctions are from the large-scale CI treatment.[109]

$$\eta_{i,AB} = -\sqrt{2}C_{i,a^2}\lambda_1\lambda_2 + \sqrt{2}C_{i,b^2}\lambda_3\lambda_4 - C_{i,ab}(\lambda_2\lambda_3 - \lambda_1\lambda_4)\,, \quad (3.59)$$

where $i = 0, 1, 2$ and λ_1, λ_2, λ_3, and λ_4 are coefficients next to the group orbitals localized in A and B parts of the molecule in HOMO and LUMO:

$$|a\rangle = \lambda_1|A\rangle - \lambda_2|B\rangle\,, \quad (3.60)$$

$$|b\rangle = \lambda_3|A\rangle + \lambda_4|B\rangle\,, \quad (3.61)$$

and C_{i,a^2}, C_{i,b^2}, and $C_{i,ab}$ are the coefficients in the expansion of correlated wavefunction for S_i ($i = 0, 1, 2$) next to the configurations that can be labeled as closed shell, singly excited (HOMO$\rightarrow$LUMO) and doubly excited (HOMO$\rightrightarrows$LUMO).

In this manner it is easy to follow development of the dominantly "covalent" character of the ground state S_0 with increasing twist angle ($\eta_{0,AB}$ large) as well as the increasing "polarity" of S_1 and S_2 states due to increasing values of η_{1,B^2} and η_{2,A^2}, respectively. An avoided crossing between S_1 and S_2 is present in the neighborhood of $\theta \sim 80°$ due to the change in sign of η_{1,A^2} and η_{2,B^2} but the direction of opposite dipole moments in S_1 and S_2 remains unchanged because η_{1,B^2} is dominant in S_1 and η_{2,A^2} prevails in S_2.

D. Complex Dative π-Bond

1. *Free Molecules*

The donor B as well as the acceptor A subunits of the system typically have locally excited states that have to be considered in addition to the charge-separated states which we have been discussing. Locally excited states of corresponding ions might also have relatively low energies. Therefore, as the starting point several zero-order states have to be considered, for example,

$$AB\,, \quad A^{\overline{\cdot}}B^{\dot{+}}\,, \quad A^*B\,, \quad AB^*\,, \quad A^{\overline{\cdot}*}B^{\dot{+}*}\,, \quad A^{\overline{\cdot}}B^{\dot{+}*}\,, \quad A^{\overline{\cdot}*}B^{\dot{+}}\,.$$

We have shown that the energy of the charge-separated state is favored for twisted geometries, but the energies of many other zero-order states are lower at the planar than at the twisted geometries, similarily to the ground-state case. Therefore, crossings of zero-order states are likely to occur along the twist, which will be strongly avoided when the interaction is allowed for. In order to consider the interaction among zero-order states, the matrix elements among all considered configurations must be constructed as we did for the two-orbital two-electron model. The interaction among charge-separated and local-excited configurations will be studied in the next section. After solving the secular equation, the wavefunctions of the ground and excited states, which can then be a mixture of considered zero-order states (configurations) of the same symmetry, are constructed. Figure 3.6 schematically shows the ground state as well as two singlets and triplets for several cases differing in relative position of locally excited and charge-separated zero-order states. In principle, several close-lying locally excited states might be present. Also, the detailed specification of locally excited states (L_a or L_b in the case of aromatic subunits) has not been made on purpose in order to emphasize a concurrence between two different types of states: locally excited and charge-separated states for mechanism of dual fluorescence. As illustrated in Fig. 3.6 going from left to right, depending on the relative position of locally excited and charge-separated states there might be two minima on S_1 energy surface at planar and twisted geometries separated or not separated by a barrier. At planar S_1 minima the corresponding wavefunctions should be of locally excited nature and at 90° twisted minimum of "dot–dot" types $A^{\overline{\cdot}} B^{\dot{+}}$ characterized by a charge separation. By lowering the relative energy of the state characterized by $A^{\overline{\cdot}} B^{\dot{+}}$ with respect to the energy of the locally excited state changing donor–acceptor subunits appropriately the chances for the minimum in S_1 at the 90° twist becomes larger (left to right of Fig. 3.6). Increasing the

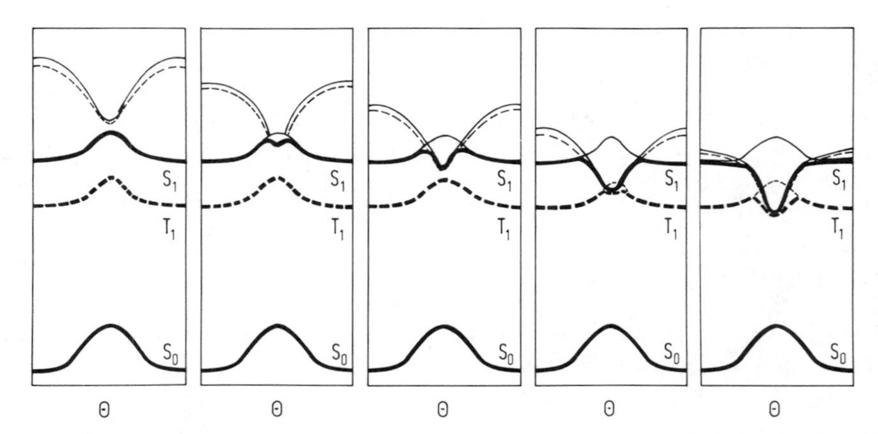

Fig. 3.6. Schematic energies of zero-order singlet states (thin solid lines) and of final states (bold solid lines) as a function of the twist angle θ. Triplet states are indicated by dashed thin and heavy lines. From left to right the energies of the singlet and triplet zero-order states are lower with respect to the locally excited states.

electron affinity of the acceptor A and decreasing the ionization potential of donor B but keeping the energy of locally excited states almost constant, the twisted minimum will develop and become deeper. This minimum can be separated by a large barrier from the minimum at planar geometry if the relative energy of charge-separated to locally excited state is large. If the choice of donor–acceptor subunits lowers substantially this relative energy, the barrier might completely disappear.

It is interesting to consider also the triplet states. We have shown that T_1 lies close to S_1 for TICT states of a simple dative bond but locally excited triplets certainly lie well above locally excited singlets. Therefore, a very strong donor–acceptor combination in the case of the complex dative bond will be needed in order to obtain minimum in T_1 for twisted geometries.

It is possible to assume that the absorption intensity originates mainly from the locally excited zero-order state since the $\cdot A^- B^+ \cdot$ zero-order state should barely contribute at all owing to the small overlap density of donor and acceptor orbitals. Therefore, the S_1 or T_1 should be reached at planar geometries by vertical excitation into the dominantly locally excited states. Only for the best donor–acceptor subunit combinations and high-lying locally excited states, a motion from planar minimum toward twisted minimum in S_1 or T_1 will be barrierless. Otherwise, a thermal activation is needed to overcome the barrier caused by crossings of zero-order states.

After the TICT minimum is reached, the transition moment between charge-separated state and the ground state represented by $A^{\overline{\cdot}} B^{\dot{+}}$ and AB:, respectively, is expected to be fairly small owing to almost no overlap between part A and B. Therefore, the fluorescence intensity will be small and significant contributions most probably stem from neighboring geometries ($\theta \neq 90°$) for which the emission from admixed locally excited states can occur. The return from S_1 or T_1 minimum to S_0, which can proceed in radiative or radiationless manner, usually does not lead to formation of *cis–trans* isomers as one would expect from the assumed energy surfaces. This is due most probably to rapid thermal *cis–trans* interconversion in the S_0 state. So far in this Section electronic properties of the free molecules have been addressed.

2. *Environmental Influence*

The important determining factor for the relative energy of the charge-separated state $A^{\overline{\cdot}} B^{\dot{+}}$ to the "hole–pair" state AB: is certainly the nature of the environment (polar solvent). Note that, although the emission from the TICT state has been observed in the gas phase, most of the observations have been made in polar solvents. We can again use the cases given in Fig. 3.6 but, now, keep the substrate fixed and change the solvent (from left to right). It is also instructive to describe the influence of polar solvent by adding to the substrate coordinate a single geometrical coordinate representing the collective effect of the orientation of the solvent molecules in the solvation shell. The range for the "solvation" variable starts with the point that represents the average solvent arrangement around the nonpolar ground state of the substrate and ends with the point representing the average solvent arrangement around the highly polar charge-separated state. The energy surface change along the solvation coordinate, shown in Fig. 3.7, is similar to the changes covering the cases from the left to the right of the Fig. 3.6. Development of the solute

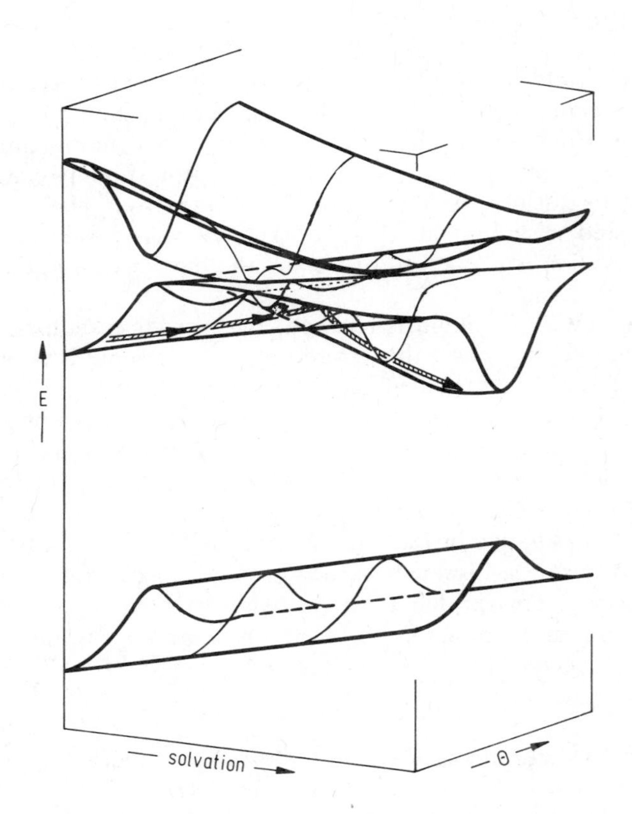

Fig. 3.7. Schematic two-dimensional energy plot for the singlet states as a function of solvation variable and the twist angle θ. The cases shown in Fig. 3.6 are represented from left to right along the solvation coordinate.

geometry toward a twisted minimum involves simultaneous twisting within the molecule as well as rotation and translation of the neighboring solvent molecules. Consequently, the rates will depend also on the specific properties of the solvent.

In summary, if the good combination of donor–acceptor subunits can be found so that the charge-separated state does not lie substantially higher than the locally excited states and if the polar solvent is adequately chosen to lower the energy of the charge-separated state, TICT state will be responsible for the anomalous fluorescence emission.

E. The Origin of the Dual Fluorescence of 9,9′-Bianthryl

Using simple principles developed previously, we take up the qualitative explanation of observed dual fluorescence for 9,9′-bianthryl[75–77] in polar solvents. In this case the solute contains two equal subunits orthogonally twisted to each other in the ground state. Both halves of the molecule have low-lying acceptor orbitals (high electron affinity) and high-lying donor orbitals (low ionization potential). The influence of solvent is necessary in order to break the symmetry so that one subunit of a molecule can act as an acceptor and the other subunit can act as a donor in a charge-separated orthogonally twisted state. For 9,9′-bianthryl the planar geometries are energetically unfavorable because of reverse steric hindrance. Let us draw the analogy to the appearance of the "sudden polarization" effect in excited singlets S_1 and S_2 of 90° twisted, slightly one-end perturbed ethylene (pyramidalized ethylene). In this case we have states characterized by two "hole–pair" polar structures. Owing to small perturbation, their energy difference is small. At the 90° twist and close to it their interaction is negligible ($\gamma_{AB} = 0$) and S_1 is dominated by one, S_2 by another VB-like "hole–pair" structure giving rise to strong polarity in the small interval of the twist angle. Deviation to planarity makes mixing between two "hole–pair" structures stronger and the polarity of S_1 and S_2 disappears gradually. For 9,9′-bianthryl two "dot–dot" structures $C_{14}H_9^{\dot{-}}$–$C_{14}H_9^{\dot{+}}$ and $C_{14}H_9^{\dot{+}}$–$C_{14}H_9^{\dot{-}}$ have equal energies in an isolated perfectly symmetrical molecule.

Now the geometrical changes and polar solvent can remove this isoenergetic situation, stabilizing one charge-separated "dot–dot" structure over the other by introducing polarity into the corresponding wavefunction. The analogy to ethylene is, of course, not complete because here we have to consider locally excited configurations of similar energies: $C_{14}H_9^*$–$C_{14}H_9$, $C_{14}H_9$–$C_{14}H_9^*$.

Consequently, in order to discuss qualitatively the nature and the energy of the first excited state S_1, mixing of four configurations schematically shown in Chart III have to be considered for the free

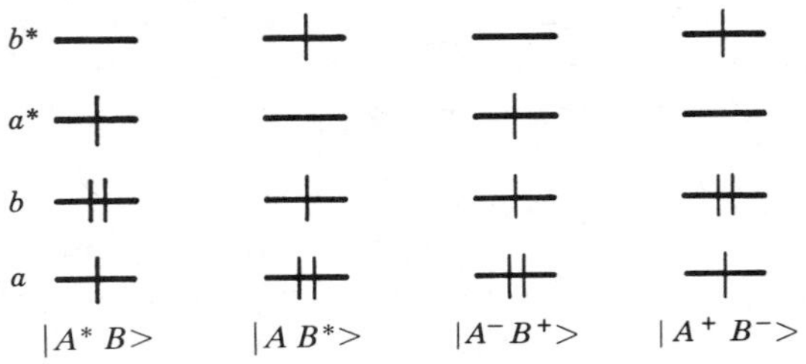

b^*, a^*, b, a

$|A^* B\rangle$ $|A B^*\rangle$ $|A^- B^+\rangle$ $|A^+ B^-\rangle$

molecule:

$$\begin{matrix} |A^*B\rangle \\ |A^{\cdot -}B^{\cdot +}\rangle \\ |A^{\cdot +}B^{\cdot -}\rangle \\ |AB^*\rangle \end{matrix} \begin{pmatrix} E_{\text{loc}} & H_3 & H_4 & H_1 \\ H_3 & E_{\text{CT}} & H_2 & H_3 \\ H_4 & H_2 & E_{\text{CT}} & H_4 \\ H_1 & H_3 & H_4 & E_{\text{loc}} \end{pmatrix} . \tag{3.62}$$

In this simple qualitative consideration we have neglected the interaction with the ground-state configuration $|AB\rangle$. Let us for simplicity assume that the overlap between the two subunits is negligible. Although both subunits are equal, we will keep notation A and B for them. The excitation within subunits A and B is labeled by A^* and B^*, respectively, and is represented by promoting one electron from the MO a to the MO a^* and from the MO b to the MO b^*, respectively. The matrix element between two locally excited configurations is

$$\begin{aligned} H_1 &= \langle A^*B|\hat{H}|AB^*\rangle = \langle {}^{1,3}\cdots b^2aa^*|\hat{H}|{}^{1,3}\cdots a^2bb^*\rangle \\ &= \tfrac{1}{2}\langle\{|b\bar{b}a\bar{a}^*| \pm |b\bar{b}a^*\bar{a}|\}|H|\{|a\bar{a}b\bar{b}^*| \pm |a\bar{a}b^*\bar{b}|\}\rangle \\ &= -\langle a(1)b(1)|\frac{e^2}{r_{12}}|a^*(2)b^*(2)\rangle + \langle a(1)a^*(1)|\frac{e^2}{r_{12}}|b(2)b^*(2)\rangle \\ &\quad \pm \langle a(1)a^*(1)|\frac{e^2}{r_{12}}|b(2)b^*(2)\rangle . \end{aligned} \tag{3.63}$$

We have written out only parts of the configurations that differ in occupation number of two electrons within subunit A and within subunit B. Positive signs refer to singlets, negative signs to triplets. The occupancy of an orbital by the spin β is indicated by an overbar.

Since we neglected overlap between orbitals of subunits, overlap densities $a(1)b(1)$ and $a^*(2)b^*(2)$ are negligible. The matrix element H_1 for the singlet state can be approximated by the electrostatic interaction between transition densities aa^* on subunit A and bb^* on subunit B:

$$H_1(\text{S}) = \langle A^*B|\hat{H}|AB^*\rangle \approx 2\langle a(1)a^*(1)|\frac{e^2}{r_{12}}|b(2)b^*(2)\rangle , \tag{3.64}$$

$$H_1(\mathrm{T}) = 0\,. \tag{3.65}$$

For orthogonal geometries $H_1(\mathrm{S}) = 0$.

The matrix element between two "dot–dot" configurations $\cdot A^- B^{\ddot{+}}$ and $\cdot A^+ B^{\ddot{-}}$ representing charge transfer to A and to B, respectively, is

$$\begin{aligned} H_2 &= \langle A^- B^+ | \hat{H} | A^+ B^- \rangle = \langle {}^{1,3} \cdots a^2 b a^* | \hat{H} |{}^{1,3} \cdots b^2 a b^* \rangle \\ &= \tfrac{1}{2} \langle \{ |a\bar{a}b\bar{a}^*| \pm |a\bar{a}a^*\bar{b}| \} | H | \{ |b\bar{b}a\bar{b}^*| \pm |b\bar{b}b^*\bar{a}| \} \rangle \\ &= \langle a(1)b(1) | \frac{e^2}{r_{12}} | a^*(2)b^*(2) \rangle\,. \end{aligned} \tag{3.66}$$

If the overlap between two subunits is negligible, the overlap densities $a(1)b(1)$ and $a^*(2)b^*(2)$ are negligible and H_2 matrix element vanishes as well.

The matrix elements between "dot–dot" and locally excited configurations are:

$$\begin{aligned} H_3 &= \langle A^- B^+ | \hat{H} | A^* B \rangle = \tfrac{1}{2} \langle \{ |a\bar{a}b\bar{a}^*| \pm |a\bar{a}a^*\bar{b}| \} | \hat{H} | \{ b\bar{b}a\bar{a}^* \pm |b\bar{b}a^*\bar{a}| \} \rangle \\ &= -\mathcal{E}(a, b) + \langle a(1)a^*(1) | \frac{e^2}{r_{12}} | b(2)a^*(2) \rangle \\ &\pm \langle a(1)a^*(1) | \frac{e^2}{r_{12}} | b(2)a^*(2) \rangle \\ &- \langle a(1)b(1) | \frac{e^2}{r_{12}} | a^*(2)a^*(2) \rangle \end{aligned} \tag{3.67}$$

and

$$\begin{aligned} H_4 &= \langle A^- B^+ | \hat{H} | A B^* \rangle = \tfrac{1}{2} \langle \{ |a\bar{a}b\bar{a}^*| \pm |a\bar{a}a^*\bar{b}| \} | \hat{H} | \{ a\bar{a}b\bar{b}^*| \pm |a\bar{a}b^*\bar{b}| \} \rangle \\ &= \mathcal{E}(a^*, b^*) - \langle a^*(1)b^*(1) | \frac{e^2}{r_{12}} | b(2)b(2) \rangle \\ &+ \langle b(1)b^*(1) | \frac{e^2}{r_{12}} | b(2)a^*(2) \rangle \\ &\pm \langle b(1)b^*(1) | \frac{e^2}{r_{12}} | b(2)a^*(2) \rangle\,, \end{aligned} \tag{3.68}$$

where $\mathcal{E}(a, b)$ and $\mathcal{E}(a^*, b^*)$ are off-diagonal elements of the Fock operator and they do not vanish since a, a^*, b, and b^* are MOs of the two subunits and not canonical orbitals of the whole system. In the expressions for H_3 and H_4 only two-electron integrals in which one of the two charge densities representing the overlap of MOs located in two subunits have been kept. It is clear that for orthogonal geometries H_3 and H_4 are relatively small quantities.

Let us now consider influence of the solvent on the energies of charge-separated configurations. If the dipole moment of the $\langle A^{\overline{\cdot}} B^{\dot{+}} \rangle$ is labeled by $\boldsymbol{\mu}_0$, and if the molecule is placed in a solvent, its dipole moment will give rise to the reaction field

$$\mathbf{F} = \boldsymbol{\mu}_0 f_\varepsilon , \tag{3.69}$$

where f_ε might be approximated as (cf. Onsager model):

$$f_\varepsilon = \frac{2}{c^3} \frac{\varepsilon - 1}{2\varepsilon + 1} , \tag{3.70}$$

where ε is the dielectric constant of the solvent and c is equilvalent to the sphere radius of the solvent cavity. The reaction field will influence energies E_{CT}, stabilizing one of them and destabilizing the other one, but the energy of locally excited states will remain almost unaffected. The Hamiltonian matrix for configurations of interest can be written as

$$\begin{pmatrix} E_{\mathrm{loc}} & H_3 & H_4 & 0 \\ H_3 & E_{\mathrm{CT}} + \mu_0^2 f_\varepsilon & 0 & H_3 \\ H_4 & 0 & E_{\mathrm{CT}} - \mu_0^2 f_\varepsilon & H_4 \\ 0 & H_3 & H_4 & E_{\mathrm{loc}} \end{pmatrix} . \tag{3.71}$$

For the molecule in the locally excited state the influence of polar solvent is negligible and the terms due to the reaction field of the surroundings are absent in the Hamiltonian matrix. Therefore, the solutions for S_1 obtained by diagonalization of matrix (3.71) and by diagonalization of an analogous matrix without $\mu_0^2 f_\varepsilon$ present should be compared in energy. If the former has lower energy, the S_1 is of charge-separated nature. $E_{\mathrm{CT}} - E_{\mathrm{loc}}$ will depend on H_3, H_4, and the strength of the polar solvent. For the 9,9′-bianthryl even if H_3 and H_4 increase for slight deviations from orthogonal geometries, the S_1 state will have charge-separated nature as long as the dipole solvation energy is sufficient to keep E_{CT} lower than E_{loc}.

These oversimplified considerations illustrate clearly that, although we do not have π-donor linked to π-acceptor, if the low-lying excited states are of polarizable nature, they can become of charge-transfer nature under the influence of a polar solvent and they are good candidates for solvent-dependent fluorescence.

At this point the question can be raised if an excited state of nonpolar excimers can also develop charge-separation nature under the influence of polar solvent as is the case of 9,9′-bianthryl. The interaction between four important configurations is again given by the matrix (3.62). The basic differences between these two cases is due to different conditions for polarizability of excited states of the systems with π-interacting subunits and the systems with σ-interacting subunits. In the case of 9,9′-bianthryl, the interaction between two π subunits is broken by twisting, while in the case of excimers two partners have to be separated by almost infinite distances in order to break the σ-interaction. But this would destroy excimers and produces two molecules as ion pairs. Consequently, the matrix element H_1 does not vanish for excimers as was the case for 9,9′-bianthryl. H_3 and H_4 are also no longer small, because they are proportional to the overlap. Therefore, the mixing of locally excited and charge-transfer configurations will be large and the dipole moments of the resulting states are not any longer 0 and $\boldsymbol{\mu}_0$. Therefore, it is to be expected that an excimer S_1 state is not of charge-separated nature even in the very strong polar solvent as long as two components interact.[116] If the excimer is destroyed by pulling apart two of the partners, H_1, H_3, and H_4 will be small. Energies of locally excited and charge separated configurations increase. Of course, the influence of a polar solvent can stabilize partially energies of the charge-separated configurations so that, in principle, an ion pair can be formed from the excimer. But, usually, the process is exothermic in the opposite direction. In other words, free radical anions $A^{\overline{\cdot}}$ and cations $B^{\dot{+}}$ diffuse together giving rise to excimers that are then detected by their fluorescence.

So far, the knowledge about the electronic structure of excited states has helped to identify the important factors responsible for dual fluorescence, but dynamical aspects of the process remain open and will be treated in Section IV.

IV. STOCHASTIC DESCRIPTION OF CHEMICAL REACTIONS

A. Introduction

This section consists of two major parts. The first part (sections B through G) is a review of the statistical description of the Markov processes and of the Langevin equation. The topics considered here are the fundamental definitions and properties that play a central role in the understanding of the theoretical models used to describe unimolecular reactions. Section

B introduces the formal and general properties of Markov processes with particular reference to the Chapman–Kolmogorov and Kolmogorov–Feller relationships governing the time evolution of the transition probability densities. Section C gives a brief discussion of the master equation. The applicatory value of the formal equations previously derived is illustrated in Section D by studying birth–death processes, especially those connected with reversible and irreversible unimolecular reactions. It is also shown how the formal equations of Section B may be useful in analyzing the Poisson process and the one-dimensional random walk. Section E gives an account of diffusion processes; Kolmogorov's backward and forward equations are established. The latter expression is most commonly known as the Fokker–Planck equation. The approximation of the master equations by Fokker–Planck equations is discussed briefly in Section F by introducing the Kramers–Moyal expansion. Section G investigates the Langevin equation and its connection with the diffusion process. Brownian motion is examined in the low- and high-friction situations. The stochastic differential equations method is introduced and extended to the derivation of the Smoluchowski and Kramers equations, which may be regarded as the bases of discussion of chemical reactions.

The second part (sections H and I) is devoted to a detailed discussion of the dynamics of unimolecular reactions in the presence and the absence of a potential barrier. Section H presents a critical examination of the Kramers approach. It is stressed that the expressions of the reaction rates in the low-, intermediate-, and high-friction limits are subjected to restrictive conditions, namely, the high barrier case and the quasi-stationary regime. The dynamics related to one-dimensional diffusion in a bistable potential is analyzed, and the exactness of the time dependence of the reaction rate is emphasized. The essential results of the non-Markovian theory extending the Kramers conclusions are also discussed. The final section investigates in detail the time evolution of an unimolecular reaction in the absence of a potential barrier. The formal treatment makes evident a two-time-scale description of the dynamics.

B. Markov Processes

1. *Definition of Stochastic Processes*

A physical system S that evolves probabilistically in time can be mathematically described by a time-dependent random variable $X(t)$. It is assumed that (1) one can measure values $x_1, x_2, x_3, \ldots, x_n$ of $X(t)$ at instants $t_1, t_2, t_3, \ldots, t_n$ representing the possible states of the system S and (2) one can define a set of joint probability distribution functions

$$F(x_1, t_1; x_2, t_2; \dots; x_n, t_n) = \Pr[X(t_1) < x_1; X(t_2) < x_2; \dots; X(t_n) < x_n] \quad (4.1)$$

that describe the system completely.[117-120]

The state space, that is, the set of possible states of S, as well as the time, can be continuous or discrete; it is usual to consider only increasing times $t_1 \leq t_2 \leq \cdots \leq t_k \leq \cdots \leq t_n$.

Two outcomes $y(t)$ and $z(t)$ of the process $X(t)$ are shown in Figs. 4.1*a* and 4.1*b* corresponding, respectively, to the discrete and continuous cases. The joint density function is obtained by differentiating Eq. (4.1) with respect to $x_1, x_2, \dots, x_n$:

$$f(x_1, t_1; x_2, t_2; \dots; x_n, t_n) = \frac{\partial^n}{\partial x_1 \, \partial x_2 \cdots \partial x_n} F(x_1, t_1; x_2, t_2; \dots; x_n, t_n) \,. \quad (4.2)$$

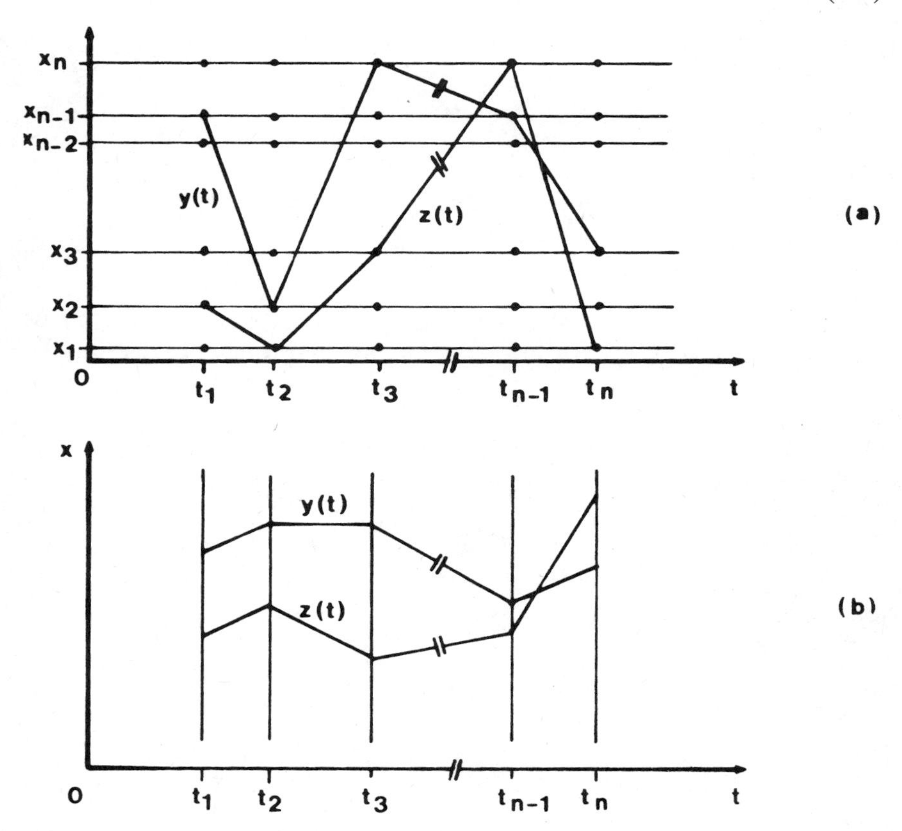

Fig. 4.1. Stochastical process $X(t)$; (*a*) discrete case; (*b*) continuous case.

More precisely, for a continuous process, one has

$$\Pr[x_1 \le X(t_1) < x_1 + \Delta x_1; x_2 \le X(t_2) < x_2 + \Delta x_2; \ldots; x_n \le X(t_n) < x_n + \Delta x_n] = f(x_1, t_1; x_2, t_2; \ldots; x_n, t_n)\, \Delta x_1\, \Delta x_2 \cdots \Delta x_n\,. \tag{4.3}$$

If one integrates the left-hand side of Eq. (4.2) with respect to certain variables, one obtains the joint density of the remaining variables, for instance,

$$f(x_1, t_1; x_2, t_2) = \int \cdots \int f(x_1, t_1; x_2, t_2; \ldots; x_n, t_n)\, dx_3\, dx_4 \cdots dx_n \tag{4.4}$$

and particularly

$$f(x_k, t_k) = \int \cdots \int f(x_1, t_1; x_2, t_2; \ldots; x_k, t_k; \ldots; x_n, t_n) \times dx_1 \cdots dx_{k-1}\, dx_{k+1} \cdots dx_n\,, \tag{4.5}$$

which is the marginal density of the random variable $X(t_k)$. The probability to find the system at time t_k in the state x_k is given by

$$\Pr[x_k \le X(t_k) < x_k + \Delta x_k] = f(x_k, t_k)\, \Delta x_k\,. \tag{4.6}$$

The future development of the process taking into account the knowledge of the past can be analyzed by conditional probability distributions and densities, for example,

$$\begin{aligned}\Pr[X(t_n) < x_n / X(t_1) = x_1; \ldots; X(t_{n-1}) = x_{n-1}] &= F(x_n, t_n / x_1, t_1; x_2, t_2; \ldots; x_{n-1}, t_{n-1}) \\ &= \frac{F(x_1, t_1; x_2, t_2; \ldots; x_n, t_n)}{F(x_1, t_1; x_2, t_2; \ldots; x_{n-1}, t_{n-1})}\end{aligned} \tag{4.7}$$

and

$$\begin{aligned}f(x_n, t_n / x_1, t_1; x_2, t_2; \ldots; x_{n-1}, t_{n-1}) &= \frac{\partial}{\partial x_n} F(x_n, t_n / x_1, t_1; \ldots; x_{n-1}, t_{n-1}) \\ &= \frac{f(x_1, t_1; x_2, t_2; \ldots; x_n, t_n)}{f(x_1, t_1; x_2, t_2; \ldots; x_{n-1}, t_{n-1})}\,.\end{aligned} \tag{4.8}$$

By repeated application of Eq. (4.8), one finds

$$f(x_1, t_1; x_2, t_2) = f(x_2, t_2/x_1, t_1)f(x_1, t_1)\,,$$
$$f(x_1, t_1; x_2, t_2; x_3, t_3) = f(x_3, t_3/x_1, t_1; x_2, t_2)f(x_2, t_2/x_1, t_1)f(x_1, t_1)\,,$$
$$\vdots$$
$$\begin{aligned} f(x_1, t_1; x_2, t_2; \ldots; x_n, t_n) = {} & f(x_n, t_n/x_1, t_1; x_2, t_2; \ldots; x_{n-1}, t_{n-1}) \\ & \times f(x_{n-1}, t_{n-1}/x_1, t_1; x_2, t_2; \ldots; x_{n-2}, t_{n-2}) \\ & \times \cdots f(x_2, t_2/x_1, t_1)f(x_1, t_1)\,. \end{aligned} \tag{4.9}$$

A stochastic process is said to be independent if

$$f(x_1, t_1; x_2, t_2; \ldots; x_n, t_n) = \prod_{i=1}^{n} f(x_i, t_i)\,, \tag{4.10}$$

which means that the value of $X(t)$ at time t is independent of its value in the past or the future.

2. *Continuous and Discrete Markov Processes*

A stochastic process $X(t)$ is called a Markov process[119,121–123] if for every n and $t_1 \le t_2 \le \cdots \le t_n$

$$\begin{aligned} \Pr\,[X(t_n) < x_n/X(t_1) = x_1; X(t_2) = x_2; \ldots; X(t_{n-1}) = x_{n-1}] \\ = \Pr\,[X(t_n) < x_n/X(t_{n-1}) = x_{n-1}]\,. \end{aligned} \tag{4.11}$$

Equivalently, the conditional density obeys the following condition:

$$f(x_n, t_n/x_1, t_1; x_2, t_2; \ldots; x_{n-1}, t_{n-1}) = f(x_n, t_n/x_{n-1}, t_{n-1})\,. \tag{4.12}$$

The Markov property states that the conditional probability of $X(t_n)$ for given values of $X(t_1)$, $X(t_2)$, . . . , $X(t_{n-1})$ depends only on the most recent value $X(t_{n-1})$.

The Markov assumption asserts that the past and the future are statistically independent when the present is known; this means that

$$f(x_1, t_1/x_2, t_2; x_3, t_3/x_2, t_2) = f(x_1, t_1/x_2, t_2)f(x_3, t_3/x_2, t_2)\,. \tag{4.13}$$

Indeed, one has

$$\begin{aligned} f(x_1, t_1; x_2, t_2; x_3, t_3) &= f(x_1, t_1/x_2, t_2; x_3, t_3/x_2, t_2)f(x_2, t_2) \\ &= f(x_3, t_3/x_1, t_1; x_2, t_2)f(x_1, t_1/x_2, t_2)f(x_2, t_2)\,, \end{aligned} \tag{4.14}$$

and taking into account Eq. (4.12), Eq. (4.13) follows.

For a continuous or a discrete process, Eq. (4.13) is equivalent to, respectively,

$$\begin{aligned}\Pr[x_1 \le X(t_1) < x_1 + \Delta x_1 / X(t_2) = x_2; x_3 \le X(t_3) < x_3 + \Delta x_3 / X(t_2) = x_2] \\ = \Pr[x_1 \le X(t_1) < x_1 + \Delta x_1 / X(t_2) = x_2] \\ \times \Pr[x_3 \le X(t_3) < x_3 + \Delta x_3 / X(t_2) = x_2] \quad (4.15)\end{aligned}$$

and

$$\begin{aligned}\Pr[X(t_1) = x_1 / X(t_2) = x_2; X(t_3) = x_3 / X(t_2) = x_2] \\ = \Pr[X(t_1) = x_1 / X(t_2) = x_2] \Pr[X(t_3) = x_3 / X(t_2) = x_2] . \quad (4.16)\end{aligned}$$

Under the Markov assumption, the hierarchy of the joint probability densities [Eqs. (4.9)] describing the evolution of the system takes the following form:

$$\begin{aligned}f(x_1, t_1; x_2, t_2) &= f(x_2, t_2 / x_1, t_1) f(x_1, t_1) , \\ f(x_1, t_1; x_2, t_2; x_3, t_3) &= f(x_3, t_3 / x_2, t_2) f(x_2, t_2 / x_1, t_1) f(x_1, t_1), \\ &\vdots \\ f(x_1, t_1; x_2, t_2; \ldots ; x_n, t_n) &= f(x_n, t_n / x_{n-1}, t_{n-1}) \\ &\quad \times f(x_{n-1}, t_{n-1} / x_{n-2}, t_{n-2}) \\ &\quad \times \cdots f(x_2, t_2 / x_1, t_1) f(x_1, t_1) .\end{aligned} \quad (4.17)$$

These results clearly show that the time development of the system is entirely determined by the transition probability densities $f(x_k, t_k / x_{k-1}, t_{k-1})$ and the initial probability density of states $f(x_1, t_1)$.

a. The Chapman–Kolmogorov Equation. Assuming $t_3 \ge t_2 \ge t_1$,

$$f(x_2, t_2; x_3, t_3 / x_1, t_1) = \frac{f(x_1, t_1; x_2, t_2; x_3, t_3)}{f(x_1, t_1)} . \quad (4.18)$$

With the aid of Eq. (4.17) this relationship can be expressed simply

$$f(x_2, t_2; x_3, t_3 / x_1, t_1) = f(x_3, t_3 / x_2, t_2) f(x_2, t_2 / x_1, t_1) , \quad (4.19)$$

and the Chapman–Kolmogorov expression follows by integrating with respect to x_2:

$$f(x_3, t_3 / x_1, t_1) = \int f(x_3, t_3 / x_2, t_2) f(x_2, t_2 / x_1, t_1) \, dx_2 . \quad (4.20)$$

In the case where the states are of discrete type, one sees that

$$f(x_3, t_3/x_1, t_1) = \sum_{x_2} f(x_3, t_3/x_2, t_2) f(x_2, t_2/x_1, t_1) . \qquad (4.21)$$

In Fig. 4.2, two issues $y(t)$ and $z(t)$ are depicted starting at time t_1 from the same state x_1, passing through $y_2 \in (x_2, x_2 + \Delta x_2)$ at time t_2, and finally arriving at two different states x_3' and x_3'' at the instant t_3; Eq. (4.20) yields

$$f(x_3, t_3/x_1, t_1) \approx f(x_3, t_3/x_2, t_2) f(x_2, t_2/x_1, t_1) \Delta x_2 . \qquad (4.22)$$

In physical systems it can happen that the transition probability densities are homogeneous in time and/or in space. A stochastic process $X(t)$ is stationary if $X(t)$ and $X(t+\tau)$ obey the same probability laws for every τ; this means that all joint probability densities verify time translation invariance

$$f(x_1, t_1; x_2, t_2; \ldots ; x_n, t_n) = f(x_1, t_1 + \tau; \ldots ; x_n, t_n + \tau) , \qquad (4.23)$$

which leads to the following special results: the initial probability density

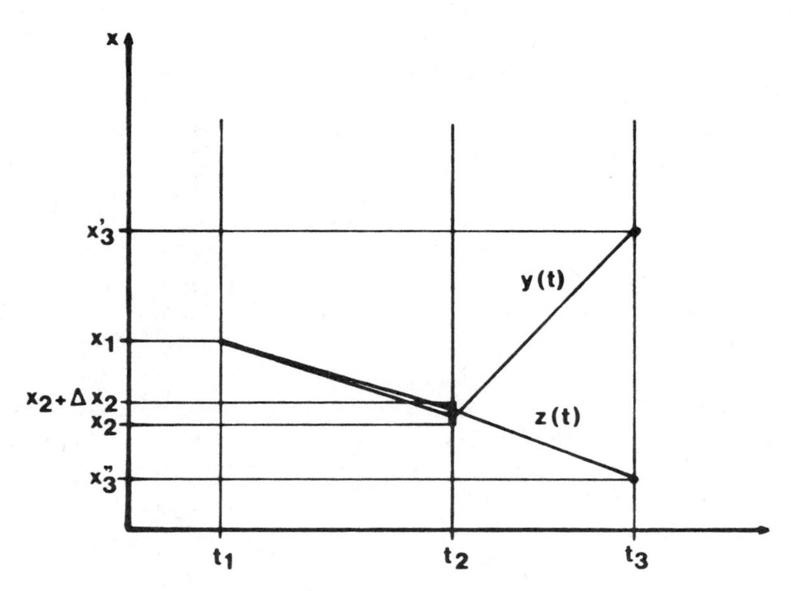

Fig. 4.2. Example of two different outcomes for the same starting point (discrete case).

of states is independent of time and the transition probability can be simply written as

$$f(x_2, t_2/x_1, t_1) = f(x_2, t_2 - t_1/x_1, 0)\,. \tag{4.24}$$

The Chapman–Kolmogorov equation then becomes

$$f(x_3, \theta + \tau/x_1) = \int f(x_3, \tau/x_2) f(x_2, \theta/x_1)\, dx_2\,. \tag{4.25}$$

Let us define the function $p(t_1, x_1; t_3, x_3)$ by

$$p(t_1, x_1; t_3, x_3) = f(x_3, t_3/x_1, t_1)\,. \tag{4.26}$$

For a stationary Markovian process that is homogeneous in space, it yields

$$p(t_1, x_1; t_3, x_3) = p(t_3 - t_1; x_3 - x_1) \tag{4.27}$$

and the Chapman–Kolmogorov equation is given by

$$p(\theta + \tau, y) = \int p(\tau, y - x) p(\theta, x)\, dx\,. \tag{4.28}$$

Note that $p(\theta + \tau, y)$ is the convolution of $p(\tau, y)$ and $p(\theta, y)$; clearly

$$p(\theta + \tau, y) = p(\theta, y) \otimes p(\tau, y)\,. \tag{4.29}$$

If $\tilde{p}(\theta, k)$ and $\tilde{p}(\tau, k)$ are the Fourier transforms of the corresponding functions, Eq. (4.29) leads to

$$\tilde{p}(\theta + \tau, k) = \tilde{p}(\theta, k) \cdot \tilde{p}(\tau, k)\,. \tag{4.30}$$

In Figs. 4.1 and 4.2, the broken lines do not represent the sample paths of the process $X(t)$, but join the outcoming states of the system observed at a discrete set of times $t_1, t_2, \ldots, t_n$. To understand the behavior of $X(t)$, it is necessary to know the transition probability. In Fig. 4.3 are given numerical simulations of a Wiener process $W(t)$ (Brownian motion) and a Cauchy process $C(t)$, both supposed one dimensional, stationary, and homogeneous. Their transitions functions are defined by[119,123–126]

$$\begin{aligned} p_W(x, t; y, t') &= p_W(y - x, t' - t) \\ &= [4\pi D(t' - t)]^{-1/2} \exp[-(y - x)^2/4D(t' - t)] \end{aligned} \tag{4.31}$$

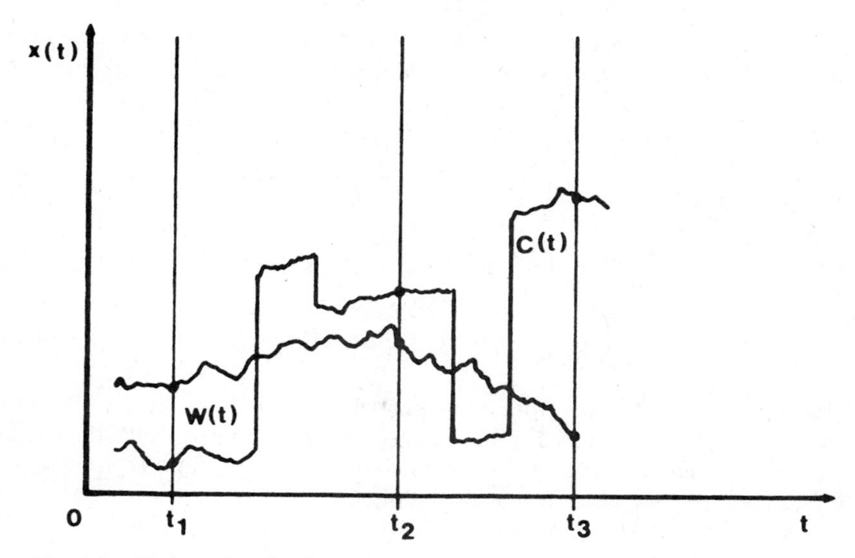

Fig. 4.3. Trajectories of a Brownian motion $W(t)$ and of a Cauchy process $C(t)$.

and

$$p_C(x, t; y, t') = \alpha \frac{t' - t}{\pi} [(y - x)^2 + \alpha^2(t' - t)^2]^{-1} . \qquad (4.32)$$

The different behaviors between the two processes are striking. In contrast to the Brownian motion, which is irregular but continuous, the Cauchy process appears greatly discontinuous.

It is easy to verify that the Fourier transforms $\tilde{p}_W(\tau, k)$ and $\tilde{p}_C(\tau, k)$ satisfy Eq. (4.30); indeed

$$\tilde{p}_W(\tau, k) = \exp(-2D\tau k^2) \qquad (4.33)$$

and

$$\tilde{p}_C(\tau, k) = \exp(-\alpha\tau|k|) . \qquad (4.34)$$

The essential difference between the two transition probability densities lies in the fact that for the gaussian distribution $p_W(\tau, x)$ the different moments $E[X^m]$, $m = 1, 2, \ldots, n$, exist, while for the Cauchy distribution $p_C(\tau, x)$ they do not exist. The Levy distributions characterized by $\tilde{p}(\tau, k) = \exp(-\alpha|k|^q\tau)$ with $0 < q < 2$[117,127,128] play a prominent role in the theory of relaxation processes.[129–133] *b. Rates of Passage and Transition.* Consider a nonhomogeneous Markovian process $\{X(t), t \geq 0\}$ with a transition probability density function

$$p(t_1, x; t_2, y) = \Pr[X(t_2) = y/X(t_1) = x] \tag{4.35}$$

and compelled to the following assumptions[119,122,134]:

$$\lim_{\Delta t \to 0} \frac{1}{\Delta t} [p(t, x; t + \Delta t, y)] = Q(t; x, y), \tag{4.36}$$

$$\lim_{\Delta t \to 0} \frac{1}{\Delta t} [1 - p(t, x; t + \Delta t, x)] = q(t; x). \tag{4.37}$$

The quantities have the following probabilistic interpretation: $q(t; x)\,\Delta t$ [respectively, $1 - q(t; x)\,\Delta t$] represents the probability that the system $X(t)$ undergoes a change (remains unchanged) in the time $(t, t + \Delta t)$ and $Q(t; x, y)\Delta t$ is the conditional probability to observe a transition to y in $(t; t + \Delta t)$ given $X(t)$ is in state x. Thus

$$\Pr[X(t + \Delta t) = x/X(t) = x] = 1 - q(t; x)\,\Delta t, \tag{4.38}$$

$$\Pr[X(t + \Delta t) = y/X(t) = x] = Q(t; x, y)\Delta t. \tag{4.39}$$

$q(t; x)$ and $Q(t; x, y)$ are called, respectively, the rates of passage and transition. It seems reasonable to require that the rate functions for every state x and time t satisfy the expression

$$q(t; x) = \sum_{y \neq x} Q(t; x, y). \tag{4.40}$$

c. Kolmogorov–Feller Equations. The transition probability density $p(s, x; t, y)$ for a discrete Markov process satisfies the following differential-difference equation[119,122,134]:

Kolmogorov–Feller backward equation

$$\frac{\partial}{\partial s} p(s, x; t, y) = q(s; x)p(s, x; t, y) - \sum_{\xi \neq x} p(s, \xi; t, y)Q(s; x, \xi), \tag{4.41}$$

Kolmogorov–Feller forward equation

$$\frac{\partial}{\partial t} p(s, x; t, y) = -q(t; y)p(s, x; t, y) + \sum_{\xi \neq y} Q(t; \xi, y)p(s, x; t, \xi). \tag{4.42}$$

Indeed, from the Chapman–Kolmogorov equation and the relations (4.38)–(4.40), one obtains

$$\begin{aligned} p(s, x; t, y) &= \sum_{\xi} p(s, x; t-\Delta t, \xi) p(t-\Delta t, \xi; t, y) \\ &= p(s, x; t-\Delta t, y) p(t-\Delta t, y; t, y) \\ &\quad + \sum_{\xi \neq y} p(s, x; t-\Delta t, \xi) p(t-\Delta t, \xi; t, y) \\ &= [1 - q(t; y)\,\Delta t]\, p(s, x; t-\Delta t, y) \\ &\quad + \sum_{\xi \neq y} Q(t; \xi, y) p(s, x; t-\Delta t, \xi)\,\Delta t \end{aligned} \tag{4.43}$$

and

$$\begin{aligned} &\frac{1}{\Delta t}\,[p(s, x; t, y) - p(s, x; t-\Delta t, y)] \\ &\quad = -\,q(t; y) p(s, x; t-\Delta t, y) + \sum_{\xi \neq y} Q(t; \xi, y) p(s, x, t-\Delta t, \xi)\,. \end{aligned} \tag{4.44}$$

Letting $\Delta t \to 0$ in this relation, one obtains the forward equation (4.42). The backward equation (4.41) can be deduced by a similar treatment.[119,122,134]

d. Survival Probability Function. It is interesting to calculate the probability $P_0(s, x; t)$ that a system found in state x at time s does not undergo a change until time t. The absence of change of the state of the system during the time interval $t - s$ results from two exclusive possibilities: (1) the system undergoes no change until time t and makes a transition in the time interval $t + \Delta t$; (2) the system presents no change until time $t + \Delta t$. So,

$$P_0(s, x; t) = q(t; x)\,\Delta t\, P_0(s, x; t) + P_0(s, x; t + \Delta t)\,. \tag{4.45}$$

It is obvious that

$$\frac{P_0(s, x; t+\Delta t) - P_0(s, x; t)}{\Delta t} = -\,q(t; x) P_0(s, x; t)\,. \tag{4.46}$$

Letting $\Delta t \to 0$, one obtains

$$\frac{\partial P_0(s, x; t)}{\partial t} = -\,q(t; x) P_0(s, x; t)\,. \tag{4.47}$$

Remarking that $P_0(t, x; t) = 1$, it is apparent that the survival probability $P_0(s, x; t)$ is related to the rate of disappearance $q(t; x)$ by

$$P_0(s, x; t) = \exp\left(-\int_s^t q(t'; x)\, dt'\right). \tag{4.48}$$

For instance, if $q(t; x) = \lambda$(constant), one easily sees that $P_0(s, x; t) = \exp[-\lambda(t - s)]$ characterizes a Poisson process (see Section IV.D.2.a).

For chemical reactions, Eq. (4.47) is interpreted as a kinetic equation, where $q(t; x)$ is identified as a time- and state-dependent rate of reaction.

C. Master Equation

Let $\{N(t), t \geq 0\}$ be a discrete Markov process defined by the rates of passage and of transition $q(t; m)$ and $Q(t; m, n)$. The version of the forward Kolmogorov equation (4.42) for which the condition (4.40) $q(t; m) = \Sigma_{n \neq m} Q(t; m, n)$ is valid, is called the master equation.[134–138]

It is clear that by substituting Eq. (4.40) into Eq. (4.42), one finds

$$\frac{\partial}{\partial t} p(s, m; t, n) = \sum_{\xi} [Q(t; \xi, n)p(s, m; t, \xi) - Q(t; n, \xi)p(s, m; t, n)]\,. \tag{4.49}$$

For a continuous Markov process, the master equation is of the form

$$\frac{\partial}{\partial t} p(s, x; t, y) = \int [Q(t; \xi, y)p(s, x; t, \xi) - Q(t; y, \xi)p(s, x; t, y)]\, d\xi\,. \tag{4.50}$$

Note that Eq. (4.49) is a differential-difference equation and Eq. (4.50) an integro-differential equation, and that $Q(t; n, n)$ [respectively, $Q(t; y, y)$] is arbitrary, since the term for which $\xi = n$ [$\xi = y$] in Eq. (4.49) [Eq. (4.50)] is equal to zero.

The probability density $P_1(t, y)$ to find at time t the system in the state y is given by [see Eqs. (4.17) and (4.26)]

$$\begin{aligned} P_1(t, y) &= \int f(x, s; y, t)\, dx \qquad \text{recalling that } t > s \\ &= \int f(y, t/x, s) f(x, s)\, dx \\ &= \int p(s, x; t, y) f(x, s)\, dx\,. \end{aligned} \tag{4.51}$$

Multiplying Eq. (4.50) by the initial condition $f(x, s)$ and integrating with respect to x, results in

$$\frac{\partial}{\partial t} P_1(t, y) = \int [Q(t; \xi, y)P_1(t, \xi) - Q(t; y, \xi)P_1(t, y)]\, d\xi\,. \quad (4.52)$$

The discrete version of the master equation takes the following form:

$$\frac{\partial}{\partial t} P_1(t, n) = \sum_{\xi} [Q(t; \xi, n)P_1(t, \xi) - Q(t; n, \xi)P_1(t, y)]\,. \quad (4.53)$$

In this form, the master equation is a gain–loss equation for the probability to find the system at time t in a given state: the first sum (or integral) on the right-hand side represents the transitions into the state n (or y) from all other states, the second sum represents transitions from the state n (or y) to all other states.

For many physical applications, modeled by a homogeneous Markov process in time and space, the rate of transition is time independent and depends only on the difference of the starting and arriving states. Therefore, one can see that the master equation is given by

$$\frac{\partial}{\partial t} P_1(t, n) = \sum_{m} [Q(m, n)P_1(t, m) - Q(n, m)P_1(t, n)]\,. \quad (4.54)$$

As a final note, it has to be stressed out that Eqs. (4.49) and (4.50) and Eqs. (4.52) and (4.53) hold for an arbitrary stochastic process. These evolution equations cannot give any information about whether or not the process is Markovian.[135] The master equation concept has been used to analyze some examples of multistate relaxation processes.[139]

D. Birth–Death Processes (or One-Step Processes)

1. *General Properties*

A class of continuous-time Markov processes with integer state space and with transitions allowed only between adjacent states plays a central role in the stochastic description of chemical kinetics.[122,134,135]

Let $\{N(t), t \geq 0\}$ be an integer Markovian process, and $Q(t; m, n)$ and $q(t; m)$, the rates of transition and passage [see Eqs. (4.36)–(4.40)] satisfying the nonhomogeneous birth–death process conditions, that is,

$$\begin{aligned} Q(t; m, n) &= 0 && \text{if } |m - n| > 1\,, \\ Q(t; m, n) &= B_m(t) && \text{if } n = m + 1\,, \\ Q(t; m, n) &= D_m(t) && \text{if } n = m - 1\,, \\ q(t; m) &= +[B_m(t) + D_m(t)]\,. \end{aligned} \quad (4.55)$$

$N(t)$ can be interpreted as the size of the population at time t. The most simple kind of birth–death process is that for which the rates of birth and death are time independent. A birth–death process $N(t)$ is called a pure birth process (respectively pure death process) if $D_m = 0$ ($B_m = 0$) for any m.

From the Kolmogorov equations (4.41) and (4.42), one obtains the difference-differential equations for the birth–death process. The backward equation is given by

$$\begin{aligned} \dot{p}_b(t; m, n) &= B_m(t)p_b(t; m+1, n) + D_m(t)p_b(t; m-1, n) \\ &\quad - [B_m(t) + D_m(t)]\, p_b(t; m, n) \\ &\ \vdots \\ \dot{p}_b(t; 0, n) &= B_0(t)p_b(t; 1, n) - B_0(t)p_b(t; 0, n) \end{aligned} \tag{4.56}$$

with $m > 0$ and $n \geq 0$; and the forward equations take the form

$$\begin{aligned} \dot{p}_f(t; m, n) &= B_{n-1}(t)p_f(t; m, n-1) + D_{n+1}(t)p_f(t; m, n+1) \\ &\quad - [B_n(t) + D_n(t)]\, p_f(t; m, n) \\ &\ \vdots \\ \dot{p}_f(t; m, 0) &= -B_0(t)p_f(t; m, 0) + D_1(t)p_f(t; m, 1) \end{aligned} \tag{4.57}$$

with $m \geq 0$ and $n > 0$.

The master equation (4.53) becomes

$$\begin{aligned} \dot{P}_1(t, n) &= B_{n-1}(t)P_1(t, n-1) + D_{n+1}(t)P_1(t, n+1) \\ &\quad -[B_n(t) + D_n(t)]P_1(t, n) \\ &\ \vdots \\ \dot{P}_1(t, 0) &= -B_0(t)P_1(t, 0) + D_1(t)P_1(t, 1)\,. \end{aligned} \tag{4.58}$$

At this point it is interesting to note that this equation can be derived directly by considering only probabilistic arguments. Indeed, consider the event $\{N(t + \Delta t) = n\}$, which is the sum of the following mutually exclusive events: (1) at time t the size of the population was $n - 1$ and a birth occurs in the interval $(t, t + \Delta t)$ with a probability $B_{n-1}(t)\,\Delta t$; (2) at time t the size was n and during $(t, t + \Delta t)$ no birth or death is observed, this latter event having a probability $\{1 - [B_n(t) + D_n(t)]\,\Delta t\}$; (3) at time t the size was $n + 1$ and in the interval $(t, t + \Delta t)$ one death occurs with a probability $D_{n+1}(t)\,\Delta t$; and (4) changes in the size of the population greater than 1 occur with a probability $O(\Delta t)$*. Thus

$$\begin{aligned}\Pr[N(t+\Delta t)=n] \equiv P_1(t+\Delta t, n) = {} & P_1(t, n-1)B_{n-1}(t)\,\Delta t \\ & + P_1(t, n)\{1-[B_n(t)+D_n(t)]\,\Delta t\} \\ & + P_1(t, n+1)D_{n+1}(t)\,\Delta t + O(\Delta t)\end{aligned} \tag{4.59}$$

and it follows that the limit of $[P_1(t+\Delta t, n) - P_1(t, n)]/\Delta t$ as Δt tends to zero equals the right-hand side of Eq. (4.58).

2. *Examples*

a. Poisson Process. A pure birth process is called a Poisson process if the birth rate is equal to a constant (independent of the time and the state): $B_n(t) = \lambda$ whatever n. The master equations reduce to

$$\begin{aligned}\dot{P}_1(t, n) &= -\lambda P_1(t, n) + \lambda P_1(t, n-1)\,, \\ \dot{P}_1(t, n-1) &= -\lambda P_1(t, n-1) + \lambda P_1(t, n-2)\,, \\ &\vdots \\ \dot{P}_1(t, 1) &= -\lambda P_1(t, 1) + \lambda P_1(t, 0)\,, \\ \dot{P}_1(t, 0) &= -\lambda P_1(t, 0)\,.\end{aligned} \tag{4.60}$$

One sees immediately, because of $P_1(0, 0) = 1$, that

$$\begin{aligned}P_1(t, 0) &= \exp(-\lambda t) \\ P_1(t, 1) &= \lambda \exp(-\lambda t) \otimes \exp(-\lambda t) = \lambda t \exp(-\lambda t)\,, \\ &\vdots \\ P_1(t, n) &= [(\lambda t)^n/n!] \exp(-\lambda t)\,.\end{aligned} \tag{4.61}$$

The forward equations (4.57) have the simple forms

$$\begin{aligned}\dot{p}_f(t; m, n) &= \lambda p_f(t; m, n-1) - \lambda p_f(t; m, n)\,, \\ \dot{p}_f(t; m, n-1) &= \lambda p_f(t; m, n-2) - \lambda p_f(t; m, n-1)\,, \\ &\vdots \\ \dot{p}_f(t; m, 0) &= -\lambda p_f(t; m, 0)\,,\end{aligned} \tag{4.62}$$

* $O(\Delta t)$ stands for an arbitrary expression with the property that $\lim_{\Delta t \to 0}[O(\Delta t)/\Delta t] = 0$.

one easily concludes that

$$p_f(t; m, n) = [(\lambda t)^{n-m}/(n-m)!] \exp(-\lambda t) \qquad \text{if } n \geq m ,$$
$$p_f(t; m, n) = 0 \qquad \text{if } n < m . \tag{4.63}$$

One notes that these relations yield spatial homogeneity and that $P_0(0, m; t) = p(t; m, m) = e^{-\lambda t}$, as predicted by Eq. (4.48).

b. Unimolecular Reactions.[137,140–143] 1. IRREVERSIBLE PROCESSES. A unimolecular reaction by irreversible transformation of reactants R into products P, characterized by a rate of reaction k,

$$R \xrightarrow{k} P ,$$

can be modeled by a linear death process $\{N(t), t \geq 0\}$ representing the number of reactants at time t: the rate of disappearance $D_n(t)$ is proportional to the population of reactants.

Defining $B_n(t) = 0$ and $D_n(t) = kn$ for k a positive constant, the master equation can be written as

$$\begin{aligned} \dot{P}_1(t, n) &= k(n+1)P_1(t, n+1) - knP_1(t, n) \\ &\vdots \\ \dot{P}_1(t, 0) &= kP_1(t, 1) \end{aligned} \tag{4.64}$$

recalling that $P_1(t, n) = \Pr[N(t) = n]$ and $P_1(0, n) = \delta_{nn_0}$. To obtain the solution of these equations it is convenient to use the generating function $G_N(t, z)$ given by

$$G_N(t, z) = \sum_{i=0}^{\infty} z^i P_1(t, i) . \tag{4.65}$$

Noting that

$$\frac{\partial}{\partial z} G_N(t, z) = \sum_{i=1}^{\infty} iP_1(t, i) z^{i-1} , \tag{4.66}$$

from Eq. (4.64) and Eqs. (4.65) and (4.66), it results that

$$\frac{\partial}{\partial t} G_N = -k(z-1) \frac{\partial}{\partial z} G_n . \tag{4.67}$$

Solving this partial differential equation with the initial condition

$G_N(0, t) = z^{n_0}$, one obtains

$$G_N(t, z) = (1 - e^{-kt} - ze^{-kt})^{n_0} , \tag{4.68}$$

which is related to the binomial distribution by

$$P_1(t, n) = \binom{n_0}{n} (1 - e^{-kt})^{n_0 - n} e^{-nkt} . \tag{4.69}$$

The expectation $E[N(t)]$ and the variance $\mathrm{Var}\,[N(t)]$ of the number of reactants have the forms of

$$E[N(t)] = \frac{\partial}{\partial z} G_N(t, z)_{z=1} = n_0 e^{-kt} , \tag{4.70}$$

$$\begin{aligned} \mathrm{Var}\,[N(t)] &= \frac{\partial^2}{\partial z^2} G_N(t, z)_{z=1} + \frac{\partial}{\partial z} G_N(t, z)_{z=1} - E^2[N(t)] \\ &= n_0 e^{-kt}(1 - e^{-kt}) . \end{aligned} \tag{4.71}$$

For the pure linear death process, the forward Kolmogorov equation [see Eq. (4.57)] is

$$\dot{p}(t; m, n) = -knp(t; m, n) + k(n + 1)p(t; m, n + 1) . \tag{4.72}$$

Noting that $p(t; m, n) = 0$ if $n > m$, then

$$\dot{p}(t; m, m) = -kmp(t; m, m) . \tag{4.73}$$

From Eqs. (4.72) and (4.73), it is easy to see that

$$p(t; m, n) = k(n + 1)e^{-knt} \otimes p(t; m, n + 1) , \tag{4.74}$$

$$p(t; m, m) = e^{-kmt} . \tag{4.75}$$

Solving this system recursively, the transition probability is given by

$$p(t; m, n) = \binom{m}{n} e^{-knt}(1 - e^{-kt})^{m-n} , \tag{4.76}$$

which demonstrates a spatial inhomogeneous process.

Remembering that [see Eq. (4.51)]

$$P_1(t, n) = \sum_m p(t; m, n)f(m, 0) \tag{4.77}$$

with the initial condition $f(m, 0) = \delta(m - n_0)$, one finds the master equation result (4.69).

The following comments are useful.

1. The time evolution of the mean value of the number of reactants [Eq. (4.70)] coincides with the one given by the deterministic kinetic equation, that is,

$$\dot{C}_R(t) = -kC_R(t), \tag{4.78}$$

$C_R(t)$ being the concentration of the reactant at time t and $C_R(0)$ equalling n_0.

2. The time development of the expectation and the variance of $N(t)$ can be directly evaluated by derivating Eq. (4.67) with respect to z and by substituting the derivatives by the basic expressions of Eqs. (4.70) and (4.71). Thus

$$E[\dot{N}(t)] = -kE[N(t)], \tag{4.79}$$

$$\text{Var}\,[\dot{N}(t)] = -2k\,\text{Var}\,[N(t)] + kE[N(t)]. \tag{4.80}$$

3. In expression (4.64) can be included some spontaneous deactivation of the reactants. Then one obtains

$$\dot{P}_1(t, n) = -k_s P_1(t, n) + k(n+1)P_1(t, n+1) - knP_1(t, n). \tag{4.81}$$

Let be $Q_1(t, n) = e^{-k_s t} P_1(t, n)$, it follows that

$$\dot{Q}_1(t, n) = k(n+1)Q_1(t, n+1) - knQ_1(t, n) \tag{4.82}$$

and

$$P_1(t, n) = e^{-k_s t} \binom{n_0}{n} (1 - e^{-kt})^{n_0 - n} e^{-nkt}. \tag{4.83}$$

4. This model can be the extended to reactions in which the rate is time dependent, so that Eqs. (4.64), (4.67), and (4.69) become

$$\dot{P}_1(t, n) = (n+1)k(t)P_1(t, n+1) - nk(t)P_1(t, n), \tag{4.84}$$

$$\frac{\partial}{\partial t} G_N = -k(t)(z-1)\frac{\partial}{\partial z} G_N, \tag{4.85}$$

$$P_1(t, n) = \binom{n_0}{n}\left[1 - \exp\left(-\int_0^t k(u)\,du\right)\right]^{n_0 - n} \exp\left(-n\int_0^t k(u)\,du\right), \tag{4.86}$$

and Eq. (4.79) has the simple form

$$E[\dot{N}(t)] = -k(t)E[N(t)] . \tag{4.87}$$

2. REVERSIBLE PROCESSES. A reversible unimolecular reaction,

$$R \underset{k_{-1}}{\overset{k_1}{\rightleftharpoons}} P ,$$

with forward and backward rate constants k_1 and k_{-1} can be analyzed as a homogeneous birth–death process $\{N(t), t \geq 0\}$. Noting that n_0 represents the total number of reactants and products, one can easily check that

$$\begin{aligned} \dot{P}_1(t, n) = {} & k_1(n+1)P_1(t, n+1) + k_{-1}(n_0 - n + 1)P_1(t, n-1) \\ & - [k_1 n + k_{-1}(n_0 - n)]P_1(t, n) \end{aligned} \tag{4.88}$$

and also that

$$\frac{\partial}{\partial t} G_N = [k_1 + (k_{-1} - k_1)z - k_{-1}z^2] \frac{\partial}{\partial z} G_N + n_0 k_{-1}(z-1)G_N . \tag{4.89}$$

Recalling that $P_1(0, n) = \delta_{nn_0}$, it is not difficult to find that

$$G_N(t, z) = \left(\frac{Ke^{-(k_1+k_{-1})t}(z-1) + K - z}{K} \right)^{n_0} , \tag{4.90}$$

$$E[N(t)] = \frac{n_0}{k_1 + k_{-1}} [k_1 \exp - (k_1 + k_{-1})t + k_{-1}] , \tag{4.91}$$

$$\text{Var}\,[N(t)] = \frac{n_0}{1+K} [1 + Ke^{-(k_1+k_{-1})t}] \left(1 - \frac{1 + Ke^{-(k_1+k_{-1})t}}{1+K} \right) , \tag{4.92}$$

where $K = k_1/k_{-1}$ represents the equilibrium constant.

Note that (1) for both reactions $E[N(t)]/\text{Var}\,[N(t)] \neq 1/E^{1/2}[N(t)]$, which means that the number of reactants is not governed by Poisson statistics; (2) the average value in the stochastic model is identical to the deterministic result.

If the solution of a deterministic reaction rate equation differs from the first moment corresponding to the solution of the master equation, it can generally be considered as a differently conditioned average of the same random variable.[144]

3. TIME-RESOLVED SOLVATOCHROMISM EFFECT. We consider the model of Rapp et al.[91] describing the following physical situation. Neutral polar solvent molecules are surrounding an excited polar molecule. The solute–solvent interaction is modeled as follows: (1) every reorientation of a solvent dipole gives rise to a decrease $\delta\tilde{\nu}_S$ of the center of gravity of the spectral transition probability; (2) the maximum allowed spectral shift is fixed to $\Delta\tilde{\nu}_S$, which implies that n_0 $(=\Delta\tilde{\nu}_S/\delta\tilde{\nu}_S)$ molecules are involved in the solvatochromism process.

Let $N(t)$ be the random number of solvent molecules at time t each of which has been reoriented by the same solute-excited molecule and let k_0 be the characteristic orientation rate constant. From the preceding discussion $N(t)$ appears to be a birth process. One easily established that

$$\dot{P}_1(t, n) = -(n_0 - n)k_0P_1(t, n) + (n_0 - n + 1)k_0P_1(t, n-1) \quad (4.93)$$

and that the solution takes the form

$$P_1(t, n) = \binom{n_0}{n}[e^{-k_0t}]^{n_0-n}[1 - e^{-k_0t}]^n . \quad (4.94)$$

Furthermore, let $S(\tilde{\nu}, \tilde{\nu}_0 - n\,\Delta\tilde{\nu}_S/n_0)$ denote the emission spectrum whose center of gravity is $\tilde{\nu}_0 - n\,\Delta\tilde{\nu}_S/n_0$ and corresponding to the reorientation of n solvent molecules. Then it follows that the time-resolved fluorescence spectrum $I(t, \tilde{\nu})$ is given by the expected value of $S(\tilde{\nu}, \tilde{\nu}_0 - N(t)\,\Delta\tilde{\nu}_S/n_0)e^{-t/\tau}$, where τ is the excited state lifetime, that is,

$$I(t, \tilde{\nu}) = \sum_{n=0}^{n_0} S\left(\tilde{\nu}, \tilde{\nu}_0 - \frac{n\,\Delta\tilde{\nu}_S}{n_0}\right) P_1(t, n)e^{-t/\tau} . \quad (4.95)$$

c. *One-Dimensional Random Walk.*[117,119,123,134,136,145] Consider a particle that takes at the end of a constant time interval τ, a step of length x_0 to the right or to the left with equal probability p $(=\frac{1}{2})$. Let $X(n\tau)$ be the distance of the particle from the origin at instant $t = n\tau$. The process $\{X(n\tau), n \geq 0\}$ can be modeled as a discrete both in time and space Markov process. Assuming that the particle leaves the origin at time $t = 0$ $(n = 0)$, one easily sees that

$$X(n\tau) = X_1 + X_2 + X_3 + \cdots + X_n , \quad (4.96)$$

where the random variables $X_1, X_2, \ldots$, and X_n are mutually independent and obey to the same probability law

$$\Pr[X_i = x_0] = \Pr[X_i = -x_0] = p . \quad (4.97)$$

Of course, from Eq. (4.59), $P_1(t, k) \equiv P_1(n\tau, kx_0)$ satisfies the recurrence relation

$$P_1[(n+1)\tau, kx_0] = pP_1[n\tau, (k-1)x_0] + pP_1[n\tau, (k+1)x_0] . \tag{4.98}$$

One can also see directly the following important results:

$$\begin{aligned} E[X(n\tau)] &= 0 , \\ \mathrm{Var}\,[X(n\tau)] &= 2pnx_0^2 = nx_0^2 . \end{aligned} \tag{4.99}$$

Assume that $X(n\tau) = kx_0$; this means that at time $t = n\tau$, one has taken m steps to the right and $n - m$ steps to the left. Hence,

$$X(n\tau) = kx_0 = mx_0 - (n-m)x_0 . \tag{4.100}$$

Clearly, $X(n\tau)$ obeys a binomial distribution defined by the parameters p, m, and n, that is,

$$\Pr\,[X(n\tau) = kx_0] = \binom{n}{m} p^n = \binom{n}{(n+k)/2} p^n \tag{4.101}$$

From the De Moivre–Laplace theorem, for n large,

$$\Pr\,[X(n\tau) = kx_0] = (\pi n x_0/2)^{-1/2} \exp\,(-k^2/2n) \tag{4.102}$$

provided that k is of the order of $n^{1/2}$.

The Wiener process or Brownian motion [see Section IV.B.2.a, Eq. (4.31) and Section IV.E.4.a] can be defined as a limiting process of the symmetrical random walk. Indeed, for $t = n\tau$, the expectation and the variance are given by

$$\begin{aligned} E[X(t)] &= 0 , \\ \mathrm{Var}\,[X(t)] &= tx_0^2/\tau . \end{aligned} \tag{4.103}$$

By keeping t constant and making x_0 and τ tend to zero such that x_0 tends to zero as $\tau^{1/2}$, the variance will remain finite with a limit equal to $2Dt$. Hence, $X(t)$ is asymptotically normally distributed:

$$\frac{d}{dx} \Pr\,[X(t) < x] = (4\pi Dt)^{-1/2} \exp\left(\frac{-x^2}{4Dt}\right) . \tag{4.104}$$

E. Diffusion Processes

1. *Definition*

A Markov process $\{X(t), t \geq 0\}$ is called a diffusion process if there exist two functions $M(t, x)$ and $S^2(t, x)$ such that for any value of $\delta > 0$ the following expressions are verified[134,135,137,138]:

$$\lim_{\Delta t \to 0} \frac{1}{\Delta t} \int_{|y-x|>\delta} p(t, x; t + \Delta t, y)\, dy = 0 , \tag{4.105}$$

$$\lim_{\Delta t \to 0} \frac{1}{\Delta t} \int_{|y-x|\leq\delta} (y - x) p(t, x; t + \Delta t, y)\, dy = M(t, x) , \tag{4.106}$$

$$\lim_{\Delta t \to 0} \frac{1}{\Delta t} \int_{|y-x|\leq\delta} (y - x)^2 p(t, x; t + \Delta t, y)\, dy = S^2(t, x) . \tag{4.107}$$

The first condition means that if the system jumps from a state x to some other state y finitely different from x during the time interval $(t, t + \Delta t)$, the transition probability tends to zero faster than Δt, as Δt goes to zero. In other words, the system cannot undergo an appreciable change during a small time interval. It is of interest to recall that for a birth–death process a change of the system in a negligible time interval does not occur with a high probability, but when it jumps, it changes by a large amount. To stress the physical meaning of the last two expressions, one considers the expectation and the second moment of the variation of $X(t)$ in the time interval $(t, t + \Delta t)$ given, respectively, by

$$E[X(t + \Delta t) - X(t)] = \int (y - x) p(t, x; t + \Delta t, y)\, dy , \tag{4.108}$$

$$E\{[X(t + \Delta t) - X(t)]^2\} = \int (y - x)^2 p(t, x; t + \Delta t, y)\, dy . \tag{4.109}$$

In light of this, one notes that $M(t, x)$ and $S^2(t, x)$ are the mean rates of change, of $X(t)$ and $X^2(t)$, respectively, and are known as the drift and diffusion coefficients.

2. *Kolmogorov's Backward Equation*

From the Chapman-Kolmogorov expression [Eqs. (4.20) and (4.26)], one finds

$$p(s-\Delta s, x; t, y) = \int p(s-\Delta s, x; s, z)p(s, z; t, y)\, dz \tag{4.110}$$

and since

$$p(s, x; t, y) = p(s, x; t, y) \int p(s-\Delta s, x; s, z)\, dz\ , \tag{4.111}$$

it follows that

$$\begin{aligned}\frac{1}{\Delta s}&[p(s-\Delta s, x; t, y) - p(s, x; t, y)]\\ &= \frac{1}{\Delta s}\int \{p(s-\Delta s, x; s, z)[p(s, z; t, y) - p(s, x; t, y)]\}\, dz\\ &= \frac{1}{\Delta s}\int_{|z-x|>\delta} \{\ \}\, dz + \int_{|z-x|\le\delta} \{\ \}\, dz\ .\end{aligned} \tag{4.112}$$

Let Δs tend to zero. The left-hand side of Eq. (4.112) exists and equals $-(\partial/\partial s)p(s, x; t, y)$ and the first member of the right-hand side, by hypothesis [see Eq. (4.105)], equals zero, so

$$\begin{aligned}-\frac{\partial}{\partial s}p(s, x; t, y) = \lim_{\Delta s\to 0}\int_{|z-x|\le\delta} &[p(s, z; t, y) - p(s, x; t, y)]\\ &\times p(s-\Delta s, x; s, z)\, dz\ .\end{aligned} \tag{4.113}$$

By Taylor's expansion, one obtains

$$\begin{aligned}-\frac{\partial}{\partial s}p(s, x; t, y) = &\lim_{\Delta s\to 0}\frac{1}{\Delta s}\int_{|z-x|\le\delta}(z-x)\\ &\times\left[\frac{\partial}{\partial x}p(s, x; t, y)\right]p(s-\Delta s, x; s, z)\, dz\\ &+\lim_{\Delta s\to 0}\frac{1}{2\Delta s}\int_{|z-x|\le\delta}[(z-x)^2 + O(z-x)^2]\\ &\times\left[\frac{\partial^2}{\partial x^2}p(s, x; t, y)\right]p(s-\Delta s, x; s, z)\, dz\ .\end{aligned} \tag{4.114}$$

By conditions (4.106) and (4.107), it follows the Kolmogorov's backward equation

$$-\frac{\partial}{\partial s}p(s, x; t, y) = M(s, x)\frac{\partial}{\partial x}p(s, x; t, y) + \tfrac{1}{2}S^2(s, x)\frac{\partial^2}{\partial x^2}p(s, x; t, y)\ . \tag{4.115}$$

3. *Kolmogorov's Forward Equation*

The transition probability $p(s, x; t, y)$ satisfies the forward equation:

$$\frac{\partial}{\partial t} p(s, x; t, y) = -\frac{\partial}{\partial y} M(t, y)p(s, x; t, y) + \frac{1}{2} \frac{\partial^2}{\partial y^2} S^2(t, y)p(s, x; t, y) . \tag{4.116}$$

In the literature, this relation is commonly called the Fokker–Planck equation. It is important and instructive to point out that the derivation of the Fokker–Planck relation requires the existence of the first two moments. For the Levy processes, there does not exist a Fokker–Planck equation.

4. *Examples*

a. Brownian Motion or Wiener Process.[133,134,146] A spatial and temporal homogeneous Markov $W(t)$ process is known as a Wiener process if the transition probability density is normal, given by

$$p(s, x; t, y) = p(t-s; y-x) = \frac{\partial}{\partial y} \Pr [W(t) < y/W(s) = x]$$

$$= [4\pi D(t-s)]^{-1/2} \exp\left[-\frac{(y-x)^2}{4D(t-s)}\right] . \tag{4.117}$$

It can easily be verified that the transition probability density verifies the forward Kolmogorov's equation (4.116), with $M(t, y) = 0$ and $\frac{1}{2}S^2(t, y) = D$, which is the familiar diffusion equation homogeneous in space and time

$$\frac{\partial}{\partial t} p = D \frac{\partial^2}{\partial y^2} p . \tag{4.118}$$

b. Ornstein–Uhlenbeck Process.[133,134,146] If the transition density function verifies the following forward and backward Kolmogorov's equations

$$\frac{\partial}{\partial t} p = D \frac{\partial^2}{\partial y^2} p + \frac{1}{\tau} \frac{\partial}{\partial y} (yp) , \tag{4.119}$$

$$-\frac{\partial}{\partial s} p = D \frac{\partial^2}{\partial x^2} p - \frac{x}{\tau} \frac{\partial}{\partial x} p , \tag{4.120}$$

which correspond to $M(t, y) = -y/\tau$ and $\frac{1}{2}S^2(t, y) = D$, it can be shown that

$$p \equiv p(s, x; t, y) = p(x; t-s, y) = (2\pi\sigma^2)^{-1/2} \exp\left[\frac{-(y-m)^2}{2\sigma^2}\right], \tag{4.121}$$

where $\sigma^2 = D\tau[1 - \exp - 2t/\tau]$ and $m = x \exp(-t/\tau)$. In contrast to the Wiener process, the Ornstein–Uhlenbeck process shows only temporal homogeneity.

F. Kramers–Moyal Expansion[16,135,147–154]

Let $N(t)$ be the random variable representing at time t, for instance, the number of reactants in a reversible chemical reaction. Each reactive act is followed by a decrease or an increase of one reactant. Furthermore, let $X(t) = N(t)/V$ denote the concentration variable, where V is the volume of the chemical system and consider $\epsilon = 1/V$. Thus, one can envisage that per reactive act $X(t)$ changes by $\pm\epsilon$. The process $\{X(t), t \geq 0\}$ may be interpreted as a one-step process characterized by the following time-independent transition rate densities:

$$\begin{aligned} \Pr[\Delta X = \epsilon / X(t) = x] &= b(x)\,\Delta t\,, \\ \Pr[\Delta X = -\epsilon / X(t) = x] &= d(x)\,\Delta t\,, \\ \Pr[\Delta X = 0 / X(t) = x] &= 1 - [b(x) + d(x)]\,\Delta t\,. \end{aligned} \tag{4.122}$$

Defining the transition density function $p(t; x, y) = \Pr[X(t) = y / X(0) = x]$, it is easy to obtain from Eqs. (4.41) and (4.42) or Eq. (4.59) the following backward and forward equations:

$$\begin{aligned} p_b(t; x, y) &= p_b(t-\Delta t; x+\epsilon, y) b(x)\,\Delta t + p_b(t - \Delta t; x - \epsilon, y) d(x)\,\Delta t \\ &\quad + p_b(t - \Delta t; x, y)[1 - b(x)\,\Delta t - d(x)\,\Delta t]\,, \\ p_f(t + \Delta t; x, y) &= p_f(t; x, y - \epsilon) b(y - \epsilon)\,\Delta t + p_f(t; x, y + \epsilon) d(y + \epsilon)\,\Delta t \\ &\quad + p_f(t; x, y)[1 - b(y)\,\Delta t - d(y)\,\Delta t]\,. \end{aligned} \tag{4.123}$$

It is obvious that

$$\begin{aligned} \frac{\partial}{\partial t} p_b(t; x, y) &= b(x)[p_b(t; x+\epsilon, y) - p_b(t; x, y)] \\ &\quad + d(x)[p_b(t; x - \epsilon, y) - p_b(t; x, y)]\,, \\ \frac{\partial}{\partial t} p_f(t; x, y) &= p_f(t; x, y - \epsilon) b(y - \epsilon) - p_f(t; x, y) b(y) \\ &\quad + p_f(t; x, y + \epsilon) d(y + \epsilon) - p_f(t; x, y) d(y)\,. \end{aligned} \tag{4.124}$$

Taking the Taylor expansion in ϵ, one obtains the backward and forward Kramers–Moyal equations:

$$\begin{aligned}\frac{\partial}{\partial t} p_b(t; x, y) &= b(x) \sum_{n=1}^{\infty} \frac{(\epsilon)^n}{n!} \frac{\partial^n}{\partial x^n} p(t; x, y) \\ &\quad + d(x) \sum_{n=1}^{\infty} \frac{(-\epsilon)^n}{n!} \frac{\partial^n}{\partial x^n} p(t; x, y), \\ \frac{\partial}{\partial t} p_f(t; x, y) &= \sum_{n=1}^{\infty} \frac{(-\epsilon)^n}{n!} \frac{\partial^n}{\partial y^n} b(y) p_f(t; x, y) \\ &\quad + \sum_{n=1}^{\infty} \frac{\epsilon^n}{n!} \frac{\partial^n}{\partial y^n} d(y) p_f(t; x, y).\end{aligned} \tag{4.125}$$

It is of importance to point out that if the right-hand side is truncated after two terms (diffusion approximation), the last relation leads to an expression similar to the familiar Fokker–Planck equation (4.116). The approximation of a master equation of a birth–death process by a diffusion equation can lead to false results. Van Kampen has critically examined the Kramers–Moyal expansion and proposed a procedure based on the concept of system size expansion.[135] It can be stated that any diffusion equation can be approximated by a one-step process, but the converse is not true.

G. Langevin Equation[134,146,155–158]

1. *Brownian Motion and Langevin Equation*

To give the physical interpretation of Brownian motion, consider a small particle immersed in a fluid and whose position is defined by the three cartesian coordinates; let $X(t)$ denote the coordinates of the particle at time t. During a time interval $(t, t + \Delta t)$, the particle suffers an enormous number of collisions by the fluid molecules ($\approx 10^{21}$ per second at room temperature). Assuming that the Brownian particle is heavier than the interacting molecules, the displacement due to a single collision is negligible, but there is an observable movement resulting from the continuous bombardment of the particle. Therefore, the variation of the position $X(t + \Delta t) - X(t)$ is the result of a very large number of infinitesimal displacements. By supposing that the random infinitesimal displacements are statistically mutually independent and by appealing to the central limit theorem, the increment $X(t + \Delta t) - X(t)$ is normal (see Section IV.E).

The Langevin approach to the description of the Brownian motion consists of postulating the stochastic Newton equation of motion, given by

$$m \frac{d^2X(t)}{dt^2} = -\zeta \frac{dX(t)}{dt} + \mathscr{F}(t)$$

or (4.126)

$$m \frac{dV(t)}{dt} = -\zeta V(t) + \mathscr{F}(t) ,$$

where m is the mass of the Brownian particle, $-\zeta\, dX(t)/dt$ is the frictional force exerted by the medium, and $\mathscr{F}(t)$ is the fluctuating force related to the collisions of the molecules of the fluid inducing the displacement of the Brownian particle. Here, we will consider only one-dimensional processes. The essential hypothesis concerning the random force $\mathscr{F}(t)$ are the following:

1. $\mathscr{F}(t)$ is a normal process with zero mean.
2. $\mathscr{F}(t)$ is a white noise, namely, its autocorrelation function is of the form of

$$E[\mathscr{F}(t)\mathscr{F}(t+\tau)] = 2k_B T\zeta\delta(\tau) , \tag{4.127}$$

where k_B is the Boltzmann constant, T is the absolute temperature; this equation is a version of the fluctuation–dissipation theorem. This expression shows how the Brownian motion is related to the thermal fluctuations occuring spontaneously in the fluid.

3. $\mathscr{F}(t)$ and $V(t) = \dot{X}(t)$ are statistically independent, and the fluctuations of $\mathscr{F}(t)$ are very much faster than those of $V(t)$.

The one-dimensional Langevin equation is easily integrated; one finds[159,160]

$$V(t) = v_0 \exp\left(-\frac{t}{\tau_v}\right) + \frac{1}{m}\int_0^t d\theta\, \mathscr{F}(\theta) \exp\left(-\frac{t-\theta}{\tau_v}\right) \tag{4.128}$$

where v_0 is the initial velocity and $\tau_v = m/\zeta$ is the relaxation time of the velocity. By appealing to assumptions (1) and (2), it is not difficult to show that

$$E[V(t) - v_0 \exp(-t/\tau_v)] = 0 ,$$

$$E\{[V(t) - v_0 \exp(-t/\tau_v)]^2\} = (k_B T/m)[1 - \exp(-2t/\tau_v)] . \tag{4.129}$$

In view of Eq. (4.128) it can be demonstrated[157,159,160] that $\tilde{V}(t) = V(t) - v_0 \exp(-t/\tau_v)$ is a normal stochastic process. Thus

$$\begin{aligned} p(0, v_0; t, v) &= (d/dv) \Pr[V(t) \le v/V(0) = v_0] \\ &= \{2\pi[1 - \exp(-2t/\tau_v)]k_B T/m\}^{-1/2} \\ &\quad \times \exp\{-[v - v_0 \exp(-t/\tau_v)]^2/[1 - \exp(-2t/\tau_v)]2k_B T/m\} . \end{aligned} \tag{4.130}$$

Before calculating the probabilistic properties of the displacement $X(t)$ of the Brownian particle, the long-time behavior of the statistical characteristics of the velocity $V(t)$ should be discussed.[156,159]

For $t/\tau_v \gg 1$, the expectation and the variance approach, respectively, zero and $k_B T/m$, the transition density function becomes independent of the initial condition v_0 and equals the Maxwellian density

$$\begin{aligned} p_{eq}(v) &= \lim_{t \to \infty} p(0, v_0; t, v) \\ &= [2\pi k_B T/m]^{-1/2} \exp[-v^2/(2k_B T/m)] , \end{aligned} \tag{4.131}$$

which is the well-known and fundamental result from equilibrium statistical mechanics.

The displacement $X(t)$ is obtained by integrating Eq. (4.128):

$$\begin{aligned} X(t) &= \int_0^t V(t')\, dt' \\ &= x_0 + v_0\left[1 - \exp\left(-\frac{t}{\tau_v}\right)\right]\tau_v + \frac{1}{m}\int_0^t dt' \int_0^{t'} d\theta\, \mathscr{F}(\theta) \exp\left(-\frac{t-\theta}{\tau_v}\right) . \end{aligned} \tag{4.132}$$

With identical arguments as previously used to show that $V(t)$ is normal, it can be demonstrated that $X(t)$ is a gaussian process with expectation and variance equal to

$$\begin{aligned} E[X(t)] &= x_0 + v_0[1 - \exp(-t/\tau_v)]\tau_v , \\ \mathrm{Var}[X(t)] &= 2(k_B T/\zeta)\{t - [1 - \exp(-t/\tau_v)]\tau_v\} . \end{aligned} \tag{4.133}$$

Thus, the transition probability density is given by

$$p(0, x_0; t, x) = (d/dx)\,\Pr[X(t) \le x/X(0) = x_0]$$
$$= [4\pi k_B T/\zeta]^{-1/2}\{t - [1 - \exp(-t/\tau_v)]\tau_v\}^{-1/2}$$
$$\times \exp(-\{x - x_0 - v_0[1 - \exp(-t/\tau_v)]\tau_v\}^2/[4k_B T$$
$$\times\{t - [1 - \exp(-t/\tau_v)]\tau_v\}/\zeta]) . \qquad (4.134)$$

At short times ($t/\tau_v \le 1$), Eqs. (4.133) have the following limits:

$$E[X(t)] = x_0 + v_0 t ,$$
$$\mathrm{Var}[X(t)] = E\{[X(t) - x_0]^2\} = v_0^2 t^2 , \qquad (4.135)$$

which make evident that at the very beginning the motion is deterministic (classical mechanics). The long-time behavior ($t/\tau_v \gg 1$) leads to the interesting limiting results

$$E[X(t)] = x_0 + v_0 \tau_v ,$$
$$E[X^2(t)] - E^2[X(t)] = (2k_B T/\zeta)t . \qquad (4.136)$$

Noting from the Einstein relation that the self-diffusion coefficient can be expressed as

$$D = k_B T/\zeta , \qquad (4.137)$$

the transition density function can be expressed simply by

$$\lim_{t\to\infty} p(0, x_0; t, x) = p_D(0, x_0; t, x)$$
$$= [4\pi Dt]^{-1/2} \exp[-(x - x_0)^2/4Dt] \qquad (4.138)$$

taking into account that $v_0\tau_v$ can be ignored compared to $x - x_0$, which increases with increasing time. Here, again, the long-time expression does not depend on the initial velocity v_0: the Brownian particle has lost the memory of its initial velocity.

At this point it has to be emphasized the links of the Langevin description with the diffusion processes. By comparing the transition density functions (4.121) and (4.130), it is clear that the Langevin equation (4.126) is equivalent to the Ornstein–Uhlenbeck process. Equation (4.130) satisfies the following one-dimensional Fokker–Planck equation

$$\frac{\partial}{\partial t} p = \frac{D}{\tau_v^2} \frac{\partial^2}{\partial v^2} p + \frac{1}{\tau_v} \frac{\partial}{\partial v} vp , \tag{4.139}$$

with $p \equiv p(0, v_0; t, v)$ and $\tau_v = m/\zeta$. But only at long times the derived process $X(t)$ is a Wiener process [Eqs. (4.117) and (4.138)]; the transition probability density $p(0, x_0; t, x)$ verifies the diffusion equation

$$\frac{\partial}{\partial t} p = D \frac{\partial^2}{\partial x^2} p . \tag{4.140}$$

2. *Brownian Motion in a Force Field*

Consider the Langevin equation for a particle in a force field:

$$\begin{aligned} \frac{d}{dt} X(t) &= V(t) , \\ m \frac{d}{dt} V(t) &= -\zeta V(t) + F[X(t)] + \mathscr{F}(t) , \end{aligned} \tag{4.141}$$

here $F(X)$ is an external driving force independent of the random force $\mathscr{F}(t)$.

Let $\mathscr{W}(t)$ denote the Wiener process and let $d\mathscr{W}(t) = \mathscr{W}(t + dt) - \mathscr{W}(t)$. Note first, according to the treatment of the stochastic integration in the sense of Itô and Stratonowitch,[118,134,135,161] the white noise $\mathscr{F}(t)$ is the formal derivative of the Wiener process:

$$\mathscr{F}(t) = \frac{d\mathscr{W}(t)}{dt} . \tag{4.142}$$

Using Eq. (4.127) one obtains

$$\begin{aligned} E[d\mathscr{W}(t)] &= 0 , \\ E[d\mathscr{W}(t)^2] &= 2k_B T\zeta \, dt . \end{aligned} \tag{4.143}$$

Setting $\mathscr{W}(t) = (2k_B T\zeta)^{1/2} W(t)$, the Langevin relation—describing the motion of the particle in the phase space—can be written as a two-variable stochastic differential equation system of the form of

$$\begin{aligned} dX(t) &= V(t) \, dt , \\ dV(t) &= \frac{1}{m} \{-\zeta V(t) + F[X(t)]\} \, dt + \frac{(2k_B T\zeta)^{1/2}}{m} \, dW(t) . \end{aligned} \tag{4.144}$$

a. High-Friction Limit. This approximation is of interest in the diffusion theory of chemical reactions and was used by Kramers[16,147] in the calculations of the reaction rate. The physical ideas justifying the results are easy to understand; for a mathematical discussion of the validity of the assumptions, the interested reader is referred to Refs. 162–164.

Let us divide the second expression of Eq. (4.141) by ζ and let m/ζ tend to zero much faster than any of the terms on the right-hand side. It follows that Eqs. (4.144) reduces to

$$dX(t) = + \frac{1}{\zeta} F[X(t)]\, dt + \sqrt{2D}\, dW(t) . \tag{4.145}$$

Therefore, the motion of the particle is determined by a one-variable stochastic differential equation. The physical contents of $(m/\zeta)dV(t)/dt \rightarrow 0$ can be related to the fact that the velocity relaxes on a time scale much shorter than the time scale characterizing variations in position.

b. Stochastic Differential Equations.[154,164–166] Consider an n-dimensional diffusion process $\mathbf{X}(t) = \{X_1(t), X_2(t), \ldots, X_n(t)\}$ defined by a multivariate stochastic differential equation of the form of

$$d\mathbf{X}(t) = \mathbf{M}[\mathbf{X}(t), t]\, dt + \mathbf{S}[\mathbf{X}(t), t]\, d\mathbf{W} , \tag{4.146}$$

where $\mathbf{W}(t) = \{W_1(t), W_2(t), \ldots, W_n(t)\}$ is an n-variable Wiener process satisfying the stochastic independence conditions

$$E[dW_i(t)\, dW_j(t)] = \delta_{ij}\, dt . \tag{4.147}$$

Note that Eq. (4.146) stands for the following set of equations ($i = 1, 2, \ldots, n$):

$$dX_i(t) = M_i[\mathbf{X}(t), t]\, dt + \sum_j S_{ij}[\mathbf{X}(t), t]\, dW_j . \tag{4.148}$$

The n-variable version of Kolmogorov's forward equation (or Fokker–Planck equation) can be written as

$$\frac{\partial}{\partial t} p = - \sum_i \frac{\partial}{\partial x_i} M_i(\mathbf{x}, t)p + \frac{1}{2} \sum_{ij} \frac{\partial^2}{\partial x_i\, \partial x_j} [S(\mathbf{x}, t)S^T(\mathbf{x}, t)]_{ij} p , \tag{4.149}$$

where $p \equiv p(t_0, \mathbf{x}_0; t, \mathbf{x})$ is the conditional probability density, $S(\mathbf{x}, t)$ $[S^T(\mathbf{x}, t)]$ is the matrix [the transpose] whose ij element is S_{ij} $[S^T_{ij} = S_{ji}]$.

1. SMOLUCHOWSKI EQUATION—HIGH-FRICTION REGIME. At first we will restrict our attention to the one-dimensional case; Eqs. (4.148) and

(4.149) become

$$dX = M(X, t)\, dt + S(X, t)\, dW \tag{4.150}$$

$$\frac{\partial}{\partial t} p = -\frac{\partial}{\partial x} M(x, t)p + \frac{1}{2} \frac{\partial^2}{\partial x^2} S^2(x, t)p\,. \tag{4.151}$$

One notes from Eq. (4.145) that $M(x, t) = F(x)/\zeta$ and $S(x, t) = (2D)^{1/2}$, then Eq. (4.151) yields

$$\frac{\partial}{\partial t} p = D \frac{\partial^2}{\partial x^2} p + \frac{1}{\zeta} \frac{\partial}{\partial x} p \frac{\partial U}{\partial x}\,, \tag{4.152}$$

noting that $F(x)$ is a field force, one has $F(x) = -dU(x)/dx$. This forward equation is often called the Smoluchowski equation.[16,147]

From Eq. (4.115) one can easily derive the corresponding backward equation

$$-\frac{\partial}{\partial t_0} p = D \frac{\partial^2}{\partial x_0^2} p - \frac{1}{\zeta} \frac{\partial U}{\partial x_0} \frac{\partial}{\partial x_0} p\,. \tag{4.153}$$

Let $P_1(t, x)$ be the probability density describing the time evolution of the particle. Hence,

$$P_1(t, x) = \int p_0(t_0, x_0) p(t_0, x_0; t, x)\, dx_0\,, \tag{4.154}$$

where $p_0(t_0, x_0)$ represents the initial condition.

For a harmonic oscillator immersed in a Brownian medium, the displacement—in the high-friction approximation—is a position Ornstein–Uhlenbeck process. With $U(x) = m\omega^2 x^2/2$, the solution of Eqs. (4.152) and (4.153) is

$$p(t_0, x_0; t, x) = p(x_0; t - t_0, x) = (2\pi\sigma^2)^{-1/2} \exp[-(x - y_0)^2/2\sigma^2]\,, \tag{4.155}$$

where

$$y_0 = x_0 \exp[-(t - t_0)/\tau_c^\omega] = E[X(t)]\,, \tag{4.155a}$$

$$\sigma^2 = D\tau_c^\omega\{1 - \exp[-2(t - t_0)/\tau_c^\omega]\} = \mathrm{Var}[X(t)]\,. \tag{4.155b}$$

The characteristic time τ_c^ω is defined by

$$\tau_c^\omega = \zeta/m\omega^2\,. \tag{4.156}$$

At this point, one simply remarks that at long times ($t \gg \tau_c^\omega$) the relation (4.155) yields the Boltzmann equilibrium distribution.

Assuming a Boltzmann equilibrium distribution for the initial condition of the oscillator, that is,

$$p_0(t_0, x_0) = (2\pi D\tau_c^\omega)^{-1/2} \exp[-x_0^2/2D\tau_c^\omega] , \tag{4.157}$$

it is not difficult to see that

$$\begin{aligned} P_1(t, x) &= (2\pi D\tau_c^\omega)^{-1/2} \exp[-x^2/2D\tau_c^\omega] \\ &= \lim_{t\to\infty} \int p(t_0, x_0'; t, x)\delta(x_0' - x_0)\, dx_0' . \end{aligned} \tag{4.158}$$

It is interesting to remark that with an initial Boltzmann distribution, the probability density $P_1(t, x)$ remains invariant over all time. Figure 4.4, in which the probability densities calculated at different times with Eq. (4.155) are reported, illustrates the temporal evolution of the expectation and variance values of $X(t)$ for a $\delta(x - x_0)$ initial condition.

2. KRAMERS EQUATION—ARBITRARY FRICTION REGIME. In the presence of inertial effects, the one-dimensional motion is determined by a bivariate Fokker–Planck equation.

Comparing Eq. (4.144) with Eq. (4.146), one concludes that

$$\begin{aligned} M_1 &= v , & S_{11} &= 0 , \\ M_2 &= [-\zeta v + F(x)]/m , & S_{22} &= (2k_B T\zeta)^{1/2}/m . \end{aligned} \tag{4.159}$$

Hence Eq. (4.149) for the probability density becomes

$$\frac{\partial}{\partial t} p + v \frac{\partial}{\partial x} p - \frac{1}{m}\frac{\partial U}{\partial x}\frac{\partial}{\partial v} p = \frac{\zeta}{m}\frac{\partial}{\partial v} vp + \frac{k_B T\zeta}{m^2}\frac{\partial^2}{\partial v^2} p , \tag{4.160}$$

where $p \equiv p(t_0, x_0, v_0; t, x, v)$ is an expression due to Kramers.[16,147] In classical mechanics the left-hand side represents the total derivative of p with respect to t and equals zero (Liouville's theorem). Therefore, the contribution of the heat bath on the motion of the particle is described by the right-hand side of Eq. (4.160).

It should be emphasized that

1. The derivation of Eq. (4.160) is based on the Markovian assumption for the bivariate stochastic process $\{X(t), V(t)\}$.

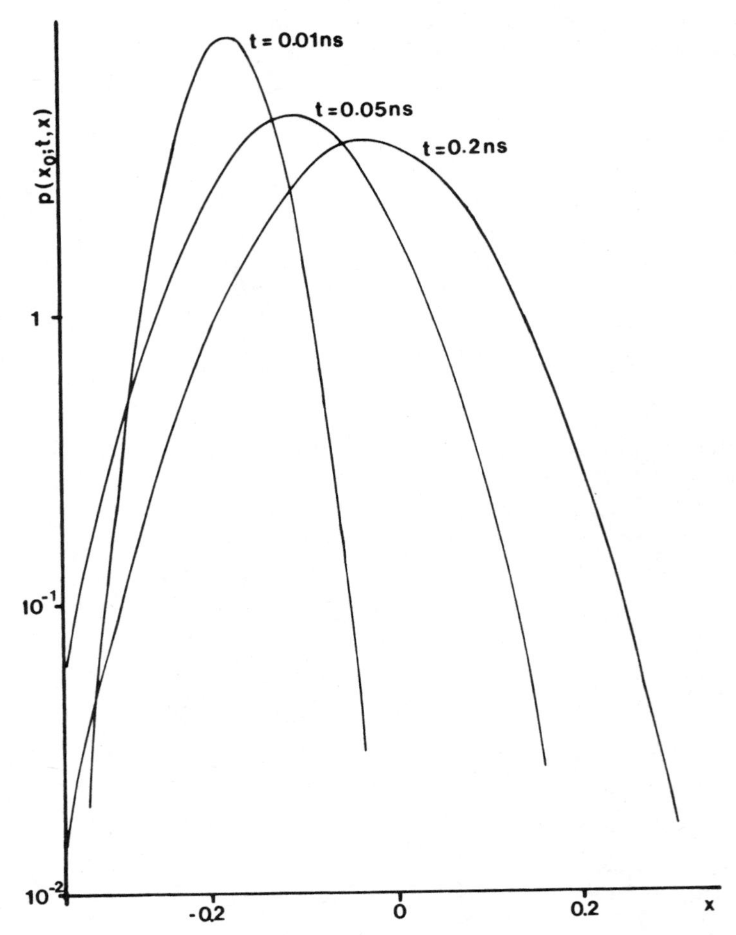

Fig. 4.4. Transition probability density $p(x_0;\ t,\ x)$ at different times [Eq. (4.155)] of a Brownian particle in the high-friction regime and in a force field deriving from an harmonic potential [initial condition $p_0(0, x_0) = \delta(x - x_0)$]: $x_0 = -0.2$ Å, $D = 0.1$ Å^2 ns^{-1}, $\tau_c^\omega = 0.1$ ns.

2. The single position process $X(t)$ (or velocity process) characterized by the marginal conditional probability density p_x (or p_v) determined from Eq. (4.160), is not in general Markovian.

3. The Smoluchowski equation (4.152) is not just a special case of the Kramers' equation (4.160) for which p is independent of v; indeed, assuming $\partial p/\partial v = 0$, Eq. (4.160) gives

$$\frac{\partial}{\partial t} p + v \frac{\partial}{\partial x} p = \frac{\zeta}{m} p \,. \tag{4.161a}$$

It is of importance to note that if p is a solution of Eq. (4.160), then $q = \int p(t_0, x_0, v_0; t, x, v)\, dv$ verifies the Smoluchowski equation.

4. Strictly speaking the Fokker–Planck equation can be derived from Eq. (4.160) by assuming that p is independent of the position and that there is no external force; Eq. (4.160) simplifies to

$$\frac{\partial}{\partial t} p = \frac{\zeta}{m} \frac{\partial}{\partial v} vp + \frac{k_B T \zeta}{m^2} \frac{\partial^2}{\partial v^2} p \,. \tag{4.161b}$$

H. Kramers' Approach to Steady-State Rates of Reaction and Its Extension to Non-Markovian Processes

1. *Kramers' Approach*

A chemical reaction, for instance, an intramolecular reaction $R \rightleftharpoons P$, is viewed as a one-dimensional motion in the phase space of a particle in a double-well potential and undergoing Markovian random forces. The dynamics of the particle are described in terms of the Kramers equation (4.160), and the rate of reaction can, in principle, be calculated from the knowledge of the probability density function $p(x_0; t, x)$.[16,147]

a. High-Friction Limit. The discussion of the range of validity of Kramers' treatment has given rise to great number of analyses.[162–164,167] Here, one examines the essential feature of the Kramers' method in the large damping limit, which is valid when the velocity relaxation is fast on the time scale characterizing the displacement of the particle and corresponding to covered distances over which the forces have changed by a large amount.

In the high-friction limit, the Smoluchowski expression (4.152) can be used to determine the time evolution of the particle and can be written as

$$\begin{aligned} \frac{\partial}{\partial t} P_1(t, x) &= \frac{\partial}{\partial x} \left(D \frac{\partial}{\partial x} P_1(t, x) + \frac{1}{\zeta} P_1(t, x) \frac{d}{dx} U(x) \right) \\ &= -\frac{\partial}{\partial x} J(t, x) \end{aligned} \tag{4.162}$$

The potential shown in Fig. 4.5 consists of two parabolas defined by

$$U(x) = \begin{cases} U_R(x) = \dfrac{m}{2} \omega_R^2 (x - x_R)^2 \,, & -\infty < x \le x_I \\ U_B(x) = Q_B - \dfrac{m}{2} \omega_B^2 (x - x_B)^2 \,, & x_I \le x < \infty \,. \end{cases} \tag{4.163}$$

The distance $x_B - x_R$ and the abscissa x_I of the inflexion point are uniquely determined by the barrier height Q_B and the angular frequencies ω_R and ω_B of the reactant well and the barrier potential.

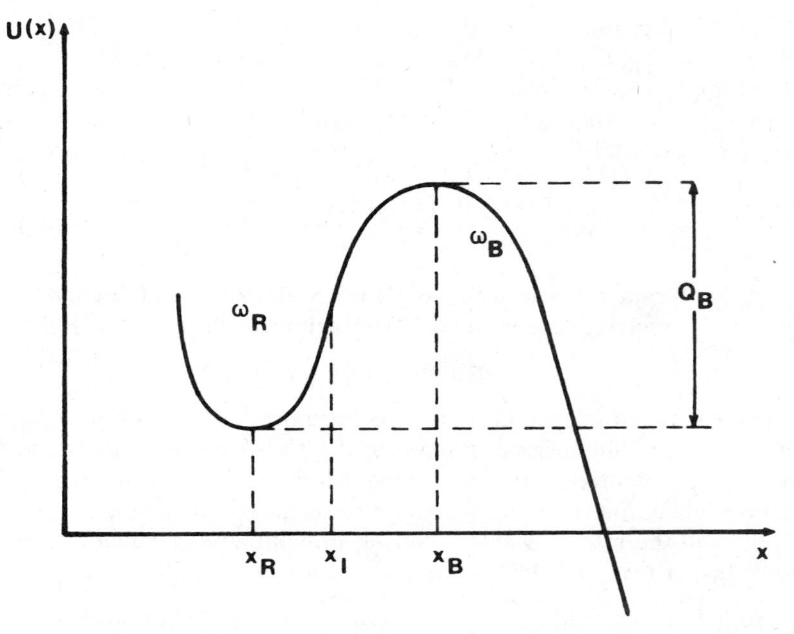

Fig. 4.5. Harmonic bistable potential [Eq. (4.163)].

The steady-state solution $P_S(x)$ is obtained by setting the left-hand side of Eq. (4.162) equal zero; hence,

$$P_S(x) = A\left[\exp - \int_{x_0}^{x} dy \, \frac{U'(y)}{k_B T}\right] \int_{x_0}^{x} dy \left[\exp \int_{x_0}^{y} dz \, \frac{U'(z)}{k_B T}\right] + B\left[\exp - \int_{x_0}^{x} dy \, \frac{U'(y)}{k_B T}\right] . \tag{4.164}$$

For the natural boundary condition $P_S(\pm\infty) = 0$, Eq. (4.164) becomes

$$P_S(x) = A\left[\exp - \int_{\infty}^{x} dy \, \frac{U'(y)}{k_B T}\right] \int_{\infty}^{x} dy \left[\exp \int_{\infty}^{y} dz \, \frac{U'(z)}{k_B T}\right] = A\left[\exp - \frac{U(x)}{k_B T}\right] \int_{\infty}^{x} dy \exp \frac{U(y)}{k_B T} . \tag{4.165}$$

Let J_0 be the stationary current, from Eq. (4.162) one sees that

$$J_0 = -D\left[\exp - \frac{U(x)}{k_B T}\right]\left[\frac{\partial}{\partial x} P_S \exp \frac{U(x)}{k_B T}\right] \tag{4.166}$$

Comparing Eqs. (4.166) and (4.165) it is obvious that $A = -J_0/D$ and

$$P_S(x) = \frac{J_0}{D}\left[\exp - \frac{U(x)}{k_B T}\right] \int_x^{\infty} dy \exp \frac{U(y)}{k_B T} . \tag{4.167}$$

It follows that

$$P_S(x_R) = \frac{J_0}{D} \int_{x_R}^{x_I} dy \exp \frac{U_R(y)}{k_B T} + \frac{J_0}{D} \int_{x_I}^{\infty} dy \exp \frac{U_B(y)}{k_B T} . \tag{4.168}$$

Under the condition of high barriers ($Q_B/k_B T \gg 1$), this expression can be approximated by

$$\begin{aligned} P_S(x_R) &\approx \frac{J_0}{D} \int_{-\infty}^{+\infty} dy \exp \frac{U_B(y)}{k_B T} \\ &\approx \frac{J_0}{D} \frac{(2\pi k_B T/m)^{1/2}}{\omega_B} \exp \frac{Q_B}{k_B T} . \end{aligned} \tag{4.169}$$

At long times the number of reactants $N(x_R)$ located at x_R is given by

$$\begin{aligned} N(x_R) &= P_S(x_R) \int_{-\infty}^{x_1} dy \exp\left(-\frac{U_R(y)}{k_B T}\right) \\ &\approx P_S(x_R) \frac{(2\pi k_B T/m)^{1/2}}{\omega_R} . \end{aligned} \tag{4.170}$$

Finally, the steady-state reaction rate k_S is calculated as the ratio of the current J_0 to the reactant population $N(x_R)$, that is,

$$k_S = \frac{m\omega_R \omega_B}{2\pi\zeta} \exp\left(-\frac{Q_B}{k_B T}\right) , \tag{4.171}$$

which is the celebrated Kramers' formula.

At this point the question that arises is: starting from an arbitrary initial distribution $P_1(0, x)$, how long does it take to the probability density function $P_1(t, x)$ to evolve to the steady-state solution $P_S(x)$?

In the foregoing treatment it is implicitly assumed that the system relaxes to the quasi-stationary distribution $P_{qs}(x, t)$ on a time scale τ_{qs} that is much shorter than the time scale $\tau_S \approx 1/k_S$ characterizing the evolution to $P_S(x)$. The following remarks on the Kramers method are of importance:

The dynamics are ignored at the very beginning.

The spatial equilibrium is assumed to be near the bottom of the reactant well:

$$P_{qs}(x, t) \approx P_S(x) P_R(t) \tag{4.172}$$

with $P_R(t) = \int P_1(t, x)\, dx$.

The result is valid only for high barriers ($Q_B/k_B T \gg 1$) [see Eq. (4.169)].

The spatial nonequilibrium is of importance in the barrier region, as can be seen in the following equation [deduced from Eq. (4.162)]:

$$\begin{aligned} J(t, x_B) &= -D\left[\exp - \frac{Q_B}{k_B T}\right]\left[\frac{\partial}{\partial x} P_1(t, x) \exp \frac{U(x)}{k_B T}\right]_{x=x_B} \\ &= -D P_S(x_B)\left[\frac{\partial}{\partial x} \frac{P_1(t, x)}{P_S(x)}\right]_{x=x_B} \end{aligned} \tag{4.173}$$

since $J(t, x_B)$ is finite and the term in brackets measures the deviation from the spatial equilibrium.

b. Low and Intermediate Frictions. 1. In situations where the inertial effects are dominating, Kramers derived,[16] under the essential assumptions of quasi-stationary regime and large barrier heights, the steady-state escape rate k_K as discussed in Sec. II.A.4.a (Eq. 2.1) given by

$$k_K = \frac{\omega_R}{4\pi\omega_B \tau_v} [(1 + (2\tau_v \omega_B)^2)^{1/2} - 1] \exp\left(-\frac{Q_B}{k_B T}\right), \tag{4.174}$$

where $\tau_v = m/\zeta$ represents the relaxation constant of the velocity of a Brownian particle [see Eqs. (4.126)].

At large damping ($\zeta \to \infty$ or better $\omega_B \tau_v \ll 1$), this expression yields Eq. (4.171), while for low friction ($\zeta \to 0$, $\omega_B \tau_v \gg 1$) it goes to the transition state theory rate k_{TST}

$$k_{TST} = \frac{\omega_R}{2\pi} \exp\left(-\frac{Q_B}{k_B T}\right). \tag{4.175}$$

This expression is expected to be valid if the time scale of the damping (τ_v) is larger than that of the motion on the barrier height ($1/\omega_B$). In contrast to the high-friction case, the barrier passage is negligibly perturbed and thus the motion of the particle to the product well is not hindered. In this case the rate is independent of the coupling between the heat bath and the particle.

2. For extremely low friction, the coupling to the thermal bath is so

weak that the quasi-stationary equilibrium cannot be maintained and energy exchange between the bath and the particle becomes rate limiting. It should be remembered that for the strong-coupling case the rate-determining process is the change of the position of the particle controlled by the friction forces.

Converting the Fokker–Planck equation (4.160) into a diffusion equation for the energy, Kramers has obtained the following approximate relation for the rate k_0 [16] (see also Eqs. 2.8 and 2.9):

$$k_0 = \frac{1}{\tau_v} \frac{Q_B}{k_B T} \exp\left(-\frac{Q_B}{k_B T}\right). \tag{4.176}$$

c. Remarks. Kramers restricted his investigation to nonreversible unimolecular reactions ($R \rightarrow P$). The particles initially are near the bottom of the reactant well, and, after passing over the barrier and moving downhill to the product well, they are not able to return. In practice, reversible reactions have to be envisaged. Modeling these situations by a bistable potential it has been shown[168–172] that the Smoluchowski escape rate k_S [Eq. (4.171)] is still correct.

Numerical results have been given,[173,174] based on the original model of Kramers, which reproduce the high-friction limit accurately. Recently, the Kramers' method has been reformulated[135,137,175] in order to emphasize the underlying assumptions, namely, the quasi-stationary (long-time) behavior of the system and the concept of a two-state system ($Q_B/k_B T \gg 1$).

The multidimensional case has been discussed for the different friction regimes.[168,176,177]

Different attempts have been elaborated[178–182] to find a general expression that accounts for the limiting forms (4.171) and (4.176) obtained in the high- and extreme-low-friction regimes. A new approach[183] made conspicuous the connection between the diverse Kramers' limits.

The applicability of the reaction-rate formula in the intermediate-friction regime to the interpretation of conformational changes of binaphtyl in solution has been discussed.[47] The intramolecular motion is governed by a bistable potential depending on a single variable, which is identified with the dihedral angle between the naphthalene moieties in binaphtyl. The influence of hydrodynamic interactions and coupling with the other coordinates of the molecules also are evoked.

d. Solvent Viscosity Dependence of the Reaction Rate in Intramolecular Conformational Dynamics. To discuss this influence, the following assumptions are made: (1) the Kramers' model applies for the high- and intermediate-friction regime; (2) the internal motion is determined by a mean bistable potential independent of the solvent properties (polarity, viscosity, etc.); (3) the temperature variation of the viscosity of the solvents is of the form

$$\eta = \eta_0 \exp Q_\eta / k_B T \; ; \tag{4.177}$$

(4) the Stokes law holds in the temperature range under investigation

$$\zeta = 6\pi a \eta \, . \tag{4.178}$$

With these expressions and $\tau_v^0 = m/\zeta_0 = (m/6\pi a \eta_0)$, Eqs. (4.171) and (4.174) can be written

$$\frac{k_S}{k_{TST}} = \omega_B \tau_v = \omega_B \tau_v^0 \exp\left(-\frac{Q_\eta}{k_B T}\right), \tag{4.179}$$

$$\frac{k_K}{k_{TST}} = \frac{1}{2k_S/k_{TST}} \left\{\left[1 + \left(2\frac{k_S}{k_{TST}}\right)^2\right]^{1/2} - 1\right\}. \tag{4.180}$$

The investigation of the isomerization process for the binaphtyl system in solution gives a reequilibration velocity time constant $\tau_v \approx 0.3$ ps for a viscosity η near to 1 cp and a value of 14 ps^{-1} for ω_B.[47]

For the most frequently studied large aromatic molecules, the angular frequency of the harmonic potential is assumed to be of the order of 10^{12}–10^{14} s^{-1} and the correlation time τ_v, which can be estimated by geometrical and hydrodynamical considerations, is usually of the order of 0.1 ps at 1 cp. Hence, it can be expected that a large number of unimolecular processes governed by intramolecular motion concern the intermediate case, and it is interesting to analyze the temperature dependence of the reaction rate k_K [Eq. (4.174)] and of the ratio k_K/k_{TST} [Eq. (4.180)] for different domains of variation of the parameter $\omega_B \tau_v$. Consider, for example, butanol solutions, whose viscosity can reasonably well be approximated by Eq. (4.177) with $\eta_0 = 10^{-3}$ cp and $Q_\eta =$ 4.6 kcal M^{-1} in the temperature range from 60°C to near the freezing point. Figure 4.6 reports for two values of $\omega_B \tau_v^0$ (5×10^4 and 10^4) corresponding, respectively, to $\omega_B \tau_v = 50$ (curves 1) and 10 (curves 2) at 1 cp, the calculated variations of k_K/k_{TST} versus the temperature and those of $(2\pi/\omega_R)k_K$ for a barrier height $Q_B = 2$ kcal M^{-1}. The straight line represents the variation of the transition-state theory rate divided by $\omega_R/2\pi$. This figure clearly illustrates that if the explored temperature range is sufficiently wide, a deviation from an Arrhenius plot should be observed. Some essential conditions for deviations from the Arrhenius

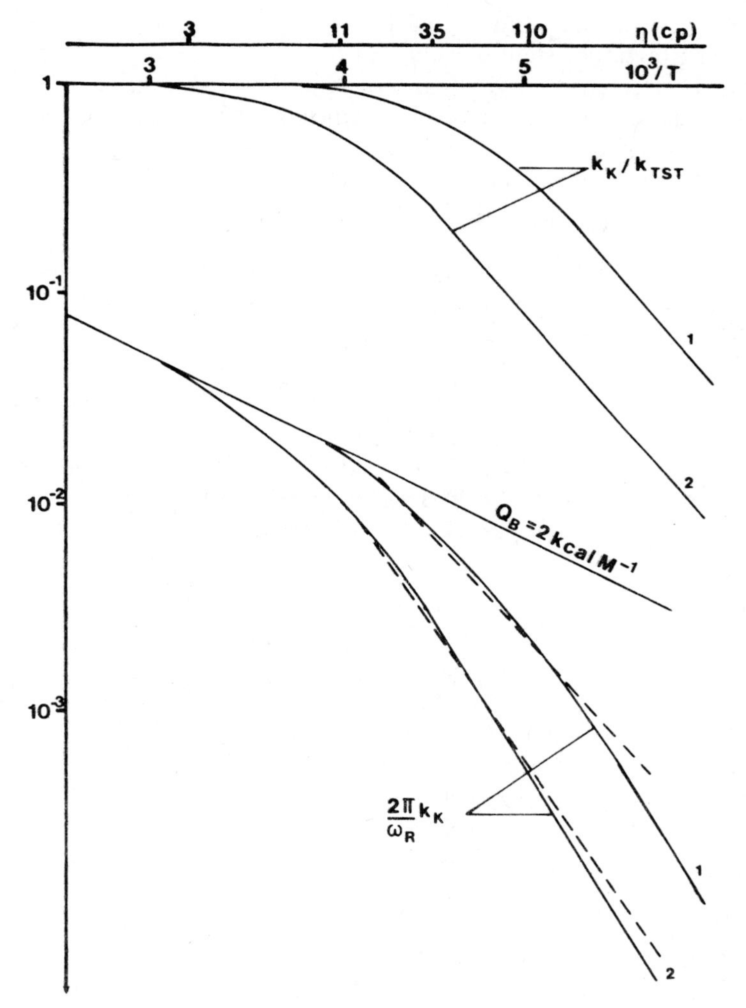

Fig. 4.6. Influence of the temperature on the Kramers' reaction rate [intermediate friction regime Eq. (4.174)] for $\omega_B \tau_v^0 = 5 \times 10^4$ (curves 1) or 10^4 (curves 2). The straight line represents $\exp(-Q_B/k_B T)$ with $Q_B = 2$ kcal M^{-1}.

plot had also been pointed out by means of Laplace transformation concerning quantum state density and transition probability functions by the late H. Labhart in 1967.[229] In our case the activation energy Q_{act} for the reaction rate, which could be deduced from an approximately exponential behavior of k_K, can be written

$$Q_{act} = Q_B + \alpha Q_\eta \,, \tag{4.181}$$

with $0 < \alpha \leq 1$, and is strongly dependent on $\omega_B \tau_v^0$ and also on the temperature (or viscosity) range where it is determined. In this example, one finds $Q_{act} \approx 4.4$ (or 5.8) kcal M^{-1} for $\omega_B \tau_v^0 = 5 \times 10^4$ (or 10^4), between $-20°C$ and $-90°C$ (dashed lines) and one can remark that the Smoluchowski limit ($\alpha = 1$) is only attained below $-80°C$ (or $-60°C$).

e. Dynamics in the High-Friction Limits. Using an exact time-dependent solution for a one-dimensional diffusion in a bistable potential, van Kampen was able to present in detail the different time scales characterizing the evolution of the particle.[184] The Smoluchowski equation is written in the form of

$$\frac{\partial}{\partial t} p = D \frac{\partial^2}{\partial x^2} p + D \frac{\partial}{\partial x} p \frac{d}{dx} W \,, \tag{4.182}$$

where $p \equiv p(x_0; t, x)$ and $W = U(x)/k_B T$. He showed that for a symmetrical square well potential (Fig. 4.7*a*) and for an initial distribution located sharply at the bottom of one well, that is, $p_0(0, x_0) = \delta(x - x_0)$, Eq. (4.182) is exactly solvable and can be expressed in terms of eigenfunctions and eigenvalues of the Schrödinger equation

$$\psi_i'' + [\lambda_i - w(x)]\psi_i = 0 \,, \qquad i = 1, 2, \ldots, \tag{4.183}$$

with $w(x) = W'^2/4 - W''/2 + w_0$, and w_0 an arbitrary constant provided that the ground state $\psi_0(x)$ and the potential W satisfy

$$W(x) = -2 \log \psi_0(x) \,. \tag{4.184}$$

Thus, the general solution is

$$p(x_0; t, x) = \psi_0^2(x) + \psi_0(x) \sum_{i=1}^{\infty} \frac{\psi_i(x_0)}{\psi_0(x_0)} \psi_i(x) \exp[-D(\lambda_i - \lambda_0)t] \tag{4.185}$$

and the set of eigenvalues verify the following interesting condition

$$\lambda_1 - \lambda_0 \ll \lambda_i - \lambda_0 \,, \qquad i > 1 \,,$$

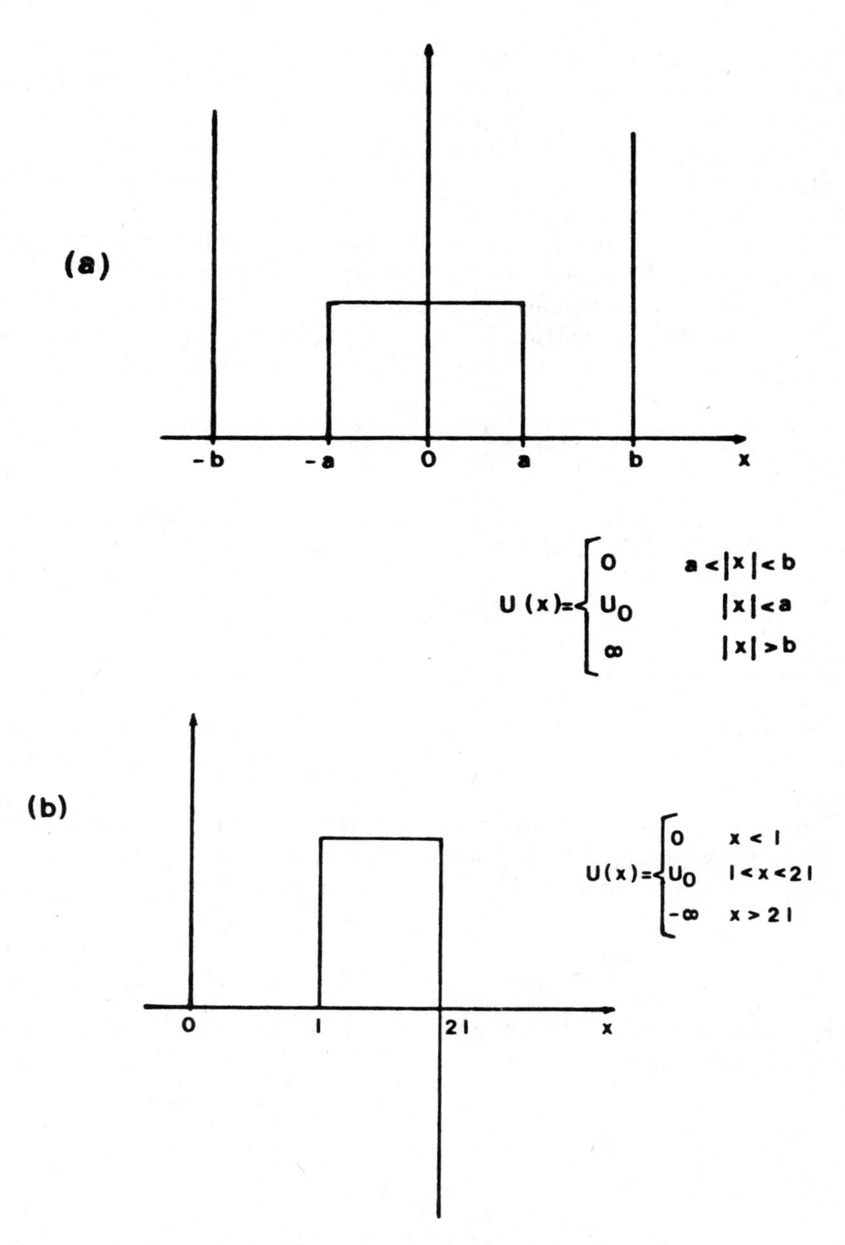

Fig. 4.7. (*a*) Rectangular potential model; (*b*) metastable potential model.

which clearly shows that (1) for times t larger than $[(\lambda_2 - \lambda_1)D]^{-1}$, Eq. (4.185) reduces to a single exponential law, and (2) the decay rate to the stationary distribution $P_S(x) \propto \psi_0^2(x)$ is given by the lowest eigenvalue.

Moreover, assuming a large barrier height ($a\sqrt{W_0} \gg 1$), van Kampen could define the time scale $t_{qs} = (DW_0)^{-1}$ on which occurs the first stage of evolution leading to quasi-stationary equilibrium.

The second stage of evolution from quasi-equilibrium (Kramers situation) takes place on a much larger time scale $1/k_S$ [Eq. (4.171)].

For a further comment it is convenient to compare the feature $(DW_0)^{-1}$ to the equivalent quantity corresponding to the case of a parabolic-shaped barrier in the linearized approximation, that is,

$$\frac{\partial}{\partial t} p = D \frac{\partial^2}{\partial x^2} p - \frac{m\omega_B^2}{\zeta} \frac{\partial}{\partial x} xp . \tag{4.186}$$

It is easy to see that $(2DW_0)^{-1}$ is replaced by the characteristic time $\tau_c^B = \zeta/m\omega_B^2 = 1/\omega_B^2 \tau_v$ and that the escape rate k_S can be written as

$$k_S = \omega_B \tau_c^B k_{TST} . \tag{4.187}$$

Keeping in mind the typical values of τ_v and ω_B, one has estimates of t_{qs} between 1 fs and some nanoseconds for viscosities from 1 to 10^3 cp.

The eigenmode expansion was also used to determine the time-dependent solution of the Smoluchowski equations for diverse bistable potentials.[185]

In particular, the analytical solution of the Smoluchowski equation for a metastable potential well depicted in Fig. 4.7b was found taking as initial condition a uniform distribution in the potential

$$p_0(0, x_0) = \begin{cases} 1/l , & 0 \le x \le l , \\ 0 , & x < 0 \quad \text{or} \quad x > l . \end{cases} \tag{4.188a}$$

The probability $P_1^W(t)$ of finding the particle in the potential well at time t has the following form:

$$P_1^W(t) = \frac{2}{\pi^2} \left[1 + \exp\left(-\frac{U_0}{k_B T} \right) \right]^{-1} \sum_{n=-\infty}^{n=+\infty} (2n + \nu)^{-2}$$
$$\times \exp\left[-\left(\frac{\pi}{2l} \right)^2 D(2n + \nu)t \right] , \tag{4.188b}$$

where $\nu = (2/\pi) \arctan \exp(-U_0/2k_B T)$.

Note that this expression is valid for any barrier height and can be approximated by a single exponential

$$P_1^{\mathrm{W}}(t) \approx \exp(-k^{\mathrm{W}}t) \tag{4.189}$$

with

$$k^{\mathrm{W}} = \frac{D}{l^2}\left(\exp\frac{U_0}{k_{\mathrm{B}}T} + \frac{1}{3}\right)^{-1} . \tag{4.190}$$

In the high barrier assumption the escape rate k^{W} becomes

$$k^{\mathrm{W}} = \frac{D}{l^2}\exp - \frac{U_0}{k_{\mathrm{B}}T} . \tag{4.191}$$

f. Time-Dependent Escape Rate (*Reaction Rate*).[186] To explain the meaning of $P_1^{\mathrm{W}}(t)$—the probability of finding the particle still in the well at time t, recalling that it was generated within it with uniform probability—let the random variable T be the time at which the particle crosses the barrier (suffers a reaction). It is obvious that the survival probability defined as the probability of the particle being in the well after time t satisfies

$$\Pr[T > t] = P_1^{\mathrm{W}}(t) . \tag{4.192}$$

One now considers the random variable $N(t)$ representing the total number of particles that are still in the well at time t. Clearly, each particle is characterized by Eq. (4.192) and $N(t)$ is taking values $0, 1, \ldots, n$. It is not difficult to show that the expected number of particles in the well at time t is

$$\begin{aligned} E[N(t)] &= n\Pr[T > t] \\ &= m(t) . \end{aligned} \tag{4.193}$$

Thus $m(t) - m(t + \Delta t)$ is the mean number of particles crossing the barrier in the time interval $(t, t + \Delta t)$. Letting $k(t)\,\Delta t$ be the ratio of the expected number of escapes in the interval $(t, t + \Delta t)$ to the mean number of particles $m(t)$ in the well at time t, one has

$$k(t) = \lim_{\Delta t \to 0} \frac{1}{\Delta t}\,\frac{m(t) - m(t + \Delta t)}{m(t)} = -\frac{1}{m(t)}\,\frac{d}{dt}\,m(t) \tag{4.194}$$

and the following kinetic equation

$$\frac{d}{dt} m(t) = -k(t)m(t) . \tag{4.195}$$

Integrating this expression and noting that $m(0) = n$, one finds

$$m(t) = n \exp\left(-\int_0^t k(t)\, dt\right) . \tag{4.196}$$

From the definition of the survival probability and Eqs. (4.192) and (4.193), it is easy to verify that

$$\begin{aligned} k(t) &= \lim_{\Delta t \to 0} \frac{1}{\Delta t} \Pr\left[t < T \le t + \Delta t / T > t\right] \\ &= -\frac{1}{P_1^{W}(t)} \frac{d}{dt} P_1^{W}(t) . \end{aligned} \tag{4.197}$$

The escape rate $k(t)$ is interpreted as the probability density that a particle crosses the barrier (undergoes a reaction) in an infinitesimal interval t, $t + \Delta t$ assuming that it was in the well (did not react) up to time t. This interpretation only holds if the same probability density $f_T(t) = -d \Pr[T > t]/dt$ characterizes the particles in the well.

If one considers the metastable potential model, by inspection of Eqs. (4.188) and (4.197) it is apparent that the rate is time dependent, which results from nonexponential decay of the survival probability. Note that if $P_1^{W}(t) = \exp(-\lambda t)$, then $k(t) = \lambda$ and $m(t) = n \exp(-\lambda t)$.

2. *Non-Markovian Theory of Activated Rate Process*

In the Kramers' approach the friction ζ models collisions between the particle and the surrounding medium, and it is assumed that the collisions occur instantaneously. There is a time-scale separation between the reactive mode and its thermal bath. The dynamics are described by the Langevin equation (4.141). The situation where the collisions do not occur instantaneously but take place on a time scale characterizing the interactions between the particle and its surrounding can be described by a generalized Langevin equation (GLE),[158,187]

$$\begin{aligned} &\frac{d}{dt} X = V , \\ &m \frac{d}{dt} V = -\int d\theta\, \zeta(t-\theta) V(\theta) + F(X) + \mathcal{F}(t) , \end{aligned} \tag{4.198}$$

where the time-dependent friction $\zeta(t)$ and the gaussian random force $\mathcal{F}(t)$ are related by the fluctuation–dissipation theorem

$$E[\mathcal{F}(t)\mathcal{F}(t+\tau)] = k_B T\zeta(\tau)$$
$$= k_B T\zeta_0 C(\tau) \tag{4.199}$$

The term $\int d\theta\ \zeta(t-\theta)V(\theta)$ implies that the determination of the properties of the process $\{X(t), V(t)\}$ at time t depends on its behavior during the interval $(0, t)$. Hence, the process $\{X(t), V(t)\}$ is not Markovian. The GLE expression reduces to the conventional Langevin equation if the friction decays on a time scale much shorter than the time development of the velocity. Indeed, the convolution $\zeta(t) \otimes V(t)$ can be written as $\zeta_0\delta(t) \otimes V(t) = \zeta_0 V(t)$. Thus, the static friction component is given by

$$\zeta_0 = \int_0^\infty \zeta(t)\, dt\ .$$

In contrast to the diffusion processes, the fluctuating force $\mathcal{F}(t)$ can no longer be derived from the Wiener process, and its spectral density defined as $2k_B T\zeta_0$ times the Fourier transform of the memory function $C(t)$ is frequency limited and has no more the white noise characteristics.

To make the meaning of the physical interpretation of the local friction complete, consider an oscillator immersed in a non-Markovian fluid. For high-frequency oscillations, the heat bath (of limited power spectrum) cannot respond to the oscillator motion; thus, on a sufficiently short time scale, the oscillator is weakly coupled to the bath. In the other extreme case when the oscillator displacements are very slow, the heat bath is able to thoroughly adjust its motion; this situation corresponds to the adiabatic regime for which the oscillator behavior is governed by the phenomenological Langevin equation. At intermediate frequencies there is a time lag between the motions of the oscillator and the surroundings.

a. Intermediate- and High-Friction Limits. Grote and Hynes[57,188] first pointed out the non-Markovian effects in the barrier dynamics. As mentioned previously, for these limits the well population is considered to be in equilibrium and the potential barrier crossing process is the rate-determining step. Two distinct treatments have been considered to solve this non-Markovian escape problem: the stable state picture[57] and the linearized harmonic approximation.[189,190] They led to the same result for the steady-state rate of reaction

$$k_{GH} = k_{TST}\,\frac{\lambda_r}{\omega_B}\ ; \tag{4.200}$$

the reactive frequency λ_r is the largest real and positive root of the self-consistent equation

$$\lambda_r^2 + \hat{\zeta}(\lambda_r)\lambda_r/m - \omega_B^2 = 0 \quad (4.201)$$

where $\hat{\zeta}(\lambda_r)$ corresponds to the Laplace transform of the time-dependent friction. It is easy to see that for $\zeta(t) = \zeta_0\delta(t)$, Eqs. (4.200) and (4.201) yield the Kramers' result [Eq. (4.174)].

The influence of the time-dependent friction is depicted in Fig. 4.8, where $\zeta(t)$ is supposed to be gaussian: $\zeta(t) = (2\zeta_0/\tau_c\pi^{1/2}) \times \exp[-(t/\tau_c)^2]H(t)$, where $H(t)$ is the Heaviside function. The ratio k_{GH}/k_{TST} is plotted verus $1/\tau_v\omega_B$ ($\tau_v = m/\zeta$) for various values of the characteristic parameter $1/\tau_c\omega_B$: one notes that the Kramers' limit is obtained for small values of the correlation time τ_c (i.e., $\omega_B\tau_c \ll 1$).

It should be stressed that for the double-well reaction model in the non-Markovian case a general result similar to the Kramers' expression (4.160) cannot be found. To evaluate the thermally activated escape rate, the motion within the barrier region is described by means of a GLE in which the potential near the barrier is linearized, that is,

$$\frac{dX}{dt} = V\,,$$
$$m\frac{d}{dt}V = -\zeta(t) \otimes V(t) - U'(X) + \mathcal{F}(t)\,, \quad (4.202)$$

with $U(x) = Q_B - \frac{1}{2}m\omega_B^2x^2$. For a critical examination of this linearization approximation [note that here the potential $U(x)$ is purely repulsive] the reader is referred to the discussion of van Kampen.[184]

Missing the random force term $\mathcal{F}(t)$, Eq. (4.202) is solved by Laplace transforms. Setting $\hat{X} \equiv \hat{X}(\lambda) = \mathcal{L}X(t)$ and $\hat{V} \equiv \hat{V}(\lambda) = \mathcal{L}V(t)$, it is easy to verify that[191a]

$$\hat{X} = \{v_0 + x_0[\lambda + \hat{\zeta}(\lambda)/m]\}[\lambda^2 + \lambda\hat{\zeta}(\lambda)/m - \omega_B^2]\,, \quad (4.203)$$

and taking the term $\mathcal{F}(t)$ into account, one has

$$X(t) = C_v(t)v_0 + C_x(t)x_0 + \int_0^t C_v(t-\tau)\mathcal{F}(\tau)\,d\tau\,, \quad (4.204)$$

where

$$C_v(t) = \mathcal{L}^{-1}[\lambda^2 + \lambda\hat{\zeta}(\lambda)/m - \omega_B^2]\,, \quad (4.204a)$$

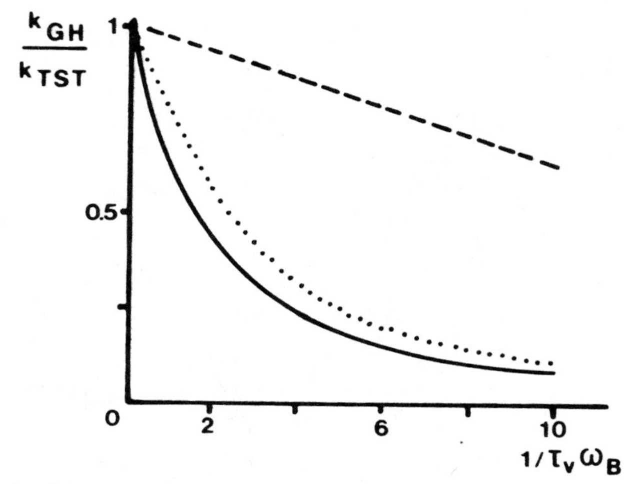

Fig. 4.8. Ratio of the rate constant (k_{GH}) for a gaussian friction to its transition-state value (k_{TST}) versus the reduced zero-frequency friction $1/\tau_v\omega_B$ for $\omega_B\tau_c = 10$ (– – –) and 1 (· · ·). The solid line corresponds to the zero-frequency friction (Kramers result).[57]

$$C_x(t) = 1 - \omega_B^2 \int_0^t C_v(\tau)\, d\tau\,. \tag{4.204b}$$

$C_v(t)$ is related to the velocity correlation function by the following expression:

$$C_v(\tau) = \frac{m}{k_B T} E[V(t)V(t+\tau)]\,. \tag{4.205}$$

Derivating Eq. (4.204) with respect to time, one sees that

$$V(t) = \dot{C}_v(t)v_0 + \dot{C}_x(t)x_0 + \int_0^t \dot{C}_v(t-\tau)\mathscr{F}(\tau)\, d\tau\,. \tag{4.206}$$

A generalized Fokker–Planck equation for the unstable state equivalent to the previous GLE (4.202) can be calculated[191b,192]:

$$\frac{\partial}{\partial t} p + v \frac{\partial}{\partial x} p - \omega^2(t)x \frac{\partial}{\partial x} p$$
$$= \frac{\tilde{\zeta}(t)}{m} \frac{\partial}{\partial x} vp + \frac{k_B T}{m^2} \tilde{\zeta}(t) \frac{\partial^2}{\partial v^2} p + \frac{k_B T}{m\omega_B^2} [\omega^2(t) - \omega_B^2] \frac{\partial^2}{\partial x\, \partial v} p\,, \tag{4.207}$$

where $\omega^2(t)$ and $\tilde{\zeta}(t)$ are defined by

$$\tilde{\zeta}(t) = -\frac{d}{dt}\ln[\det \Delta(t)]\,, \tag{4.207a}$$

$$\omega^2(t) = \det \bar{\Delta}(t)/\det \Delta(t)\,, \tag{4.207b}$$

with

$$\Delta(t) = \begin{vmatrix} \dot{C}_v(t) & \dot{C}_x(t) \\ C_v(t) & C_x(t) \end{vmatrix}, \qquad \bar{\Delta}(t) = \begin{vmatrix} \ddot{C}_v(t) & \ddot{C}_x(t) \\ C_v(t) & C_x(t) \end{vmatrix}. \tag{4.207c}$$

In the Markovian limit $[\zeta(t) = \zeta_0\delta(t),\ \omega(t) = \omega_B]$ Eq. (4.207) reduces to the Kramers' equation (4.160). It is of importance to note that the long-time behavior of the solution does not necessarily approach the Markovian result.

Hänggi and Mojtabai[189] based their treatment on the generalized Fokker–Planck expression and demonstrated that the steady-state escape rate is given by the Grote and Hynes relationship (4.200) where the reactive frequency λ_r is defined as the long-time limit given by

$$\lambda_r = \lim_{t\to\infty}\left[\left(\omega^2(t) + \left|\frac{\tilde{\zeta}(t)}{2m}\right|^2\right)^{1/2} - \frac{\tilde{\zeta}(t)}{2m}\right]. \tag{4.208}$$

Moreover, it has been established that λ_r is the root of Eq. (4.201).

b. General Expression of the Steady Escape Rate. Carmeli and Nitzan emphasized the inconsistency of the Kramers' escape rate k_K [Eq. (4.174)] which goes to the rate k_{TST} [Eq. (4.175)] calculated in the transition-state theory as the friction tends to zero and not to k_0 [Eq. (4.176)] evaluated in the extreme low-friction condition. Their recent treatment[51,52,193] extended to the double-well non-Markovian model yields a reaction rate expression k_{CN} valid throughout the whole friction domain that reduces to the extreme low- and high-friction limits.

The transmission coefficient k_{CN}/k_{TST} (Fig. 4.9) has been evaluated under the following conditions:

1. The non-Markovian heat bath presents an exponential correlation function characterized by τ_c.
2. The double-well potential is symmetrical featured by ω_R, ω_B, and $Q_B = 10.7\ \text{kcal M}^{-1}$. The variations of the ratio k_{CN}/k_{TST} are plotted as a function of $\log 1/\omega_B\tau_v$ for three different values of ω_R/ω_B and for various $\omega_B\tau_c$. Note the case $\omega_B\tau_c = 0$ corresponds to the Markovian situation.

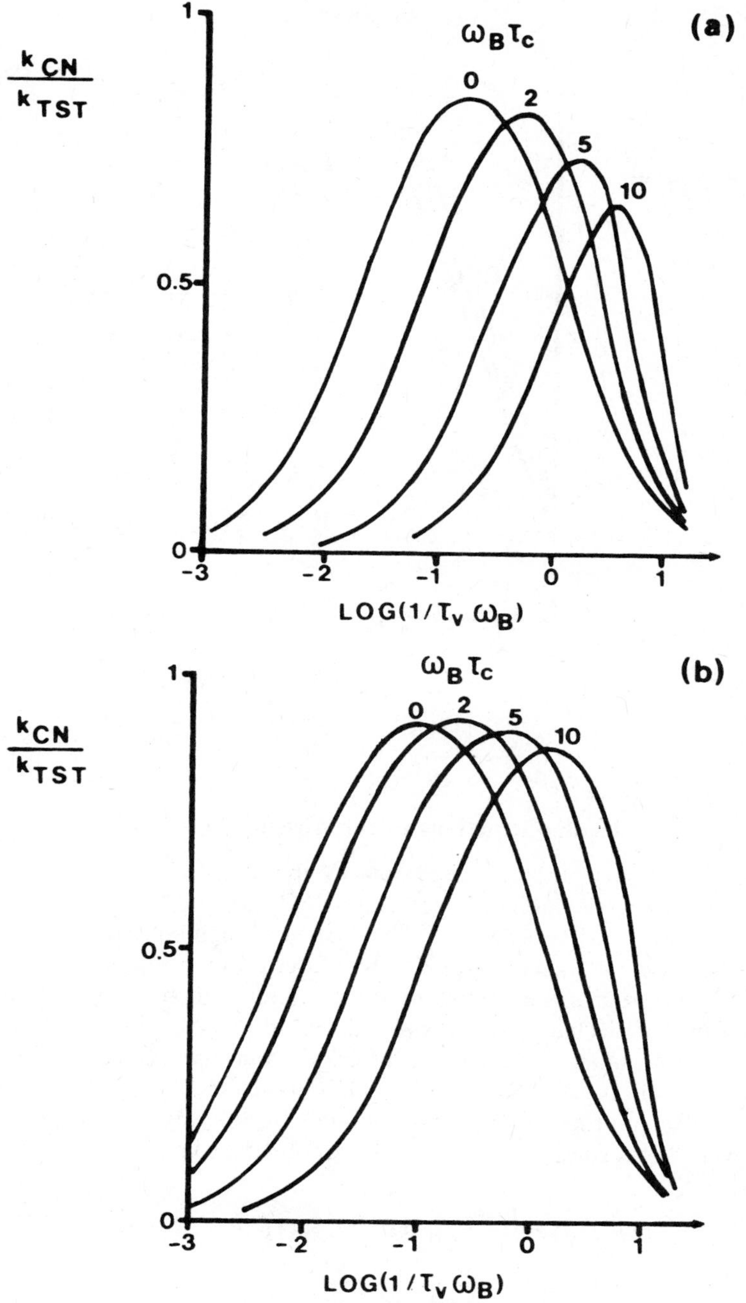

Fig. 4.9. The transmission coefficient k_{CN}/k_{TST} as a function of $1/\tau_v \omega_B$ for a symmetrical double-well potential with $Q_B = 10.7$ kcal M^{-1}. (*a*) $\omega_R/\omega_B = 5$; (*b*) $\omega_R/\omega_B = 1$; (*c*) $\omega_R/\omega_B = 0.2$.[52]

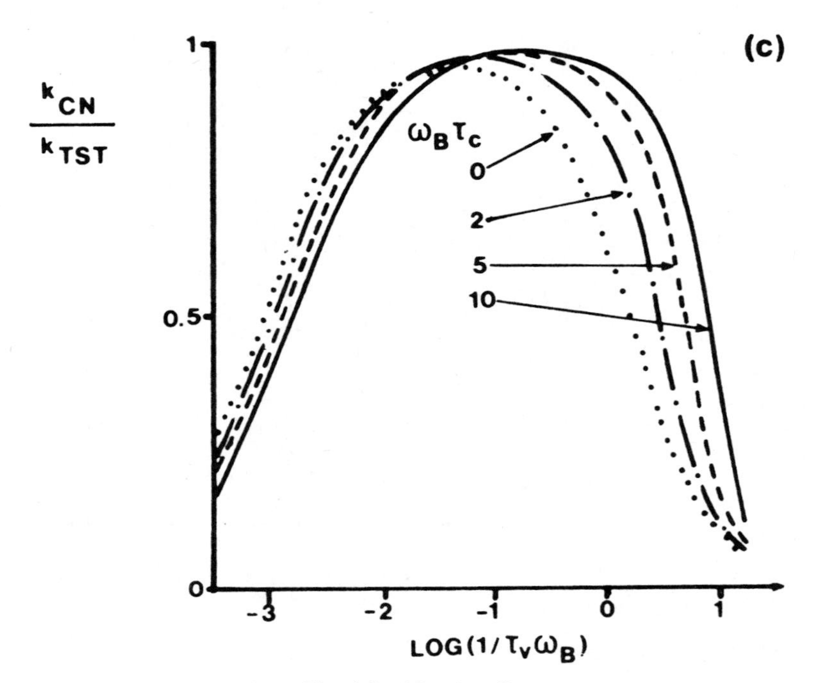

Fig. 4.9 (*Continued*).

I. Unimolecular Reactions in the Absence of a Potential Barrier

Kramers' theory and its extension to the non-Markovian case breaks down if the barrier height becomes low ($Q_B/k_BT \leq 1$) and a zero barrier model has been developed.[194,195] The cases of high and low viscosity have been treated by Bagchi et al.[195] This description concerns, for example, photoisomerization reactions in solution for which the intramolecular conformational change is in competition with the solvent motion. Here the unimolecular reaction—the passage from the reactant well to the product well in the Kramers' treatment—is modeled by a sink term depending on the reaction coordinate. In the high-viscosity case the motion is governed by the modified Smoluchowski equation:

$$\frac{\partial}{\partial t} P - D \frac{\partial^2}{\partial x^2} P - \frac{m\omega^2}{\zeta} \frac{\partial}{\partial x} xP = -S(x)P - k_r P + \delta(x + |x_0|)\delta(t) , \tag{4.209}$$

where $P \equiv P_1(t, x; x_0)$ represents the probability density of finding a particle at x at time t, starting at the beginning $t = 0$ from the position x_0. The left-hand-side expression corresponds to the motion of a one-dimensional Brownian particle subject to a force from a harmonic potential. The right-hand side includes the sink function $S(x)$, the spontaneous decay of rate k_r, and, finally, the initial condition.

If the sink corresponds to a pinhole, it is of the form

$$S(x) = k_{nr}\delta(x - x_1)\,, \tag{4.210}$$

where k_{nr} is related to the microscopic rate constant that determines the effectiveness of the reaction occurring at the position x_1.[79] If a reaction takes place whenever the particle reaches x_1 ($k_{nr} \to \infty$), P satisfies the so-called Smoluchowski equation:

$$P_1(t, x_1; x_0) = 0\,. \tag{4.211}$$

For a finite value $k_{nr} \neq 0$, P verifies the following boundary condition:

$$D\,\frac{\partial}{\partial x}\,P\bigg|_{x=x_1} = k_{nr}P|_{x=x_1}\,, \tag{4.212}$$

if $k_{nr} = 0$, Eq. (4.209) describes the motion of a Brownian particle undergoing spontaneous disappearance.

The analytical solutions of Eq. (4.209) subject to the preceding boundary conditions are known. In particular, for a totally efficient sink at the origin ($k_{nr} \to \infty$, $x_1 = 0$), the solution is of the following form:

$$P_1(t, x; x_0) = \frac{1}{\sqrt{2\pi}\sigma}\left\{\exp\left[-\frac{(x + y_0)^2}{2\sigma^2}\right] - \exp\left[-\frac{(x - y_0)^2}{2\sigma^2}\right]\right\}\exp(-k_r t)\,, \tag{4.213}$$

where

$$y_0 = |x_0|\exp(-t/\tau_c^{\omega})\,, \tag{4.214}$$

$$\sigma^2 = D\tau_c^{\omega}[1 - \exp(-2t/\tau_c^{\omega})]\,, \tag{4.215}$$

and

$$\tau_c^{\omega} = \zeta/m\omega^2 = 1/\omega^2\tau_v\,. \tag{4.216}$$

The spatial distributions at various times are represented in Fig. 4.10. Note that at long times $P_1(t, x; x_0)$ is approximated by the Rayleigh distribution:

$$P_1(t, x; x_0) = \sqrt{\frac{2}{\pi}} \frac{|x_0|}{\sqrt{D\tau_c^\omega}} \frac{x}{D\tau_c^\omega} e^{-x^2/2D\tau_c^\omega} e^{-t/\tau_c^\omega} e^{-k_r t} . \qquad (4.217)$$

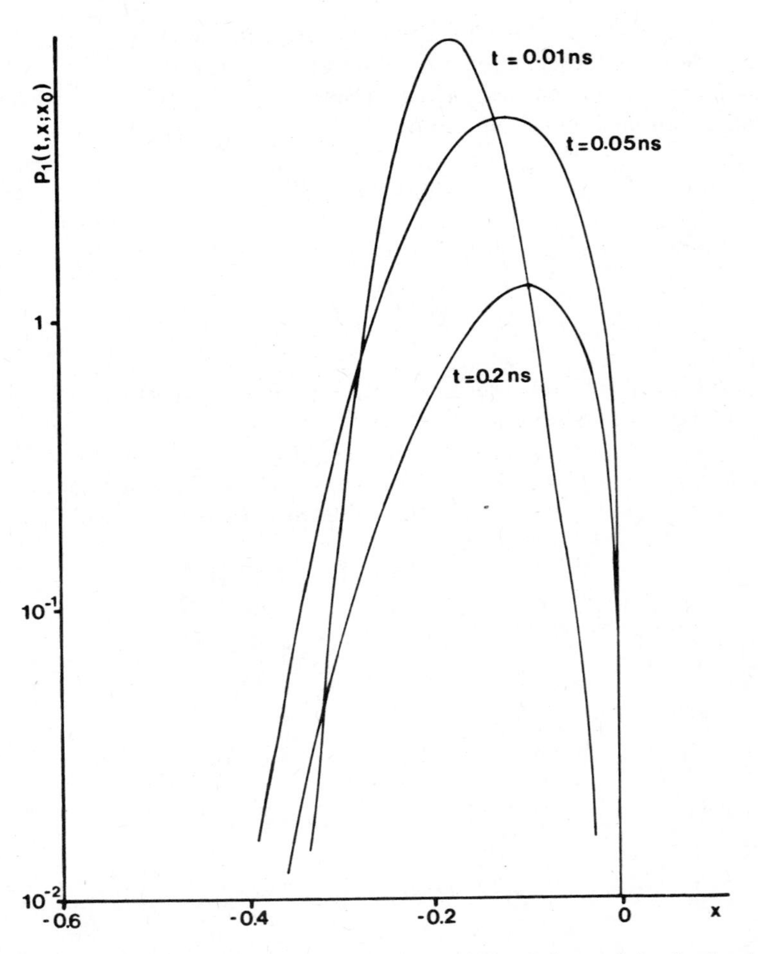

Fig. 4.10. Variation with time of the density probability $P_1(t, x; x_0)$ in the barrierless model with a totally efficient sink at $x = 0$, and $p_0(0, x_0) = \delta(x + |x_0|)$; $x_0 = -0.2$ Å, $D = 0.1$ Å^2 ns^{-1}; $\tau_c^\omega = 0.1$ ns.

Let $p(t)$ be the survival probability of the particle, that is,

$$p(t) = \int_{-\infty}^{0} P_1(t, x; x_0)\, dx = \operatorname{erf}(y_0/\sigma\sqrt{2}) \exp(-k_r t) , \qquad (4.218)$$

with

$$\frac{y_0^2}{2\sigma^2} = \left(\frac{2\tau_c^0}{\tau_c^\omega}\right) \frac{\exp(-2t/\tau_c^\omega)}{1 - \exp(-2t/\tau_c^\omega)} \qquad (4.219)$$

and

$$\tau_c^0 = x_0^2/4D . \qquad (4.220)$$

It is instructive to note that

$$2\tau_c^0/\tau_c^\omega \equiv x_0^2/2D\tau_c^\omega = U(x_0)/k_B T , \qquad (4.221)$$

where $U(x_0) = m\omega^2 x_0^2/2$ is the initial potential energy.

It is interesting to examine the behavior in the following extreme situations:

1. $t/\tau_c^\omega \ll 1$: for short times or for high friction and/or small characteristics frequencies ω, one can write

$$p(t) \approx e^{-k_r t} \operatorname{erf}\left(\frac{\tau_c^0}{t}\right)^{1/2} . \qquad (4.222)$$

This expression is identical to that calculated for $\omega = 0$: at short times, the nonexponential decay is the same as that of a free motion particle in the presence of a pinhole sink.

2. If $t/\tau_c^\omega \gg 1$, which corresponds to small friction and/or large frequencies ω or to the long-time limit, $p(t)$ becomes

$$p(t) \approx \frac{2}{\sqrt{\pi}} \left(\frac{2\tau_c^0}{\tau_c^\omega}\right)^{1/2} e^{-k_r t} e^{-t/\tau_c^\omega} , \qquad (4.223)$$

which predicts a single exponential decay. Assuming the validity of the Stokes law, the rate of reaction is proportional to the recriprocal of the viscosity.

One is in the presence of a two-time-scale description. In the short part, the system relaxes nonexponentially to a quasi-stationary state with the characteristic time τ_c^0 [Eq. (4.220)] depending on the initial condition, and the spatial distribution shows large variations. Then, on a longer-time scale, it evolves exponentially with a decay time τ_c^ω [Eq. (4.216)], whereas the shape of the spatial distribution remains unchanged and equals a Rayleigh distribution [Eq. (4.217)]. As stated for the long-time behavior of Markovian processes, the dynamics of the system are independent of the initial condition.

Furthermore, it is instructive to consider the time evolution of the dynamic variables $E[X(t)]$ and $\mathrm{Var}\,[X(t)]$ representing the expectation and the variance of the oscillator displacements. Noting that $P_1(t, x; x_0)/p(t)$ represents the probability density function associated to the random function $X(t)$, one sees that

$$E[X(t)] = \int_{-\infty}^{0} x \,\frac{P_1(t, x; x_0)}{p(t)}\, dx = -y_0/\mathrm{erf}\,(y_0/\sigma\sqrt{2})\,,$$

$$\mathrm{Var}\,[X(t)] = \int_{-\infty}^{0} x^2 \,\frac{P_1(t, x; x_0)}{p(t)}\, dx - E^2[X(t)]\,,$$

$$= \left[y_0\sigma\sqrt{\frac{2}{\pi}}\, \exp\left(-\frac{y_0^2}{2\sigma^2}\right)\Big/ \mathrm{erf}\,\frac{y_0}{\sigma\sqrt{2}} \right] + (y_0^2 + \sigma^2) - \frac{y_0^2}{\mathrm{erf}^2\,(y_0/\sigma\sqrt{2})}\,. \tag{4.224}$$

A general remark: y_0 is the mean trajectory and σ is the associated dispersion of the oscillator in the absence of the pinhole sink [see Eq. (4.215)].

One recognizes here the previously mentioned time-scale separation.

1. $t/\tau_c^\omega \ll 1$:

$$E[X(t)] \approx -|x_0|\,, \qquad \mathrm{Var}\,[X(t)] \approx 2Dt\,; \tag{4.225}$$

2. $t/\tau_c^\omega \gg 1$:

$$E[X(t)] \approx -(\pi D\tau_c^\omega/2)^{1/2}\,, \qquad \mathrm{Var}\,[X(t)] \approx (2 - \pi/2)D\tau_c^\omega\,. \tag{4.226}$$

From Eqs. (4.225) it results that at very short times the behavior of the oscillator may be approximated by a Brownian motion. On the longer time scale the relations (4.226) may be identified as the expectation and the variance of the Rayleigh distribution corresponding to the spatial equilibrium distribution (4.217).

In summary, one may stress that the two-time-scale description on which the Kramers' approach is based (see previously) clearly appears here in the time and spatial domains. During the first stage, the system relaxes rapidly and nonexponentially on a time scale $t_{qs} \approx \tau_c^0$ and behaves as if there is no external force. On the longer time scale τ_c^ω, the system is characterized by the well-defined spatial equilibrium distribution, developed equilibrium values for the dynamical variables, and relaxes exponentially.

To readily make clear the conditions of the time-scale separation in the previously described dynamics, it is suggested that one examine the variation of the survival probability $p(\theta)$ in the absence of spontaneous decay ($k_r = 0$) versus $\theta = t/\tau_c^\omega$ for various values of the parameter $W = 2\tau_c^0/\tau_c^\omega = U(x_0)/k_B T$ [see Eq. (4.221)]. From the curves in Fig. 4.11 one sees that for $W > 1$ no time-scale separation is apparent, and if $W \ll 1$, the time scales are clearly distinguishable. These conclusions are corroborated by examining the curves on Fig. 4.12 where the plots of $\ln p(\theta)$ versus θ are shown for typical values of W and different values of k_r compared to $1/\tau_c^\omega$.

1. *Time-Dependent Reaction Rate*

Using the definitions of Section H.1.f, one has

$$k(t) = -\frac{d}{dt} \ln \operatorname{erf}(y_0/\sigma\sqrt{2})$$

$$= \frac{1}{\sqrt{\pi}} \frac{2D}{\sigma^2} (y_0/\sigma\sqrt{2}) \frac{\exp(-y_0^2/2\sigma^2)}{\operatorname{erf}(y_0/\sigma\sqrt{2})} . \tag{4.227}$$

The essential feature is that $k(t)$ starts from zero at $t = 0$ and equals $1/\tau_c^\omega$ for t tending to ∞. The variations of $\tau_c^\omega k(t)$ as a function of $\theta = t/\tau_c^\omega$ are depicted in Fig. 4.13 for a set of values of $W = 2\tau_c^0/\tau_c^\omega$ characterizing the initial potential energy [Eq. (4.221)]. For high values of W, $k(t)$ increases slowly to $1/\tau_c^\omega$, whereas a maximum is observed for $W < 1$: if W is diminished, the amplitude of the maximum is enhanced and the time at which it takes place is decreased.

2. *First Passage Time*[186,194,196–198]

For clarity we briefly recall the definitions established in Section H.1.f. Let us define T the random variable representing the instant at which the particle reaches the sink and let us consider the survival probability $p(t)$; we have

$$\Pr[T > t] = p(t) . \tag{4.228}$$

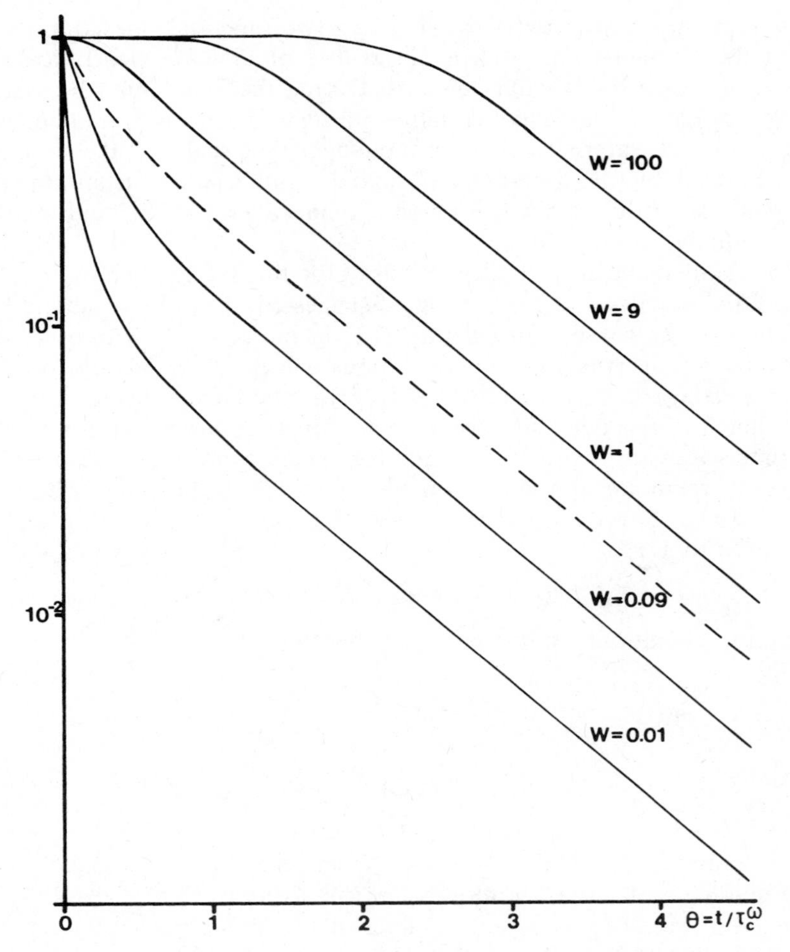

Fig. 4.11. Variation versus $\theta = t/\tau_c^{\omega}$ of the survival probability $p(\theta)$ for reaction in the absence of a barrier and spontaneous decay ($k_r = 0$). Solid lines: δ-shaped initial condition characterized by the initial potential energy $W = U(x_0)/k_B T$ [see Eqs. (4.218), (4.219), and (4.221)]. Dashed line: Boltzmann initial distribution [see Eq. (4.238)].

The probability that the particle is absorbed at time t is given by

$$\Pr[t < T < t + dt] = q(t)\, dt , \tag{4.229}$$

where

$$q(t) = -\frac{d}{dt} p(t) , \tag{4.230}$$

which is also called the first-passage-time distribution; $p(t)$, $q(t)$, and the rate $k(t)$ satisfy the following relationship:

$$k(t)p(t) = q(t) . \tag{4.231}$$

Remember that $k(t) = \lim_{\Delta t \to 0} (1/\Delta t) \Pr[t < T \leq t + \Delta t / T > t]$.

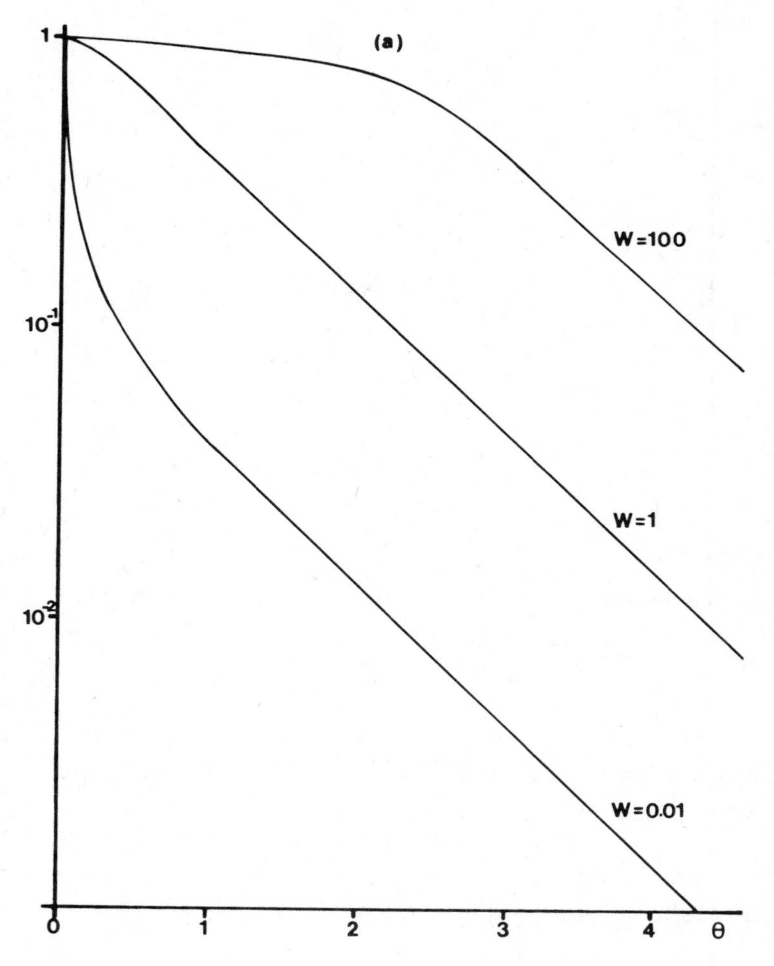

Fig. 4.12. Influence of the spontaneous decay (k_r) on the survival probability $p(\theta)$ for δ-shaped initial condition (see also legend of Fig. 4.11). (*a*) $k_r \tau_c^{\omega} = 0.1$; (*b*) $k_r \tau_c^{\omega} = 1$; (*c*) $k_r \tau_c^{\omega} = 10$.

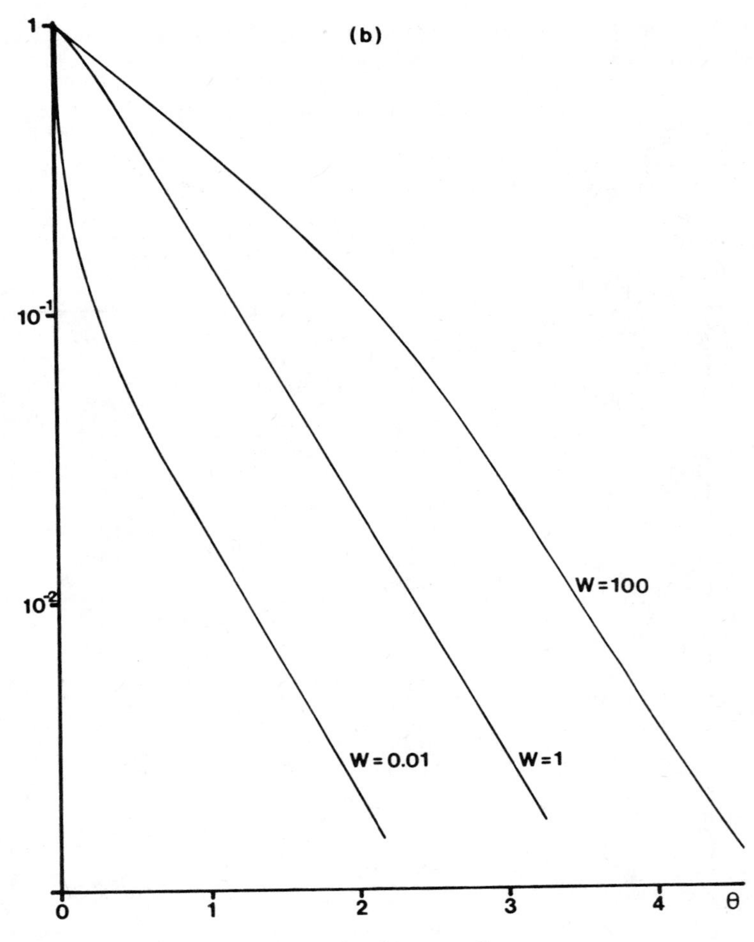

Fig. 4.12 (*Continued*).

The expectation of the first passage time may be written as

$$E[T]=\int_0^\infty tq(t)\,dt=-\int_0^\infty t\,\frac{dp(t)}{dt}\,dt\,. \tag{4.232}$$

Integrating by parts we get

$$E[T]=\int_0^\infty p(t)\,dt\,, \tag{4.233}$$

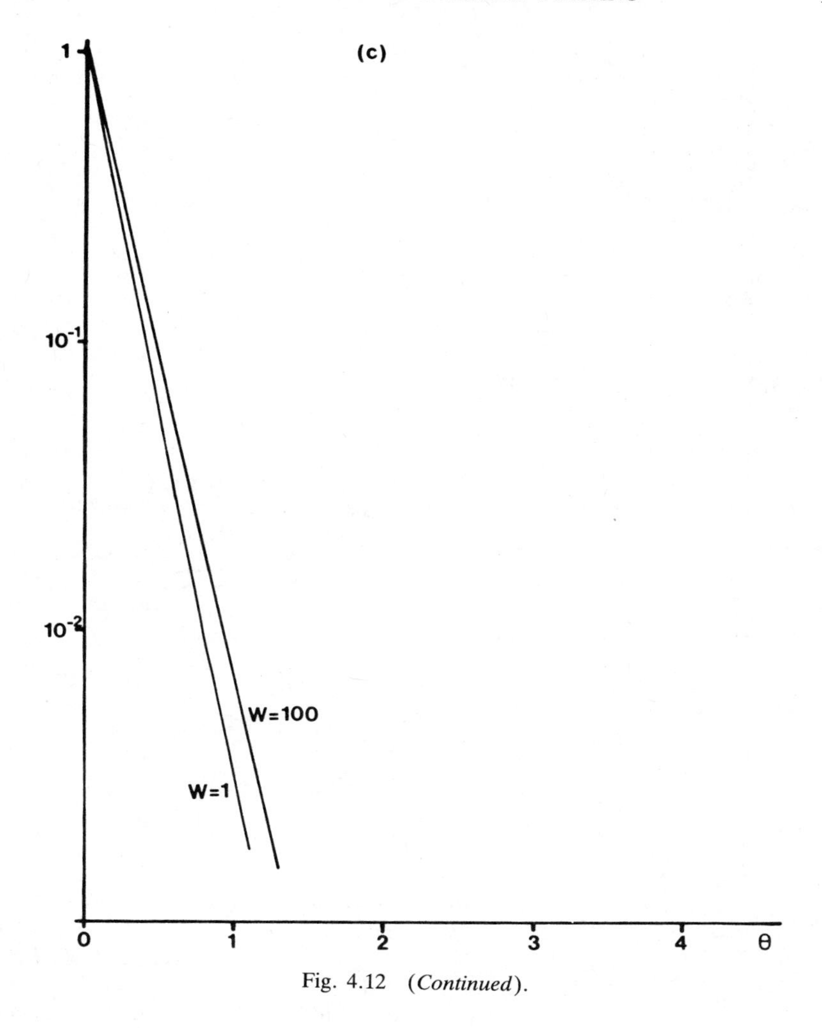

Fig. 4.12 (*Continued*).

which represents the reciprocal of the average rate of relaxation of the process.

For the particular case $\omega = 0$, $E[T]$ can be calculated explicitly; we find

$$E[T] = \int_0^\infty \operatorname{erf}(\tau_c^0/t)^{1/2} \exp(-k_r t)\, dt$$
$$= (1/k_r)\{1 - \exp[-2(k_r \tau_c^0)^{1/2}]\} . \qquad (4.234)$$

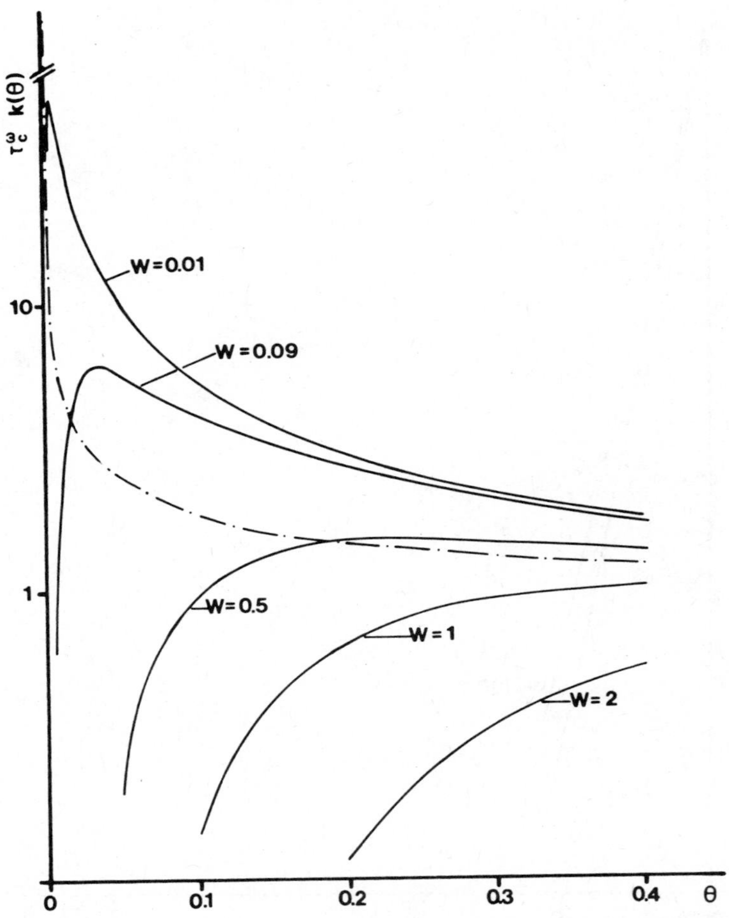

Fig. 4.13. Variation of $\tau_c^\omega k(\theta)$—the reaction rate $k(\theta)$ in the barrierless model compared to the spontaneous decay rate $1/\tau_c^\omega$—versus $\theta = t/\tau_c^\omega$. Solid lines [Eq. (4.227)]: initial condition $p_0(0, x_0) = \delta(x + |x_0|)$ for different values of x_0 defining the potential energy $W = U(x_0)/k_B T$. Dashed line [Eq. (4.241)]: Boltzmann initial distribution.

We suppose that $k_r \tau_c^0 \ll 1$, thus

$$E[T] = 2(\tau_c^0 \tau_r)^{1/2}, \tag{4.235}$$

with

$$\tau_r = 1/k_r,$$

Let us define the relative quantum yield Y (ratio of yields with and without reaction) by

$$Y=\frac{E[T]}{\tau_r}=2\left(\frac{\tau_c^0}{\tau_r}\right)^{1/2}. \tag{4.236}$$

With $\tau_c^0=x_0^2/4D$, one sees that Y should vary with temperature as $D^{-1/2}$, that is, as $(\eta/T)^{1/2}$ [see Eqs. (4.137) and (4.178)], giving an apparent[229] activation energy of $Q_\eta/2$.

3. *Influence of Initial Conditions*

A quantity relevant to the discussion of the relaxation in the presence of a sink is the broadness of the probability density function of the initial condition. The following case where the system is made reactive ($k_{nr}\neq 0$) at times $t\geq 0$, but is unreactive ($k_{nr}=0$) for $t<0$—corresponding to physical situations—can be solved easily. Indeed, the initial distribution $p_0(x_0)$ equals the Boltzmann equilibrium distribution [see also Eq. (4.157)], that is,

$$p_0(x_0)=(2\pi D\tau_c^\omega)^{-1/2}\exp(-x_0^2/2D\tau_c^\omega), \tag{4.237}$$

and the survival probability $\bar{p}(t)$ has the form[194]

$$\bar{p}(t)=\int p(t)p_0(x_0)\,dx_0$$
$$=\frac{2}{\pi}[\arcsin\exp(-t/\tau_c^\omega)]\exp(-k_r t). \tag{4.238}$$

This expression can be expanded in Taylor series

$$\bar{p}(t)=\frac{2}{\pi}\sum_{k=0}^{\infty}2k![2^{2k}(k!)(2k+1)]^{-1}e^{-(2k+1)t/\tau_c^\omega}e^{-k_r t}, \tag{4.239}$$

and clearly shows that for times greater than τ_c^ω the decay appears exponential. This is evident in Fig. 4.11 where the variation of $\bar{p}(\theta)$ (dashed curve) is compared to that of $p(\theta)$ corresponding to $p_0(x_0)$ defined by a δ-function.

For $\tau_r\gg\tau_c^\omega$, an explicit expression of the relative quantum yield can be obtained:

$$Y=\frac{\tau_c^\omega}{\tau_r}\ln 2 \tag{4.240}$$

predicting that Y is proportional to the viscosity [see Eqs. (4.216) and (4.178)].

The reaction rate $k(t)$ can be expressed as

$$k(t) = \frac{1}{\tau_c^\omega} \frac{\exp(-t/\tau_c^\omega)}{[1 - \exp(-2t/\tau_c^\omega]^{1/2} \arcsin \exp(-t/\tau_c^\omega)}, \qquad (4.241)$$

and the temporal variation of $\tau_c^\omega k(t)$ versus $\theta = t/\tau_c^\omega$ is represented in Fig. 4.13 (dashed curve). One can remark that $k(t) \to \infty$ for $t \to 0$, which is different from the case of a δ-shaped initial distribution for which it begins with zero value.

V. EXPERIMENTAL RESULTS AND INTERPRETATION

A. Dynamics of DMABN in the Excited State

1. *Dynamics of the B* State Deactivation*

For a detailed analysis of the temporal evolution of the locally excited B* state, we will consider results obtained in propanol solution, since, contrary to the aprotic solvents, the alcohols enable one to explore a wide domain of viscosity and, also, the backward reaction is much less important in alcohols. Figures 5.1*a* and 5.1*b* show experimental decay curves $i_B(t)$ recorded, for DMABN/propanol solution, in the 350 nm band (B* emission) with resolutions near 30 ps and 0.3 ns, respectively (ps laser pulses excitation; detection by streak camera or by single-photon counting).[79] The temperature was varied between −50°C and −115°C, that is, the viscosity increases from 20 to 3×10^3 cp.[199] It appears clearly that the B* decay is not exponential: in the considered temperature range, the reverse reaction B* ← A* is negligible and cannot be responsible for the deviation from an exponential law. Identical behavior has also been observed in other solvents.[78] From this observation, implying a time-dependent reaction rate for the deactivation of the B* state by the TICT state formation and especially from the fact that the variation of the fluorescence quantum yield of B* versus $1/T$ is nearly the same as or slower than that of the solvent mobility $1/\eta$ (see Section II.B and Ref. 88), it was concluded that the charge-transfer reaction in DMABN takes place without any intramolecular potential barrier. Indeed, the Kramers' approach (Ref. 16 and see Section IV.H.1) using a bistable potential with a high barrier ($E_B \gg k_B T$), and assuming that the system relaxes to the quasi-stationary distribution on a time scale very short compared to the reaction characteristic time, leads to a steady-state reaction rate whose variation with temperature should give an observed activation energy

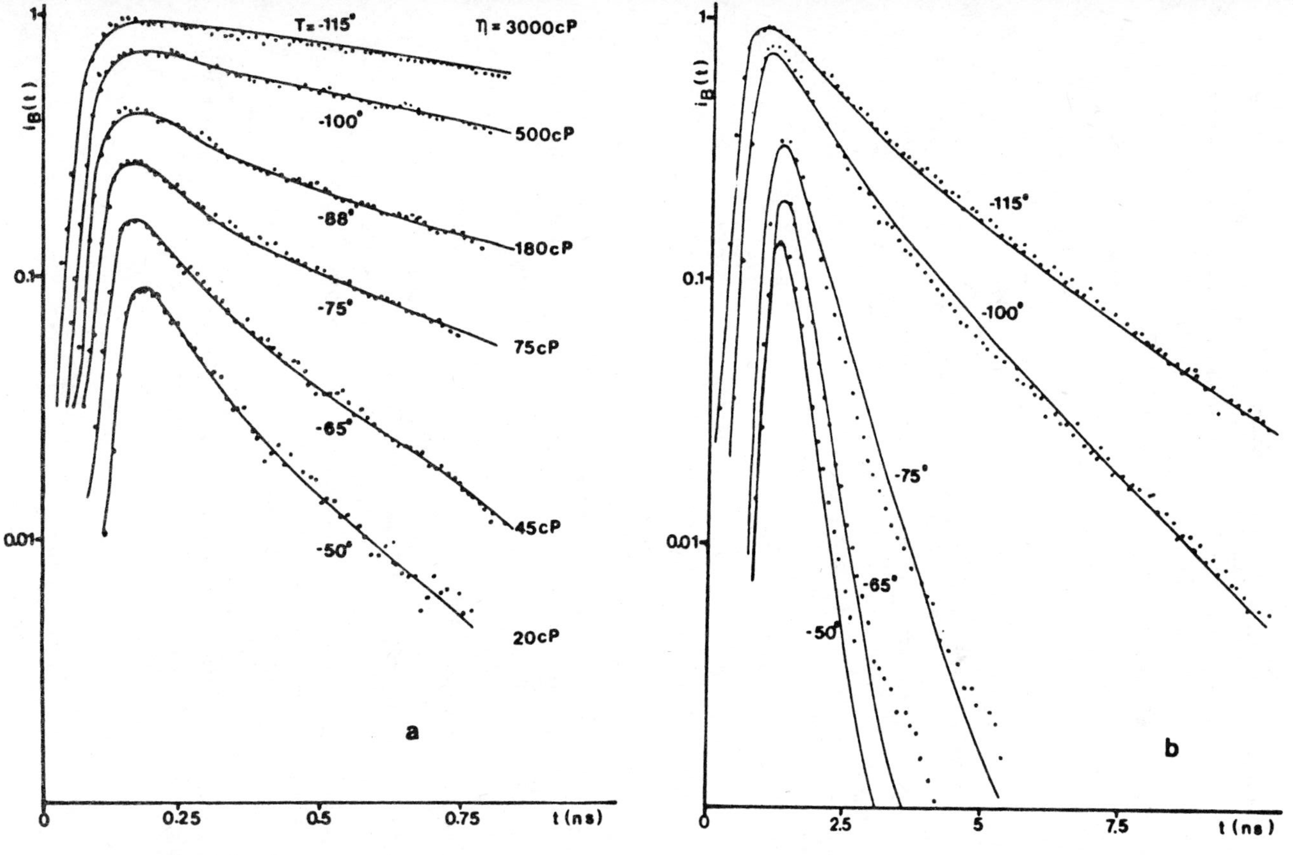

Fig. 5.1. Decays of the B* emission for DMABN/propanol solution at various temperatures (viscosities): (· · ·) dotted lines, experimental curves; solid lines, best fit decays (instrumental response function taken into account). (*a*) Streak camera data; (*b*) nanosecond resolution decay curves.

$E_{obs} = E_B + \alpha E_\eta$ with $\alpha < 1$. (E_η is the activation energy for the solvent mobility supposed to follow an Arrhenius law.) At low viscosities, E_{obs} would be near the intrinsic barrier E_B. Elsewhere numerical solution of the Smoluchowski equation with the bistable potential has made evident an important time dependence of the reaction rate for small values of E_B.[200]

Hence, the experimental results on DMABN have been discussed within the framework of the theoretical model of a barrierless electronic relaxation where the reaction is modeled by a pinhole sink on the minimum of a harmonic potential (see Section IV.I).

From the theoretical expressions (4.218) and (4.219) valid for a totally efficient sink and a δ-shaped initial distribution, it comes out that the general feature of the B* emission, proportional to the survival probability $p(t)$, is a nonexponential decay with two limit behaviors characterized by [see Eqs. (4.222) and (4.223)]

$$p(t) \approx e^{-k_r t} \operatorname{erf}\left(\frac{\tau_c^0}{t}\right)^{1/2} \qquad \text{for } t \ll \tau_c^\omega , \tag{5.1}$$

$$p(t) \approx \frac{2}{\sqrt{\pi}} \left(\frac{2\tau_c^0}{\tau_c^\omega}\right)^{1/2} e^{-k_r t} e^{-t/\tau_c^\omega} \qquad \text{for } t \gg \tau_c^\omega . \tag{5.2}$$

The experimental decays $i_B(t)$ of the 350 nm band have been compared with curves calculated (solid lines in Fig. 5.1) by adjusting the parameters τ_c^ω and τ_c^0 in Eqs. (4.218) and (4.219); the spontaneous decay rate k_r has been approximated by the value $k^B = k_f^B + k_0^B$ measured in a nonpolar solvent. It should be noted that with the photon-counting detection method the investigation of the fast initial nonexponential decay is hindered at low viscosity by poor resolution and only the exponential part of the decay is observable. At high viscosities ($\eta > 100$ cp) the deviation from an exponential law is clearly visible. For the streak camera measurements the observations are opposite to those previously mentioned: at high viscosities the semilogarithmic plot of $i_B(t)$ appears linear, whereas at low viscosities the decay shows nonexponential behavior. In Fig. 5.2 are represented the actual B* decays calculated with the best fit values of the two relaxation times τ_c^0 and τ_c^ω. Their variation with the temperature has also been examined: Fig. 5.3 shows that they follow well those of η/T and η, respectively, as expected from the expressions (4.216) and (4.220):

$$\tau_c^0 = x_0^2/4D = x_0^2\zeta/4k_B T \tag{5.3}$$

$$\tau_c^\omega = \zeta/m\omega^2 , \tag{5.4}$$

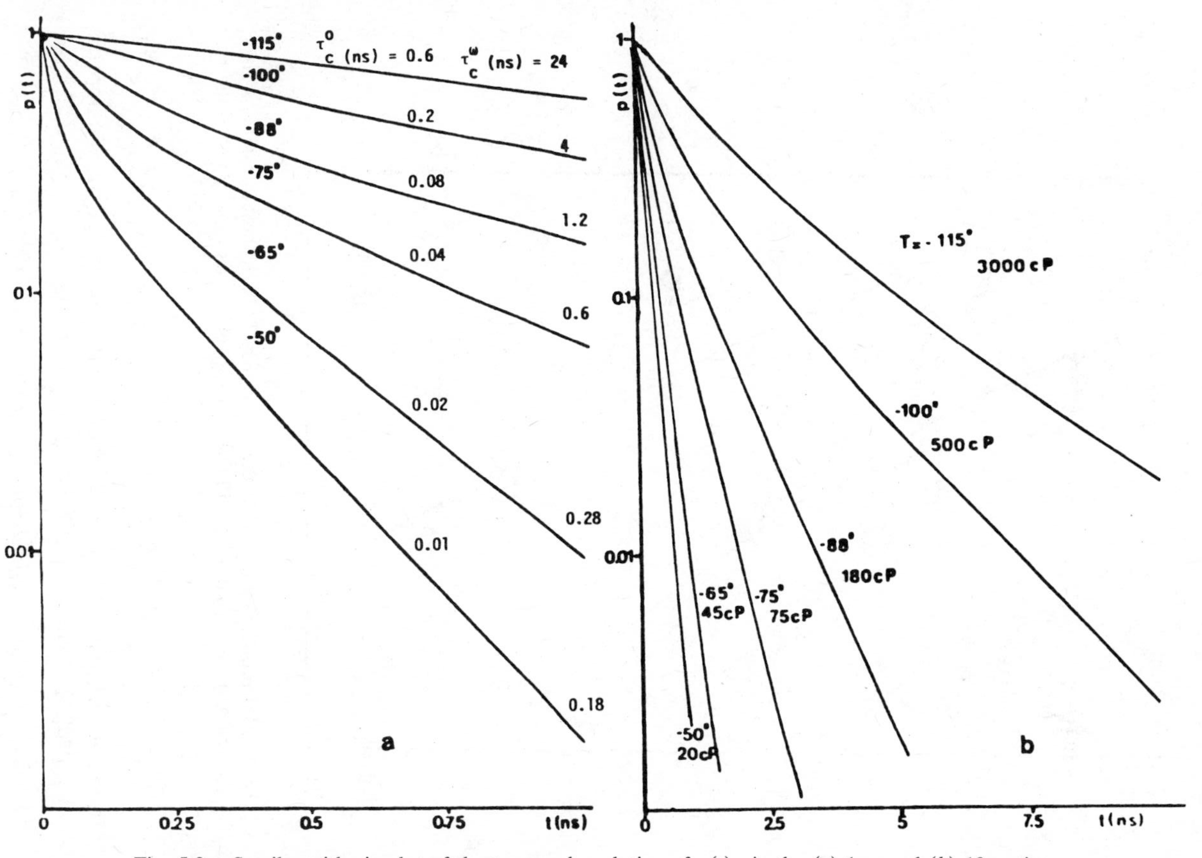

Fig. 5.2. Semilogarithmic plot of the temporal evolution of $p(t)$—in the (a) 1 ns and (b) 10 ns time range—calculated with the best fit values τ_c^ω and τ_c^0.

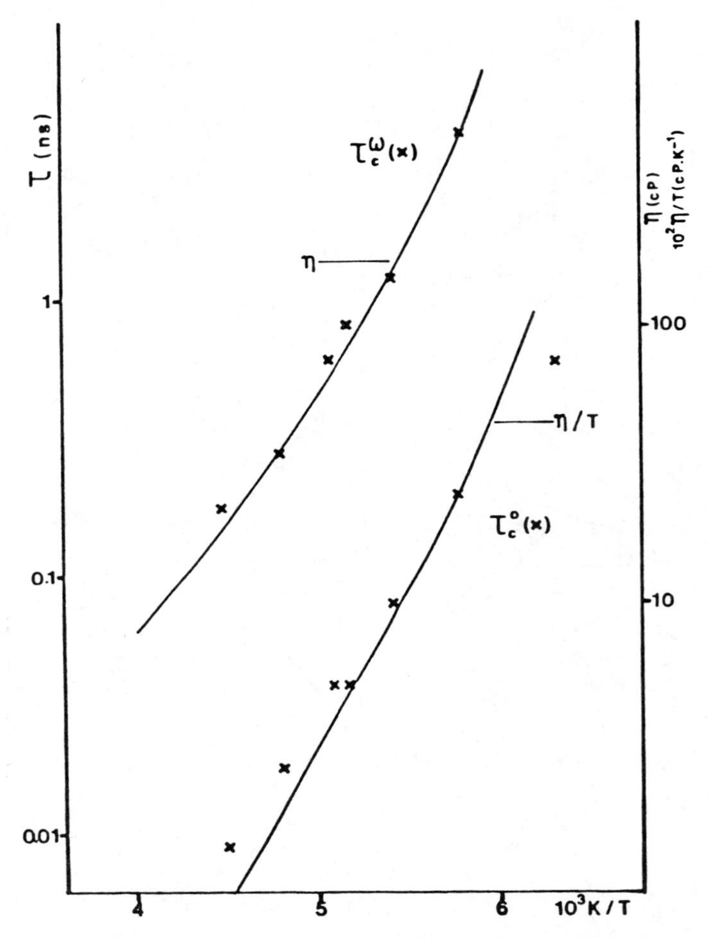

Fig. 5.3. DMABN/propanol solution. Variations versus T^{-1} of the best fit values (×) of τ_c^ω and τ_c^0 compared with the variations (solid lines) of the solvent viscosity η and of the ratio η/T.

with the Einstein relation $D = k_B T/\zeta$, and with the hydrodynamic approximation ζ proportional to η (Stokes law).

2. *Temperature Dependence of the B* State Quantum Yield*

It has been verified that the model used to analyze the dynamical characteristics of the B* deactivation also accounts for the observation that the activation energy E_{obs} for the B* quantum yield is $\leq E_\eta$. In the absence of a general analytical expression for the quantum yield $\phi \propto \int p(t)\, dt$, two extreme situations will be considered:

1. For small viscosities with $t/\tau_c^\omega \gg 1$ practically from the beginning of the emission, integration of Eq. (5.2) gives

$$\phi \propto \left(\frac{2\tau_c^0}{\tau_c^\omega}\right) \int_0^\infty e^{-k^B t} e^{-t/\tau_c^\omega}\, dt = \left(\frac{2\tau_c^0}{\tau_c^\omega}\right)^{1/2} \frac{1}{k^B + 1/\tau_c^\omega}, \qquad (5.5)$$

which shows that with $1/\tau_c^\omega > k^B$ the quantum yield is proportional to $(\tau_c^0 \tau_c^\omega)^{1/2}$ and hence to $(\eta/T^{1/2})$: the apparent activation energy should then be near E_η.

2. For very high viscosities such that $\tau_c^\omega \gg 1/k^B$ meaning that the overall emission takes place in the limit $t/\tau_c^\omega \ll 1$, we have

$$\phi \propto \int_0^\infty e^{-k^B t} \operatorname{erf}\left(\frac{\tau_c^0}{t}\right)^{1/2} dt = \frac{1}{k^B}\,[1 - e^{-2\sqrt{k^B \tau_c^0}}]. \qquad (5.6)$$

With $k^B \tau_c^0 \ll 1$, which is always true for the studied system, this expression predicts ϕ proportional to $\sqrt{\tau_c^0}$ and hence to $(\eta/T)^{1/2}$ that should give $E_{obs} \approx E_\eta/2$.

This simple analysis shows that on going from high to very low temperature, the variation of $\log \phi$ versus T^{-1} follows at first that of $\log \eta$, then it becomes less rapid: the observed activation energy for the reaction rate depends on the viscosity range.

The solid line *A* in Fig. 5.4 represents the relative variation versus $1/T$ of the quantum yield ϕ_B of the B* fluorescence calculated for the DMABN/propanol solution by numerical integration of the exact expression of $p(t)$ with the corresponding best fit values of the parameters τ_c^0 and τ_c^ω. This variation is compared to that of the viscosity (curve *B*). The apparent activation energy that can be deduced for the quantum yield ($E_{obs} \approx 3.5$ kcal M^{-1}) is significantly less than that of the viscosity in the considered temperature range ($E_\eta \approx 5$ kcal M^{-1}).

Another interesting observation is that at room temperature (small viscosity condition) and for the different alcoholic solutions (where the B* emission due to the backward reaction $B^* \leftarrow A^*$ is negligible), a straight line with slope very close to unity has been found (Fig. 5.5) for the curve $\log \phi$ versus $\log \eta$, where ϕ is the relative fluorescence quantum yield of the B* state deduced from stationary spectra with $\phi_B = 1$ in cyclohexane.

In summary, the barrierless model accounts for an activation energy less than E_η at high viscosity, the difference decreasing when the solvent becomes less viscous as experimentally observed.

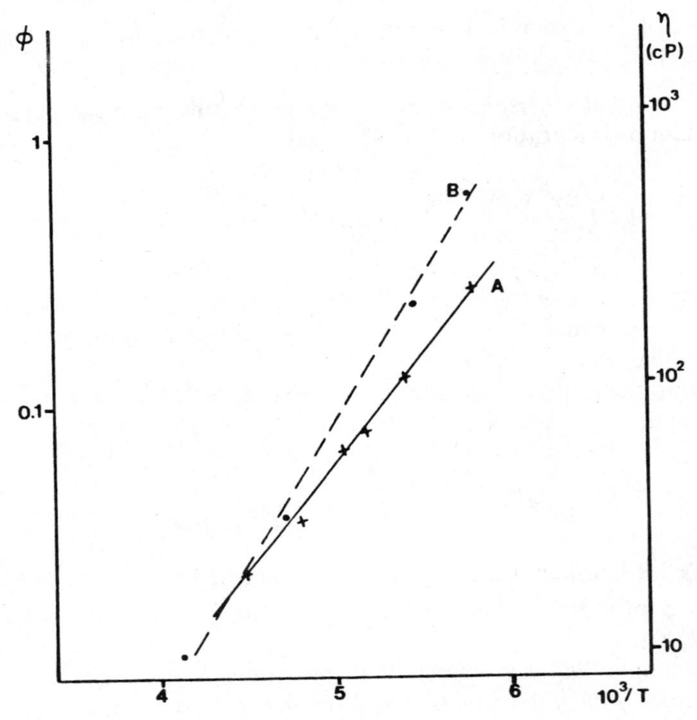

Fig. 5.4. DMABN/propanol solution. Variation with T^{-1} of the viscosity (curve B) and of the relative quantum yield of the B* state calculated with the best fit values of τ_c^ω and τ_c^0 (curve A).

3. *TICT State Formation*

Because of the spectral relaxation due to the appearance of a high dipole moment in the charge-transfer state, the dynamics of the TICT state formation has been studied by following the fluorescence rise in the whole A* band. In Fig. 5.6 are plotted, in the 10 ns time range, the experimental curve $i_A(t)$ at $-110°C$ in propanol ($\eta = 1.5 \times 10^3$ cp) and the decay of the B* emission at 350 nm. The solid curve representing the evolution of the TICT state expected in a constant reaction rate scheme shows a slower risetime with respect to that of the recorded A* emission. To interpret the experimental $i_A(t)$ curves, the time dependence of the reaction rate $k_{BA}(t)$ should be taken into account. From the coupled differential equations for the populations $n_B(t)$ and $n_A(t)$ of the B* and A* states (remembering that the reverse reaction $B^* \leftarrow A^*$ is negligible at low temperatures):

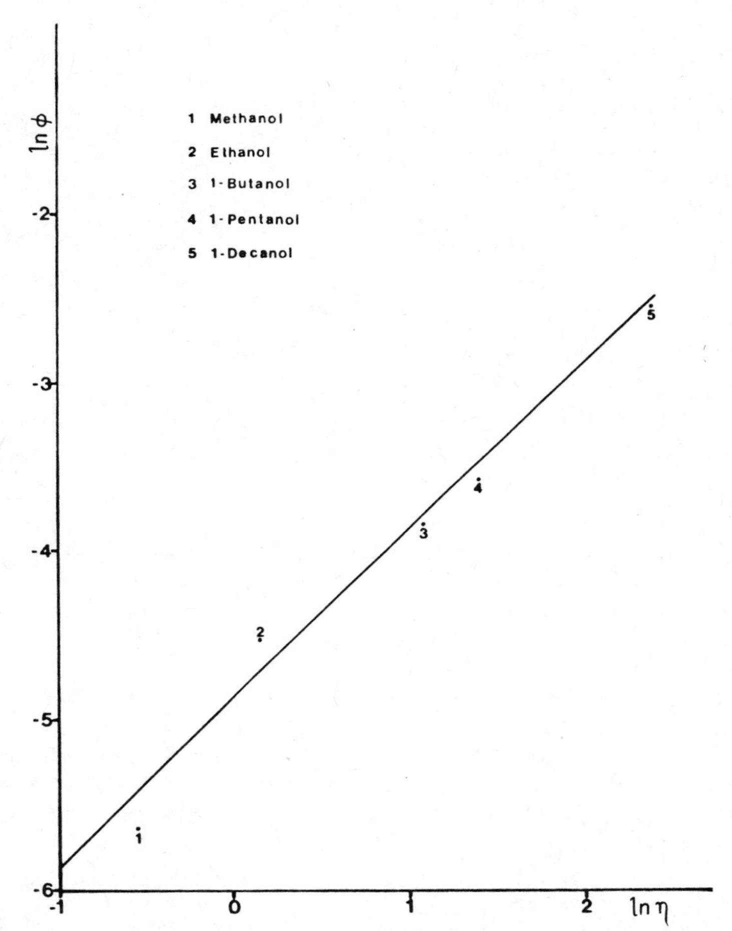

Fig. 5.5. Log–log plot of the relative B* fluorescence quantum yield ($\phi_B = 1$ in cyclohexane) versus the solvent viscosity for DMABN in various alcohols at room temperature.

$$\dot{n}_B(t) = -k^B n_B(t) - k_{BA}(t) n_B(t) , \tag{5.7}$$

$$\dot{n}_A(t) = -\tau_A^{-1} n_A(t) + k_{BA}(t) n_B(t) , \tag{5.8}$$

it results that

$$n_A(t) \propto k_{BA}(t) n_B(t) \otimes e^{-t/\tau_A} \tag{5.9}$$

with $k_{BA}(t) = -\dot{p}(t)/p(t)$ given by Eq. (4.227).

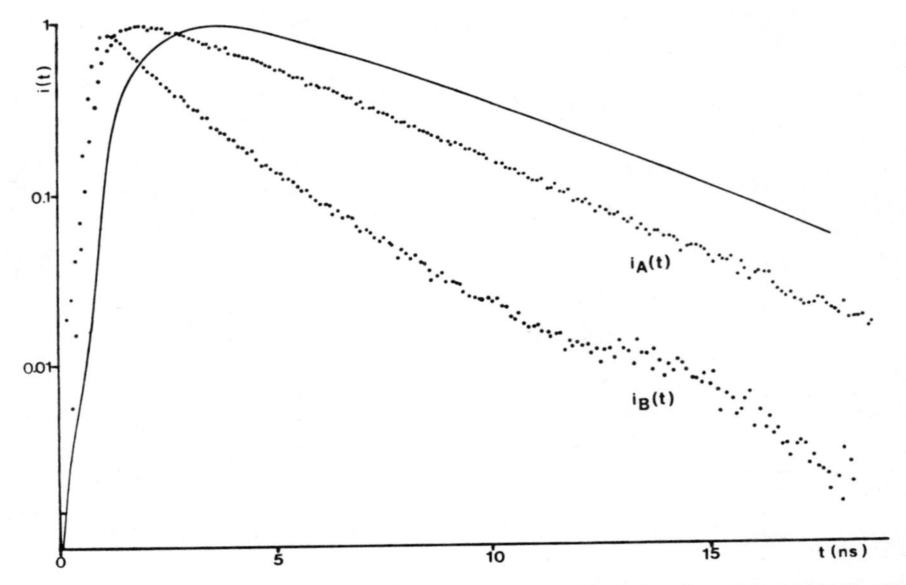

Fig. 5.6. Single-photon-counting decays (· · ·) $i_B(t)$ and $i_A(t)$ of the planar B* and TICT A* states of DMABN in propanol at −110°C. The solid line represents the convolution $i_B(t) \otimes e^{-t/\tau_A}$.

A good fit (Fig. 5.7) was obtained between the streak camera experimental data $i_A(t)$ [collected in propanol in the same temperature range as $i_B(t)$] and curves calculated with Eq. (5.9) by using the values of τ_c^0 and τ_c^ω that fit $i_B(t)$.

4. *Discussion*

The expression of $p(t)$ used to fit the experimental results corresponds to a totally efficient pinhole sink situated at the bottom of the harmonic potential ($x_1 = 0$). It was also examined how other conditions would modify the decay law and the following have been shown[79]:

1. For $x_1 > 0$ and the microscopic rate constant $k_{nr} \to \infty$, one has

$$p(t) = e^{-k_r t} \operatorname{erf}\left(\frac{x_1 + y_0}{\sigma\sqrt{2}}\right), \tag{5.10}$$

with $y_0 = |x_0| \exp(-t/\tau_c^\omega)$ and $\sigma^2 = D\tau_c^\omega(1 - e^{-2t/\tau_c^\omega})$.

The short- and long-time behaviors are

$$t/\tau_c^\omega \ll 1: \qquad p(t) \approx e^{-k_r t} \operatorname{erf}\left(\frac{\sqrt{\tau_c^0} + \sqrt{x_1^2/4D}}{\sqrt{t}}\right); \tag{5.11}$$

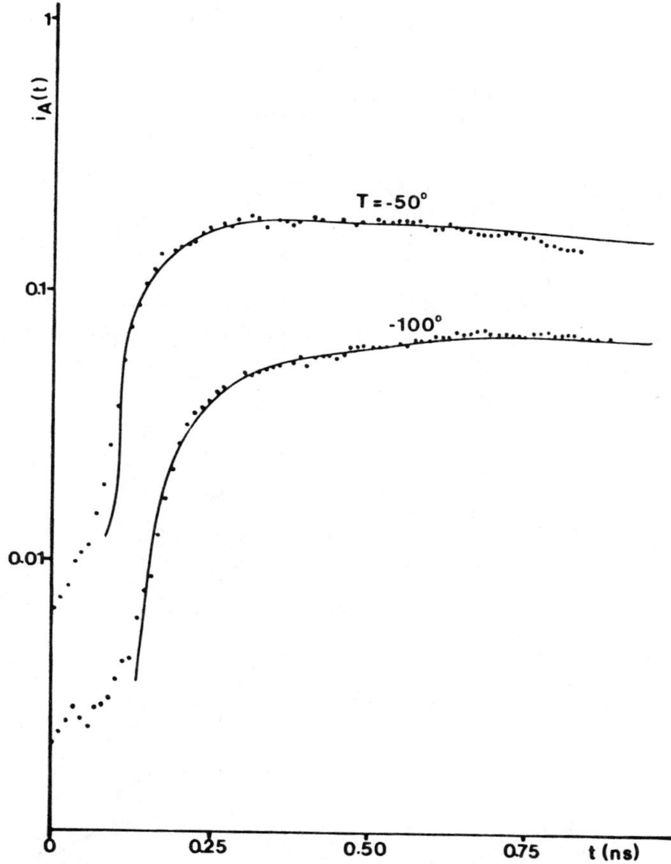

Fig. 5.7. Streak camera decay curves $(\cdots)$ $i_A(t)$ of the TICT state (propanol solution). The solid line curves are calculated with $i_A^c(t) = k_{BA}(t)n_B(t) \otimes e^{-t/\tau_A} \otimes r(t)$. Lower curve: $T = -100°C$, $\tau_c^\omega = 4$ ns, and $\tau^0 = 0.08$ ns; upper curve: $T = -50°C$, $\tau_c^\omega = 0.2$ ns, and $\tau_c^0 = 0.01$ ns. $r(t)$ is the experimental response function.

t such that $x_1 > y_0$ and $t \gg \tau_c^\omega$:

$$p(t) \approx e^{-k_r t} \operatorname{erf}\left(\frac{x_1}{\sqrt{2D\tau_c^\omega}}\right). \tag{5.12}$$

2. For radiation boundary condition ($k_{nr} < \infty$) at $x_1 = 0$, the solution is

$$p(t) = e^{-k_r t}\left\{\operatorname{erf}\left(\frac{y_0}{\sigma\sqrt{2}}\right) + e^{k_{nr}y_0/D + k_{nr}^2\sigma^2/2D^2} \operatorname{erfc}\left(\frac{y_0}{\sigma\sqrt{2}} + \frac{k_{nr}\sigma}{D\sqrt{2}}\right)\right\}, \tag{5.13}$$

and the limits are

$$t/\tau_c^\omega \ll 1: \qquad p(t) \approx e^{-k_r t}\left[e^{k_{nr}|x_0|/D}\,\mathrm{erfc}\left(\frac{\tau_c^0}{t}\right)^{1/2} + \mathrm{erf}\left(\frac{\tau_c^0}{t}\right)^{1/2}\right]; \tag{5.14}$$

$$t/\tau_c^\omega \gg 1 \text{ and } y_0 < \frac{k_{nr}\sigma^2}{D}: \qquad p(t) \approx e^{-k_r t} e^{k_{nr}^2 \tau_c^\omega/2D}\,\mathrm{erfc}\left[k_{nr}\left(\frac{\tau_c^\omega}{2D}\right)^{1/2}\right]. \tag{5.15}$$

Expressions (5.12) and (5.15) show that at long times the decrease of $p(t)$ will follow, in both cases, a single exponential law characterized by k_r

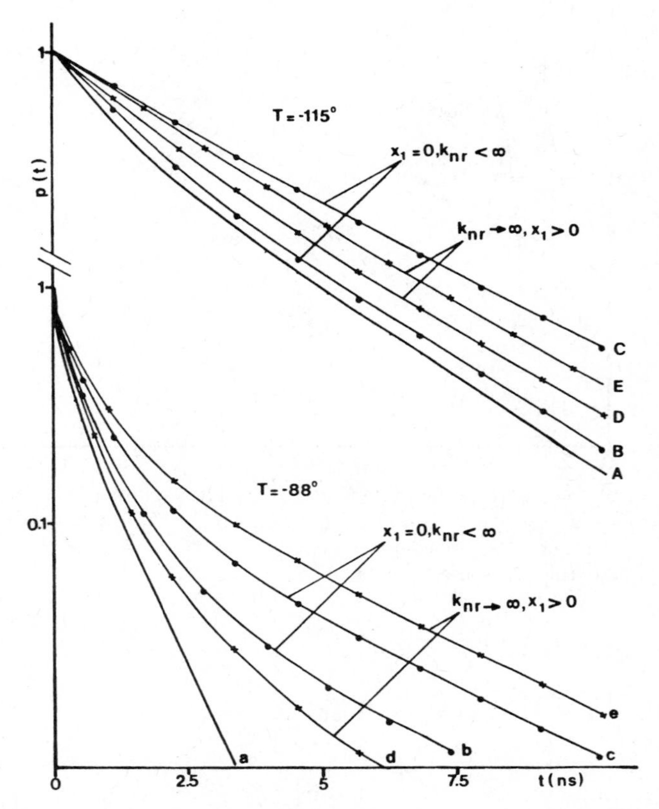

Fig. 5.8. Influence of the efficiency and the position of the pinhole sink on the shape of $p(t)$: Upper part: $T = -115°\mathrm{C}$, $\tau_c^\omega = 24$ ns, $\tau_c^0 = 0.6$ ns. A, $x_1 = 0$, $k_{nr} \to \infty$; B, $k_{nr}x_0/D = 6$; C, $k_{nr}x_0/D = 0.3$; D, $x_1/x_0 = 0.5$; E, $x_1/x_0 = 1$. Lower part: $T = -88°\mathrm{C}$, $\tau_c^\omega = 1/2$ ns; $\tau_c^0 = 0.08$ ns. a, $x_1 = 0$, $k_{nr} \to \infty$; b, $k_{nr}x_0/D = 6$; c, $k_{nr}x_0/D = 3$; d, $x_1/x_0 = 0.1$; e, $x_1/x_0 = 0.5$.

and, hence, is viscosity independent, contrary to the case $k_{nr} \to \infty$ and $x_1 = 0$. This is illustrated in Fig. 5.8: the curves have been evaluated with $k_r = k^B$ and the previous fit values of τ_c^0 and τ_c^ω at $T = -88$ and -115°C for various $k_{nr}|x_0|/D$ with $x_1 = 0$ and for different $x_1/|x_0|$ with $k_{nr} \to \infty$. The comparison of the curves a and A used to fit the experimental data with the other curves in Fig. 5.8 clearly demonstrates that no satisfactory adjustment has been possible for values of the parameters $k_{nr}|x_0|/D$ and $x_1/|x_0|$ characterizing a reasonably nonefficient pinhole sink far from the origin.

The important result of this study is the existence of two time scales in the deactivation of the B* excited state: at short times ($t < \tau_c^\omega$) the relaxation is mainly governed by the initial condition leading to a nonexponential behavior with a characteristic time $\tau_c^0 = x_0^2/4D$; and for $t \gg \tau_c^\omega$ the decay is exponential with a constant τ_c^ω proportional to the viscosity (as long as $\tau_c^\omega < 1/k^B$).

From the fitted τ_c^ω and τ_c^0 values and approximate ζ/m, the angular frequency ω of the harmonic potential has been estimated to be of the order of 10^{12} s^{-1}, which corresponds to a relatively shallow potential surface, and the parameter x_0 characterizing the initial configuration has been evaluated to be near 0.02 nm. The fact that the experimental data are accounted with a δ-shaped initial distribution can be related to the existence of a steep minimum in the ground-state potential curve, which presents a high barrier between the planar and perpendicular geometries.

B. Kinetics of Other Dialkylanilines

In the following, a comparison among the fluorescence and kinetics of two benzonitriles (BN), and the corresponding benzoic acid ethyl esters (BEE) will be made. The steady-state fluorescence spectra of the two pairs of nitriles and esters in a nonpolar and a polar solvent are shown in Figs. 5.9 and 5.10.[33] The ester always shows a higher ratio of long-wavelength (TICT) to short-wavelength (F_B) fluorescence. The ester and nitrile compounds with the dimethylpyrrolidino group (DMPYR) (Fig. 5.10) also show a strongly increased TICT component when compared to those with the dimethylamino (DMA) substituent (Fig. 5.9). Part of this increase is due to the increased donor strength of DMPYR as compared to DMA.[87] In DMPYRBEE (Fig. 5.10b), the TICT state can become so favorable energetically that even in the nonpolar solvent n-hexane dual fluorescence occurs with the TICT band as the major component. The corresponding nitrile DMPYRBN also shows an indication of a minor TICT component in n-hexane (broadened fluorescence spectrum as compared to DMABN) (see Fig. 5.10a).

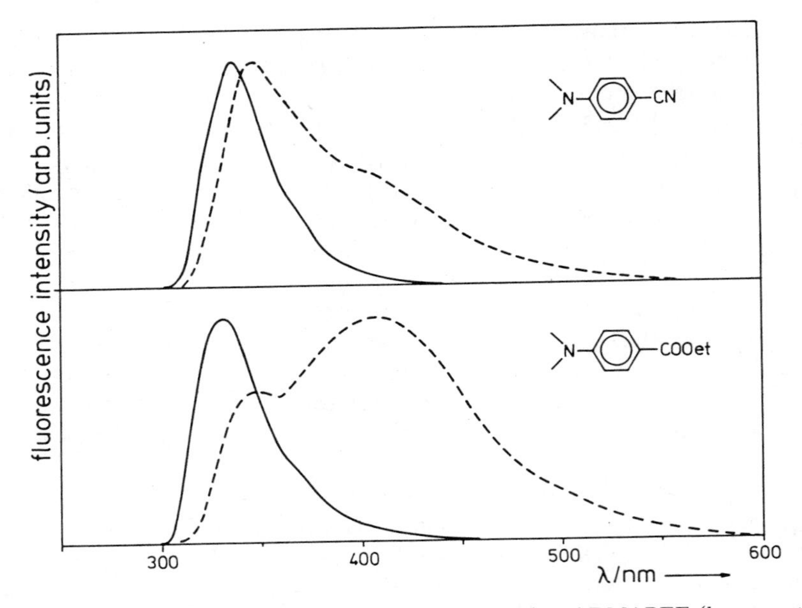

Fig. 5.9. Fluorescence spectra of DMABN (upper part, *a*) and DMABEE (lower part, *b*) in *n*-hexane (——) and *n*-butyl chloride (– – –) at room temperature.[33]

Figure 5.11 shows the behavior of the ratio of the dual fluorescence quantum yields in polar *n*-butyl chloride in a large temperature range. It can be seen that for both DMA compounds, T_m is situated at around 200 K, whereas for both DMPYR compounds, it lies above 330 K. With respect to temperature, the behavior for ester and nitrile with a similar amino substituent is comparable, but the increased importance of the TICT band for the esters reflects as an upward shifting of the ln ϕ_A/ϕ_B curves without horizontal displacement.

It can be asked why the esters exhibit this increased tendency toward the TICT state. The answer from steady-state spectra is that although ϕ_A increases for the esters by a factor of roughly 2, the main reason is a strong increase of the TICT formation rate k_{BA}.[33]

This is astonishing, since the low-temperature slopes in Fig. 5.11 are not different for esters and nitriles. Thus, the apparent Arrhenius activation energy $E_{BA} = E_{obs}$ is equal for both compounds, the difference in k_{BA} arising only from the preexponential factor. This is a further indication that E_{obs} is governed by solvent properties (E_η) and not by solute barriers (E_0) along the reaction pathway (barrierless transition in both cases).

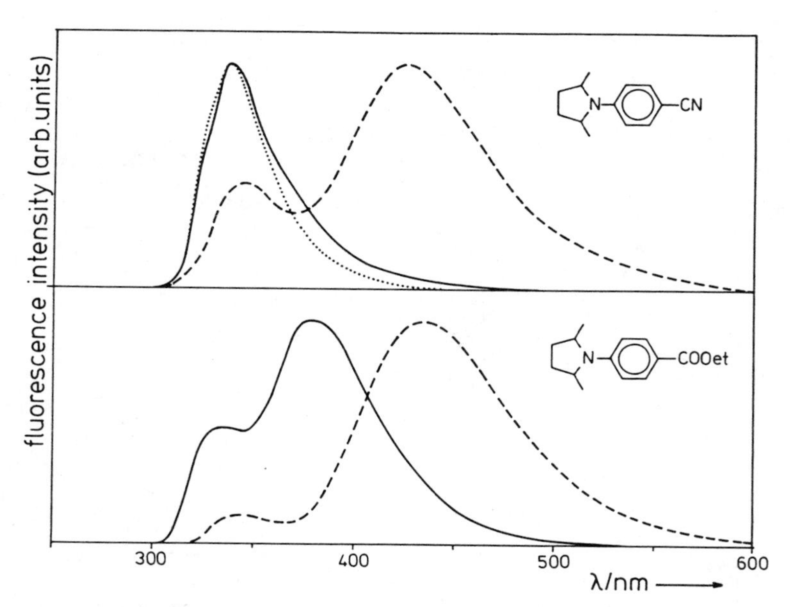

Fig. 5.10. Fluorescence spectra of DMPYRBN (upper part, *a*) and its ester DMPYRBEE (lower part, *b*) in *n*-hexane (——) and *n*-butyl chloride (---) at room temperature; the fluorescence spectrum of DMABN in *n*-hexane (···) is shown for comparison.[33]

The direct determination of k_{BA} for DMABN and its ester DMABEE in low-temperature *n*-butyl chloride by time-resolved spectroscopy is shown in Fig. 5.12. The ester is seen to decay considerably faster, and this faster decay is connected with a faster risetime of the TICT band (Table III).[50,201] From these values and from the low-temperature lifetime limit, rate constants k_{BA} given in Table III have been calculated. This table also contains the corresponding data for DMPYRBN (Fig. 2.17). The increase of k_{BA} for the ester amounts to roughly a factor of 6 in *n*-butyl chloride between -120 and -100°C as compared to the corresponding nitrile.[50]

A qualitative interpretation of this fact can be given in the framework of the stochastic theory of barrierless transitions. The comparison of the observed decays of DMABN in *n*-butyl chloride with those at similar viscosity in propanol suggests that the relaxation can be ascribed to the long-time limit $\tau_c^{\omega} = \zeta/m\omega^2$, and that the nonexponential part related to τ_c^0 is very fast and hidden by the convolution with the instrument response function. An increase of the valley frequency ω (driving force) is the most likely factor to decrease τ_c^{ω} for the ester (and thus increase k_{BA}).

This may be connected with the fact that the excited-state hypersurface

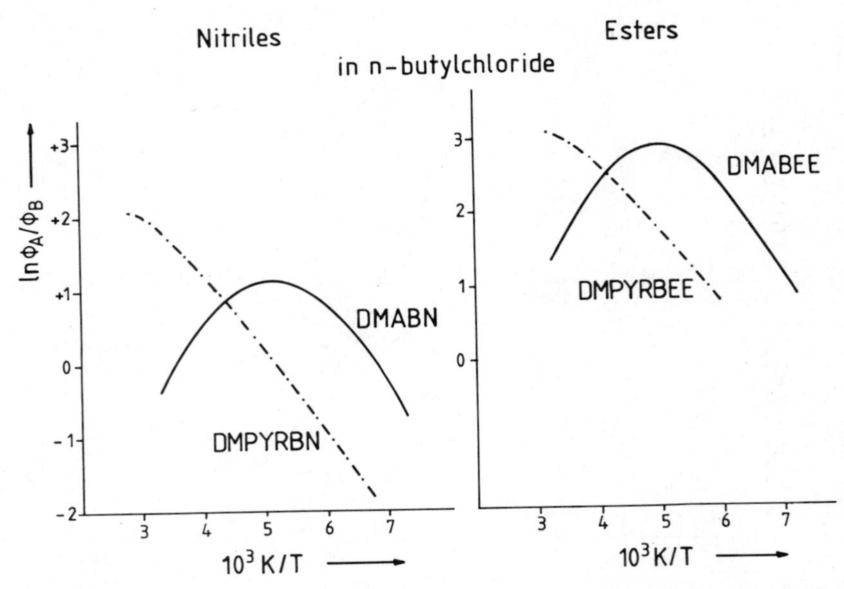

Fig. 5.11. Temperature dependence of the ratio of the dual-fluorescence quantum yields for the nitriles DMABN and DMPYRBN (left) and the esters DMABEE and DMPYRBEE (right) in *n*-butyl chloride.[33]

for DMABEE shows a qualitative difference to that of DMABN. The degree of fluorescence polarization p as given in Fig. 2.4[6] is consistent with a transition moment in the long molecular axis for the F_A (TICT) band of both nitrile and ester, as well as for the F_B band of the ester (positive values of p approaching 0.5). In contrast, the fluorescence polarization degree p for the nitrile shows small to negative values in the short-wavelength tail of the F_B band consistent with a transition moment perpendicular to the long molecular axis. Thus, for the nitrile, an (avoided) crossing of excited states of different symmetry occurs along the reaction coordinate, which is absent for the ester.[33]

TABLE III
Decay and Rise Times (in ns) of F_B and F_A Bands in *n*-Butyl Chloride[50,201]

Compound	Temp (°C)	τ_{decay} (F_B)	τ_{rise} (F_A)	$k_{BA}(10^7\ s^{-1})$
DMABN	−120	0.44	0.44	186
DMABEE	−120	<0.1	<0.1	>772
DMPYRBN	−108	1.04	0.91	75

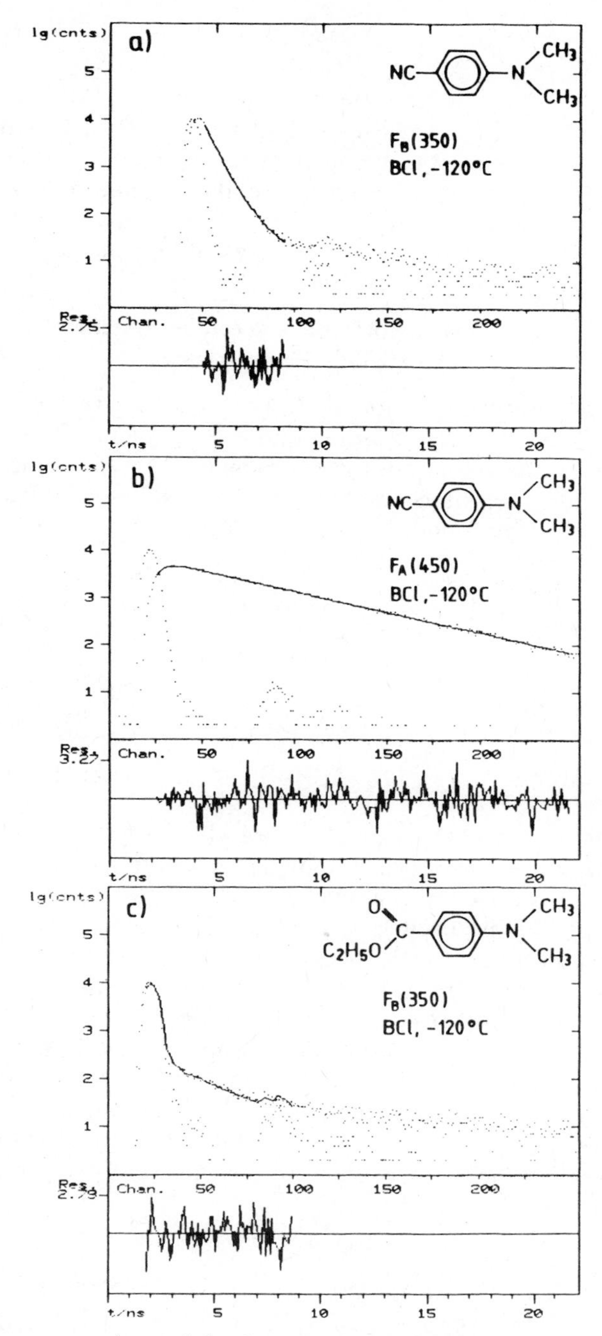

Fig. 5.12. Fluorescence decay of the short-wavelength F_B band (350 nm) of DMABN (a) and DMABEE (c) in *n*-butyl chloride at −120°C after excitation by synchrotron radiation. (b) shows the rise of the A fluorescence (450 nm) of DMABN (τ rise = 0.44 ns) which is similar to the decay time τ_1 of F_B (0.44 ns). τ_1 of DMABEE is significantly shorter (<0.1 ns), and the rise time of F_A is shortened correspondingly.[201] The very long tail (channels 100 to 256) in the F_B fluorescence has not been fitted. It may be a solvent trace impurity not visible in the spectra (<1% of the total fluorescence).

Comparing the F_B fluorescence of DMABN to DMPYRBN (Table III), the latter decays slower even at an increased temperature (reduced viscosity). If ω is assumed to be unchanged for the two nitriles, then the longer τ_c^{ω} for the compound with the bulkier amino group (increased "radius" a) suggests an increase of $\zeta = 6\pi\eta a$ that outweighs the increase of the effective particle mass m. In other words, an increased friction for DMPYRBN slows down the dynamics in the excited state.

C. Some Extended Donor–Acceptor Systems with Anomalous Fluorescence

As already discussed in Section II, larger systems like the diphenylsulfones also exhibit a charge-transfer fluorescence, and the donor can be shown not to be the amino but the entire anilino group. In the case of the anilino-substituted anthracene ADMA this has directly been demonstrated by the bridged model compound ADMAB.

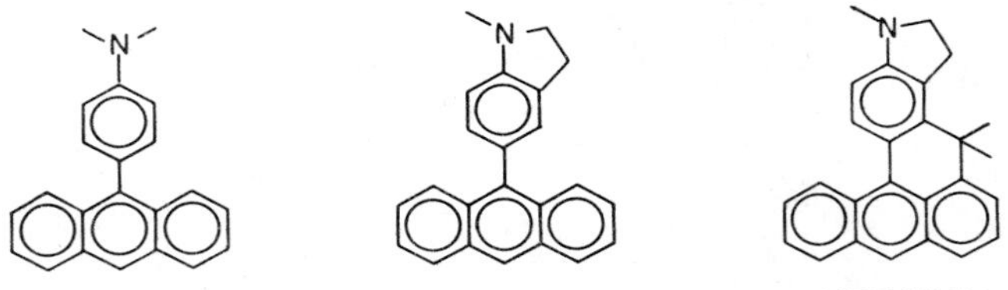

ADMA ADMAB ADMABB

Both compounds ADMA and ADMAB show a strongly solvent-dependent TICT fluorescence, although the latter is unable to reach a conformation with a twisted amino group.[202–205] Thus, the charge-separated character of the excited state is due to twisting around the C–C bond. A rigorous proof for the necesssity of C–C bond twisting for TICT state formation could be given by investigation of the doubly bridged model compound ADMABB. Like the bridged DMABN model compounds, it is expected not to exhibit the anomalous fluorescence band.

Michler's ketone MK also shows a dual fluorescence in alcohols, and, in this case, the bridged model compound MKB is available which clearly demonstrates that (1) a blocking of the phenyl group rotations (around the C–C bonds) blocks the channel toward the TICT state, and (2) the

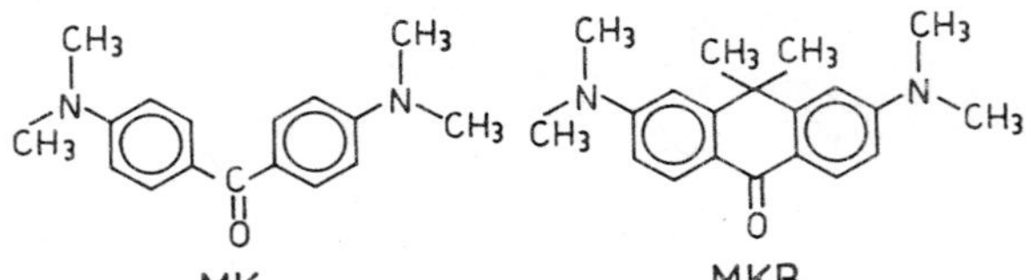

CH_3 N CH_3 C O CH_3 N CH_3 CH_3 N CH_3 CH_3 CH_3 O CH_3 N CH_3

MK MKB

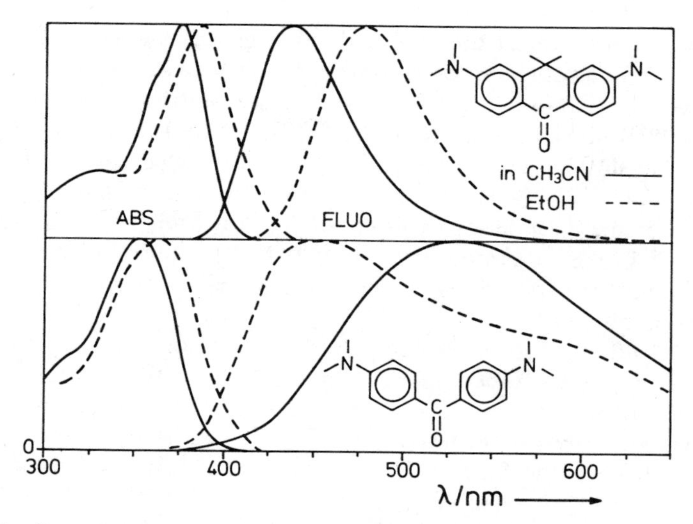

Fig. 5.13. Room-temperature absorption and fluorescence spectra of MKB (upper part) and Michler's ketone MK (lower part) in acetonitrile (——) and ethanol (– – –).

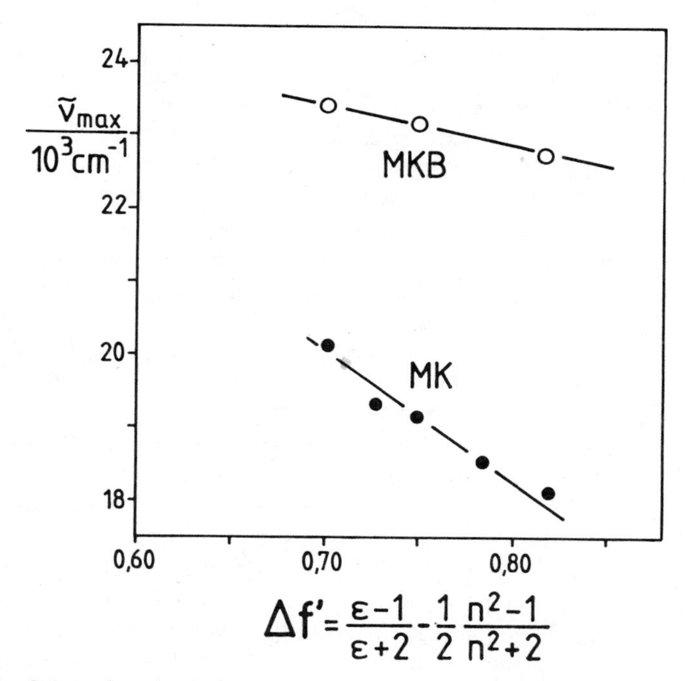

Fig. 5.14. Solvatochromic plot for the fluorescence maximum of MKB and MK in a series of homologous nitrile solvents.

possibility for rotation of the dimethylamino groups alone (as it occurs in DMABN) is not sufficient to produce TICT fluorescence in the case of MKB, even in strongly polar solvents like acetonitrile. This is shown in Fig. 5.13. MK shows a dual fluorescence in ethanol, and in acetonitrile only the redshifted TICT component remains, whereas MKB shows short-wavelength F_B fluorescence in both solvents. The charge-separated character for the anomalous band of MK is shown in the solvatochromic plot, Fig. 5.14, as described in Section II.B. The fluorescence of MKB depends much less on solvent polarity than that of MK, in a series of homologous nitriles.

How does the occurrence of two fluorescing states for MK fit into the dynamic picture developed in Section IV? The observed temperature dependence of the fluorescence quantum yield of MK in ethanol[206] yields direct evidence that in this case, also, $E_{BA} < E_\eta$. Recent time-resolved measurements at the Berlin Electron Storage Ring for Synchrotron Radiation (BESSY)[207] support this argument: The viscosity dependence of the decay of the short-wavelength fluorescence band in ethanol is consistent with an apparent value $E_{BA} \simeq 0.5E_\eta$. Moreover, the decay is nonexponential, as would be expected for a barrierless relaxation. The lifetime of the TICT state (exponential decay) is 0.65 ns in acetonitrile at room temperature, that is, it is unusually short.

D. Donor–Acceptor Systems without Anomalous Fluorescence Band

It is important to have an idea how widespread the relaxation mechanisms discussed previously are present in chemistry and biology. In a recent review article[2] many more compounds that emit an additional, anomalous TICT fluorescence have been presented, ranging from a variety of naphthalene and anthracene derivatives to biologically important compounds like indoles and purine derivatives. The question remains whether other systems which do *not* show the anomalous fluorescence band might nevertheless be understood on the same basis. It may be the case that either the two fluorescence bands are superimposed or that the product fluorescence has not yet been observed because of quenching processes or because of a red shift into the infrared region.

In the last two cases, we would still expect a behavior of the short-wavelength F_B band similar to that of DMABN: There, the k_{BA} channel toward the TICT state leads to efficient fluorescence quenching of F_B. This can be seen directly in the low-temperature range, for example, in Fig. 2.1 (Section II.A), below 180 K, where the quantum yield of F_B fluorescence increases upon cooling because the horizontal radiationless transition k_{BA} is slowed down. The simplified Kinetic Scheme II, also valid for any other monomolecular quenching process, describes this situation. The corresponding fluorescence quantum yield ϕ_f is given by

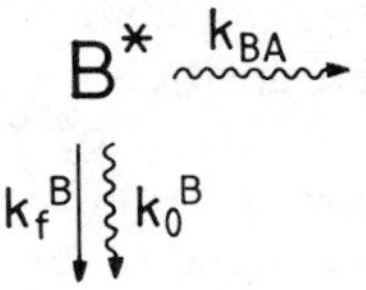

$$B^* \xrightarrow{k_{BA}} \quad \downarrow k_f^B \quad \downarrow k_0^B$$

Kinetic Scheme II

$$\phi_f = \frac{k_f^B}{k_f^B + k_0^B + k_{BA}} . \tag{5.16}$$

If the quenching process occurs toward a TICT state, the value of k_{BA} is influenced, of course, by the energetics of the adiabatic photoreation (exothermic/exergonic or not). Consider, as an example, the case of DMABN: When the good donor group $N(CH_3)_2$ is replaced by the weaker donor NH_2 (or even $NHCH_3$), the compound looses its horizontal radiationless transition channel toward the TICT state. Then the second emission band is absent, and the quantum yield of F_B fluorescence strongly increases. This also holds for the inverse case, that is, a change of the acceptor properties: Dimethylaniline shows a single but strong F_B fluorescence band (TICT states are too high in energy) but with increasing the acceptor strength by the cyano substituent, the energy of the lowest TICT state is sufficiently lowered, and the well-known dual fluorescence of DMABN with weak F_B intensity results.

The first systems without anomalous fluorescence discussed in the context of TICT states were *coumarine laser dyes*. Jones and coworkers[208–210] demonstrated that dialkylaminocoumarines like **7C** showed an increase of the nonradiative decay rate in strongly polar solvents. This effect is dramatically increased if the acceptor properties of the coumarine skeleton are stronger, for example, in **F7C**. On the other

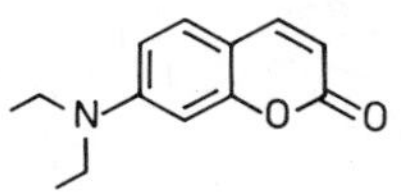

7C

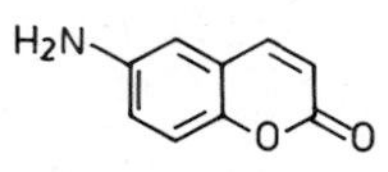

H_2N 6C

CF_3 F7C

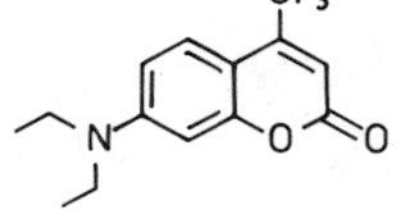

CF_3 F7CB

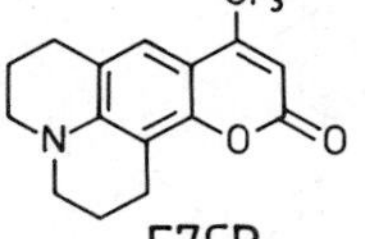

hand, the bridged derivative **F7CB** did not show the fluorescence quenching. Jones et al. ruled out the population of a triplet state as possible mechanism and suggested the transition toward a TICT state as a possible quenching mechanism.[208] In the case of **7C** and **F7C**, the TICT state seems to be nonemissive, possibly because of a very small energy gap to the ground state. The involvement of a TICT state in the quenching mechanism has recently been supported by the direct observation of weak TICT fluorescence for a similar coumarine system **6C**.[38] But it should be pointed out that the analysis of electronic transitions of coumarine dyes is dramatically impeded by a competing photoreaction in the presence of water, namely the ring-opening and formation of corresponding cinnamic acid derivatives.[231]

A comparable approach could be used to rationalize the intramolecular fluorescence quenching in *triphenylmethane* (TPM) dyes.[211] TPM dyes are structurally related to Michler's ketone, discussed in the preceding section, but they are generally charged systems. The most well-known TPM dyes, crystal violet (CV) and malachite green (MG), have been studied extensively using both steady-state and nanosecond-to-femtosecond time-resolved methods.[211–223] A complete account would be beyond the scope of this section. The important facts are that only the normal fluorescence band (F_B) is observable and that a nonradiative

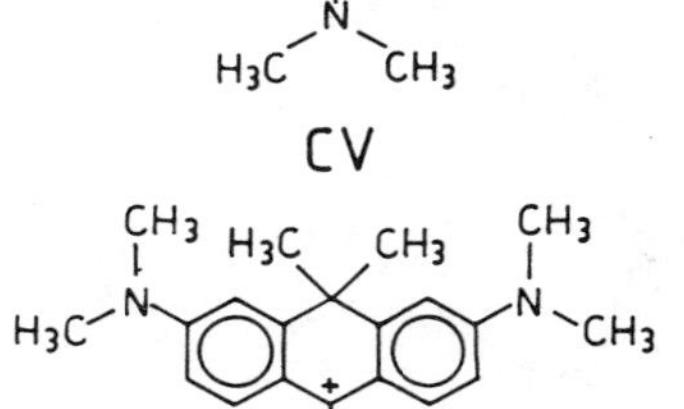

CV

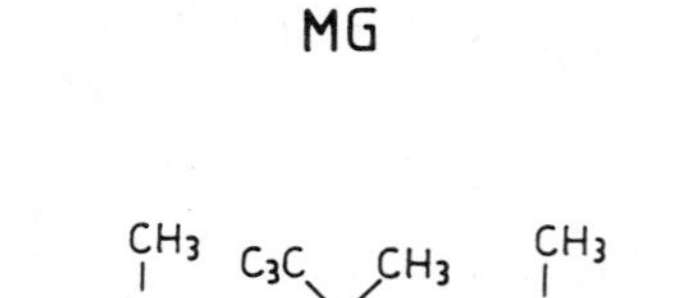

MG

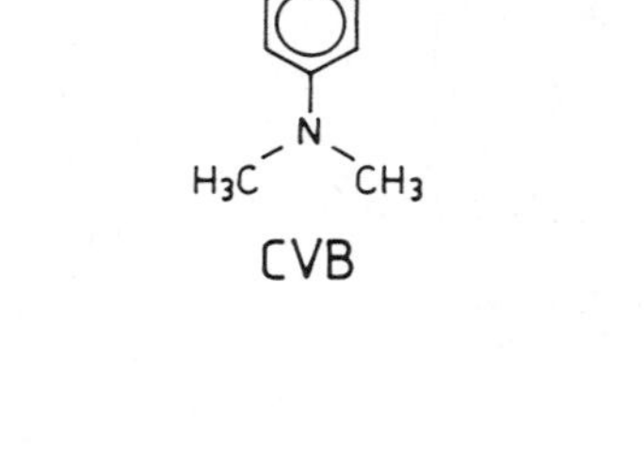

CVB

MGB

quenching process is present, the temperature (and viscosity) dependence of which is given by $E_{obs} \sim 0.7E_{\eta}$.[212] This quenching process is generally believed to be linked to a twisting relaxation of the anilino groups around the C–C bonds. Förster and Hoffman[212] showed that this relaxation occurs under the action of a driving force.

In order to gain more insight into the question of the driving force, bridged TPM model compounds CVB and MGB, where only one ring can rotate, were investigated.[211–214]

Steady-state fluorescence spectra showed that the fluorescence quantum yield ϕ_f for CVB and MGB differ by an order of magnitude for a given intermediate viscosity, but that at very high viscosities, similar high ϕ_f values and lifetimes in the range of 3–4 ns were obtained.[211] This is shown in Fig. 5.15. The comparison of CVB and MGB is instructive: If the driving force for rotation of the flexible groups were of steric origin, it should be similar for CVB and MGB (similar steric hindrance to planarity). Because of the bulkier rotating dimethylanilino group in CVB as compared to a rotating phenyl group in MGB, the former would be expected to show the slower rotation and a diminished intramolecular fluorescence quenching. The contrary is the case.

If, on the other hand, the driving force is influenced by electronic factors, we can correlate it with donor properties (for equal acceptor): in CVB the rotating group is a good donor (dimethylanilino group), but in MGB it is a poor donor (phenyl group). Thus, the observation that the quenching rate k_{BA} is much faster for CVB than for MGB suggests that donor–acceptor interactions are of crucial importance for k_{BA}: The compound CVB with the better donor possesses the stronger driving force.

It is to be expected that an increase of the acceptor properties in TPM dyes (for equal donor groups) would also cause stronger fluorescence quenching. This has indeed been found for the bridged model compound MGB (weak acceptor) when compared with the corresponding bridged TPM dye without any amino groups (strong acceptor), which shows a considerably reduced fluorescence quantum yield.[214]

Even for unbridged TPM dyes with three flexible phenyl groups, recent time-resolved data directly revealed the strong increase of the nonradiative decay rate[50,201] if one of the $N(CH_3)_2$ groups in CV is replaced by hydrogen (MG) or a cyano acceptor group (MGCN). The kinetic results are given in Table IV. They can be explained by assuming that k_{BA} (Scheme II) is increased if the energy of charge-separated states (with twisted donor–acceptor geometry) is lowered. The energies of these states might lie close to the energy of the ground state S_0, as judged from their nonemissive properties[215] and from the picosecond ground-state repopulation observed for CV and MG in low-viscosity conditions.[216–218]

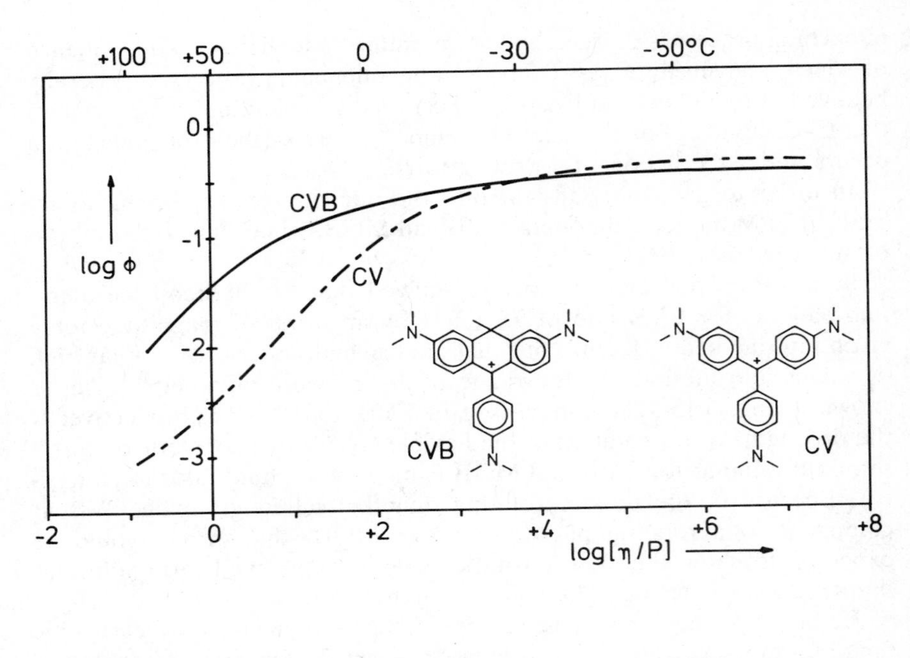

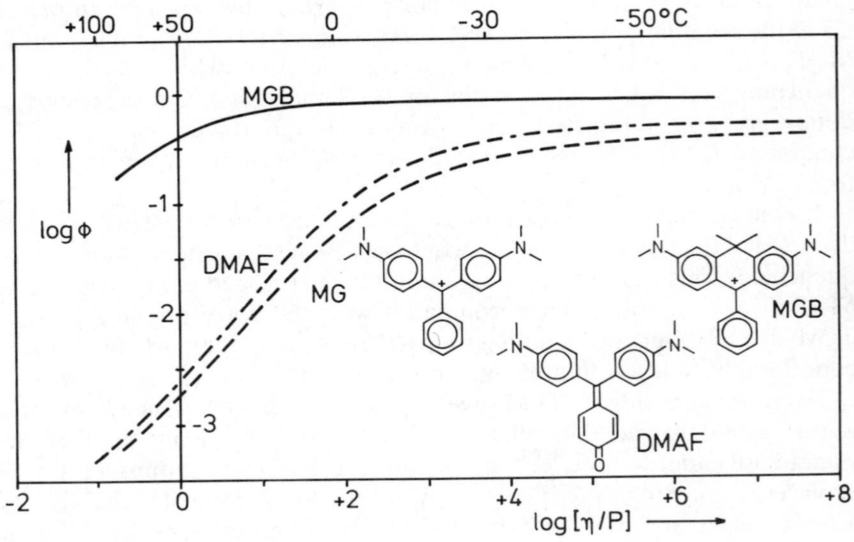

Fig. 5.15. Fluorescence quantum yields of bridged and unbridged triphenylmethane dyes in glycerol as a function of temperature/viscosity.[211] Not only cationic dyes show the strong fluorescence quenching for lower viscosities, but it is also present for neutral species as shown by the fuchsone DMAF.

TABLE IV
Decay Parameters (in ns) of Various TPM Dyes in Glycerol at −11.5°C Fitted by a Biexponential Model[50]

Compounds	τ_1	τ_2	$A_1/(A_1 + A_2)$	τ (77 K)
CV	0.82	1.88	0.53	2.92
MG	0.77	2.05	0.57	3.85
MGCN	0.67	1.65	0.68	3.16

This linkage of fluorescence quantum yields and decay rates with donor–acceptor strengths can be used as a predictive guideline for the synthesis of new fluorescent dyes.[211] As an application, the amino-rhodamine ARh has been synthesized. Instead of the usual rhodamines used as laser dyes, like rhodamine 3B (Rh-3), it possesses a rotatable phenyl group with an amino donor instead of a carboxylate acceptor. It is thus expected to show intramolecular fluorescence quenching due to a low-lying charge-separated state with a twisted dimethylanilino group as the donor. If, however, the donating $N(CH_3)_2$ substituent is blocked by protonation in acidic solvents, ARh/H^+ is formed with an $N(CH_3)_2H^+$ acceptor substituent, and the nonradiative decay should be blocked. Indeed, the experiments reveal that Rh-3 and ARh/H^+ are good fluorescers, whereas ϕ_f for ARh is diminished by a factor $>10^2$ in ethanol at room temperature.[215]

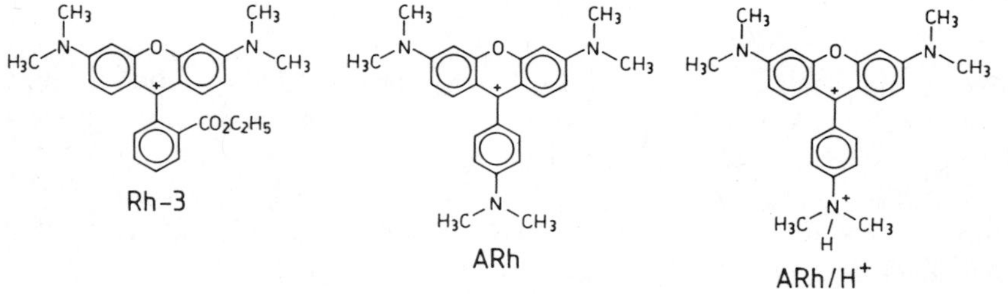

Rh-3 ARh ARh/H^+

The stochastic description of barrierless relaxations by Bagchi, Fleming, and Oxtoby (Ref. 195 and Section IV.I) was first applied by these authors to TPM dyes to explain the observed nonexponential fluorescence decay and ground-state repopulation kinetics. The experimental evidence of an activation energy $E_{obs} < E_\eta$ is also in accordance with a barrierless relaxation model. The data presented in Table IV are indicative of nonexponential decay, too. They were obtained by fitting the experiment to a biexponential model, but it can be shown[50] that a fit of similar quality can be obtained with the error-function model of barrierless relaxations. Thus, τ_1 and τ_2 are related to τ_c^0 and τ_c^ω, but, at present, we can only

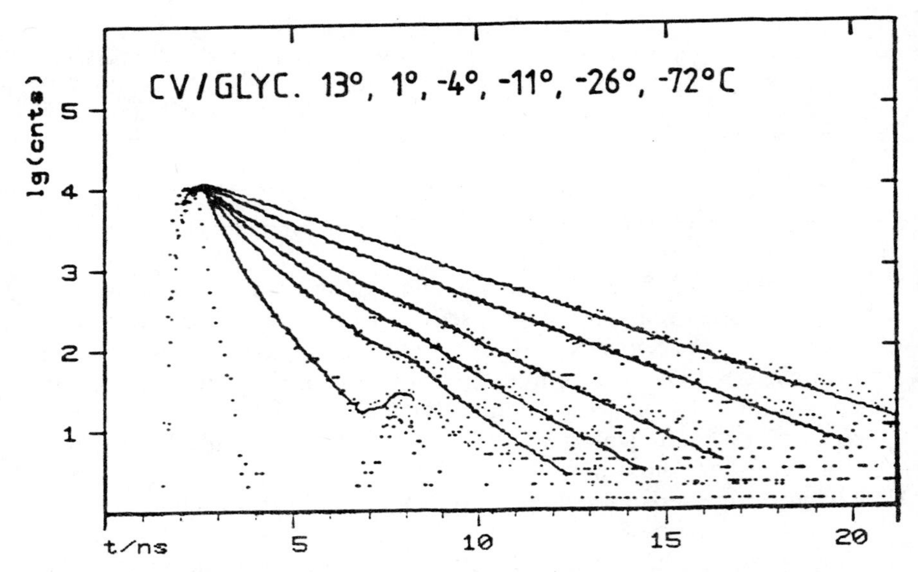

Fig. 5.16. Nonexponential decay curves of crystal violet (excited with synchrotron radiation from BESSY) in glycerol at six different temperatures. Within experimental error limits, the decays can be fitted by a biexponential model. The two lifetime components at 13°C are 330 and 850 ps, respectively. Upon reducing the temperature, both fitted decay times increase, as well as the relative weight of the slower decay component. At −72°C, a monoexponential decay is observed (2.71 ns).

discuss the general trends because a direct fit to the error-function model is still in development. As both τ_1 and τ_2 are shortened for increasing acceptor strength, τ_c^{ω} must also become shorter. Thus, quite similar to the case of the comparison of DMABN with its ester DMABEE, the shorter fluorescence decay can be traced back to an increase of ω, that is, of the driving force for the relaxation. This relaxation can be completely stopped with increasing viscosity, for example, in the case of low-temperature glycerol solutions. As shown in Fig. 5.16, the fluorescence decay is then purely monoexponential. Only by raising the temperature, the barrierless relaxation sets in, and the decay becomes increasingly nonexponential.

VI. Summary

In this article, the formation and characteristics of a new class of excited states, the "twisted intramolecular charge transfer" (TICT) states, have been dealt with from various theoretical and experimental viewpoints. The main electronic features accompanying TICT state formation are internal twisting towards a new geometrical arrangement of the molecular subunits and a large charge separation in the low-lying excited state of "charge-transfer biradicaloid" nature. These types of states appear most often as a consequence of the breaking of a π-dative bond. The energy of such highly polar states depends sensitively on a proper donor–acceptor combination of the molecular subunits as well as on the polarity of the environment, if present. Polar solvents can even polarize low-lying zwitterionic states of twisted species containing two equal subunits. In the case that the TICT state represents the S_1 surface close to the orthogonal twist, it can be responsible for the observation of an additional anomalous fluoresence band.

The TICT model is backed up by several key experiments:

1. Model compounds for DMABN, where the twisting of the dialkylamino group away from the near-planar conformation is blocked by molecular bridging, show no tendency for the emission of an anomalous fluorescence band. Model compounds for which the planar conformation is made unavailable by steric hindrance, and which exist in the perpendicular conformation already in the ground state, show an increased tendency for the "anomalous" fluorescence or display it as the only emission. The same behavior has been observed for larger systems using molecular bridging; thus internal twisting towards TICT states is likely to be a rather common phenomenon for large and flexible organic compounds.
2. Electrooptical emission measurements in the gas phase[12] show that the TICT state with its anomalous emission can be reached independently of any solvent environment (thus its formation is a true intramolecular phenomenon), and that the observed excited state dipole moment is consistent with a large degree of intramolecular charge separation.
3. Most recently the TICT model has been verified by transient absorption spectroscopy,[234] that is, by time-resolved studies of vertical transitions which start from different points on the horizontal reaction pathway.

Dynamical description of intramolecular reactions involving conformational changes is based on the Langevin–Fokker–Planck equation, founded on the properties of Markovian stochastic processes. Kinetic studies of chemical reactions are often investigated with the Kramers model, which yields results valid for high internal potential barriers and in the quasi-stationary regime (constant reaction rates). In fact, only for long times, as postulated by the Markovian process, the sytem ignores the initial conditions and there is an exponential relaxation of the initially excited state population.

Time-dependent reaction rates have been made evident for the TICT state formation in DMABN and some related compounds. This evidence is due to the observation of the nonexponential decay of the B* state (in the absence of the backward reactions) and of a faster rise-time of the A* emission than that predicted by the constant reaction rate scheme. The dynamical behavior can be accounted for quantitatively by the Smoluchowski equation in the presence of a harmonic potential without any intramolecular barrier and with a position-dependent sink.

In addition, recent investigations should be mentioned that deal with the breakdown of the Kramers theory as a problem of correct modeling[232] and its limited ability to describe photophysical processes.[233]

Acknowledgement

The first three authors acknowledge financial support by the Deutsche Forschungsgemeinschaft, the Bundesministerium für Forschung und Technologie (project 05 314 FA), and the Fonds der Chemischen Industrie, and especially W. Rettig wishes to thank the Heisenberg foundation for a stimulating grant.

References

1. E. Lippert, C. A. Chatzidimitriou-Dreismann, and K.-H. Naumann, *Adv. Chem. Phys.* **57**, 311 (1984).
2. W. Rettig, *Angew. Chem.* **98**, 969 (1986); *Angew. Chem. Int. Ed. Engl.* **25**, 971 (1986).
3. E. Lippert, *Organic Molecular Photophysics*, Vol. 2, Chap. I, J. B. Birks (ed.). Wiley, London, 1975.
4. T. Kindt and E. Lippert, in *Excited States in Organic Chemistry and Biochemistry*, B. Pullmann and N. Goldblum (eds.). D. Reidel, Dordrecht, 1977.
5. E. Lippert, W. Lüder, and H. Boos, in *Advances in Molecular Spectroscopy*, A. Mangini (ed.). Pergamon Press, Oxford, 1962, p. 443.
6. W. Rettig, G. Wermuth, and E. Lippert, *Ber. Bunsenges. Phys. Chem.* **83**, 692 (1979).
7. Z. R. Grabowski, K. Rotkiewicz, A. Siemiarczuk, D. J. Cowley, and W. Baumann, *Nouv. J. Chimie* **3**, 443 (1979).
8. E. Lippert, A. A. Ayuk, W. Rettig, and G. Wermuth, *J. Photochem.* **17**, 237 (1981).
9. W. Rettig and E. Lippert, *J. Mol. Struct.* **61**, 17 (1980).
10. G. Wermuth, *Z. Natforsch.* **38a**, 641 (1983).
11. K. Rotkiewicz, K. H. Grellmann, and Z. R. Grabowski, *Chem. Phys. Lett.* **19**, 315 (1973).
12. H. Bischof, W. Baumann, N. Detzer, and K. Rotkiewicz, *Chem. Phys. Lett.* **116**, 180 (1985).
13. J. Dobkowski, E. Kirkor-Kamińska, J. Koput, and A. Siemiarczuk, *J. Lumin.* **27**, 339 (1982).
14. V. Bonačić-Koutecký, J. Koutecký, and J. Michl, *Angew. Chem.* **99**, 216 (1987); *Angew. Chem. Intern. Ed. English* **26**, 170 (1987).

15. L. Pauling, *Proc. Nat. Acad. Sci. USA* **25**, 577 (1939).
16. H. A. Kramers, *Physica* **7**, 284 (1940).
17. T. Kihara, *Intermolecular Forces*. Wiley, Chichester, 1978, Chap. 7.
18. E. Lippert, B. Prass, W. Rettig, and D. Stehlik, *BESSY Jahresbericht 1983*, pp. 50–53.
19. K. Kimura, *Adv. Chem. Phys.* **60**, 172 (1985).
20. K. Rotkiewicz, Z. R. Grabowski, A. Krowczyński, and W. Kühnle, *J. Lumin.* **12/13**, 877 (1976).
21. Z. R. Grabowski, K. Rotkiewicz, W. Rubaszewska, and E. Kirkor-Kamińska, *Acta Phys. Polonica* **A54**, 767 (1979).
22. Z. R. Grabowski, K. Rotkiewicz, and A. Siemiarzuk, *J. Lumin.* **18/19**, 420 (1979).
23. J. Lipiński, H. Chojnacki, Z. R. Grabowski, and K. Rotkiewicz, *Chem. Phys. Lett.* **70**, 449 (1980).
24. O. S. Khalil, R. H. Hofeldt, and S. P. McGlynn, *Chem. Phys. Lett.* **17**, 479 (1972).
25. O. S. Khalil, R. H. Hofeldt, and S. P. McGlynn, *J. Lumin.* **6**, 229 (1973).
26. O. S. Khalil, R. H. Hofeldt, and S. P. McGlynn, *Spectrosc. Lett.* **6**, 147 (1973).
27. W. Rettig, K. Rotkiewicz, and W. Rubaszewska, *Spectrochim. Acta* **40A**, 241 (1984).
28. G. Wermuth and W. Rettig, *J. Phys. Chem.* **88**, 2729 (1984).
29. W. Rettig and R. Gleiter, *J. Phys. Chem.* **89**, 4676 (1985).
30. G. Wermuth, W. Rettig, and E. Lippert, *Ber. Bunsenges. Phys. Chem.* **81**, 64 (1981).
31. K. Rotkiewicz and W. Rubaszewska, *Chem. Phys. Lett.* **70**, 444 (1980).
32. K. Rotkiewicz and W. Rubaszewska, *J. Lumin.* **27**, 221 (1982).
33. W. Rettig and G. Wermuth, *J. Photochem.* **28**, 351 (1985).
34. S. Dähne, W. Freyer, K. Teuchner, J. Dobkowski, and Z. R. Grabowski, *J. Lumin.* **22**, 37 (1980).
35. Z. R. Grabowski and J. Dobkowski, *Pure and Appl. Chem.* **55**, 245 (1983).
36. N. J. Turro, J. McVey, V. Ramamurthy, and P. Lechtken, *Angew. Chemie* **31**, 597 (1979).
37. J. B. Birks, *Chem. Phys. Lett.* **54**, 430 (1978).
38. W. Rettig and A. Klock, *Can. J. Chem.* **63**, 1649 (1985).
39. V. Bonačić-Koutecký, P. Bruckmann, P. Hiberty, J. Koutecký, C. Leforestier, and L. Salem, *Angew. Chem.* **87**, 599 (1975).
40. V. Bonačić-Koutecký and J. Michl, *J. Am. Chem. Soc.* **107**, 1765 (1985).
41. P. Suppan, *J. Mol. Spectr.* **30**, 17 (1969).
42. G. Hohlneicher and B. Dick, *J. Photochem.* **27**, 215 (1984).
43. J. Saltiel, J. D'Agostino, E. D. Megarity, L. Metts, K. R. Neuberger, M. Wrighton, and O. C. ZaFiriou, *Organ. Photochem.* **3**, 1 (1973).
44. H. Görner and D. Schulte-Frohlinde, *J. Mol. Struct.* **84**, 227 (1982).
45. F. Dietz and S. K. Rentsch, *Chem. Phys.* **96**, 145 (1985).
46. J. Hicks, M. Vandersall, Z. Babarogic, and K. B. Eisenthal, *Chem. Phys. Lett.* **116**, 18 (1985).
47. J. S. McCaskill and R. G. Gilbert, *Chem. Phys.* **44**, 389 (1979); D.P. Miller and K.B. Eisenthal, *J. Chem. Phys.* **83**, 5076 (1985).
48. B. Wilhelmi, *Chem. Phys.* **66**, 351 (1982).
49. G. R. Fleming, D. H. Waldeck, K. M. Keery, and S. P. Velsko, in *Applications of*

Picosecond Spectroscopy to Chemistry, K. B. Eisenthal (ed.). Reidel, Dordrecht, 1984, p. 67.

50. W. Rettig, M. Vogel, E. Lippert, and H. Otto, *Chem. Phys.* **103**, 381 (1986).
51. B. Carmeli and A. Nitzan, *Physical Review* **A29**, 1481 (1984).
52. B. Carmeli and A. Nitzan, *J. Chem. Phys.* **80**, 3596 (1984).
53. A. Nitzan, *J. Chem. Phys.* **82**, 1614 (1985).
54. V. Sundström and T. Gillbro, *Ber. Bunsenges. Phys. Chem.* **89**, 222 (1985).
55. J. M. Dawes and M. G. Sceats, *Chem. Phys.* **96**, 315 (1985).
56. J. M. Dawes and M. G. Sceats, *Ber. Bunsenges. Phys. Chem.* **89**, 233 (1985).
57. R. F. Grote and J. T. Hynes, *J. Chem. Phys.* **73**, 2715 (1980).
58. B. Carmeli and A. Nitzan, *Chem. Phys. Lett.* **106**, 329 (1984).
59. J. B. Birks, *Photophysics of Aromatic Molecules*. Wiley, London, 1970.
60. E. Lippert, *Z. Naturforsch.* **10a**, 541 (1955).
61. E. Lippert, *Ber. Bunsenges. Phys. Chem.* **61**, 962 (1957).
62. N. Mataga, Y. Kaifu, and M. Koizumi, *Bull. Chem. Soc. Japan* **29**, 115, 465 (1956).
63. W. Rettig, *J. Mol. Struct.* **84**, 303 (1982).
64. K. D. Nolte and S. Dähne, *Adv. Mol. Relax. Interact. Processes* **10**, 299 (1977).
65. S. Hünig and O. Rosenthal, *Liebigs Ann. Chem.* **592**, 161, 180 (1955).
66. N. G. Bakhshiev, *Opt. Spectrosc.* **6**, 646 (1958); N. G. Bakhshiev, V. P. Volkov and A. V. Altaiska, *Opt. Spectrosc.* **28**, 26 (1970).
67. L. U. Bilot and A. Kawski, *Z. Naturforsch.* **A17**, 621 (1962).
68. K. Yamasaki, K. Arita, O. Kajimoto, and K. Hara, *Chem. Phys. Lett.* **123**, 24 (1986); O. Kajimoto, K. Yamasaki, K. Arita, and K. Hara, *Chem. Phys. Lett.* **125**, 184 (1986).
69. M. Zander and W. Rettig, *Chem. Phys. Lett.* **110**, 602 (1984).
70. W. Liptay, in *Excited States*, Vol. 1, E. C. Lim (ed.). Academic Press, New York, 1974.
71. W. Liptay, *Z. Naturforsch.* **A20**, 1441 (1965).
72. Y. Ooshika, *J. Phys. Soc. Japan* **9**, 594 (1954).
73. L. Onsager, *J. Am. Chem. Soc.* **58**, 1486 (1936).
74. W. Rettig and E. A. Chandross, *J. Amer. Chem. Soc.* **107**, 5617 (1985).
75. F. Schneider and E. Lippert, *Ber. Bunsenges. Phys. Chem.* **72**, 1155 (1968); **74**, 624 (1970).
76. N. Nakashima, M. Murakawa, and N. Mataga, *Bull. Chem. Soc. Japan* **49**, 854 (1976).
77. W. Rettig and M. Zander, *Ber. Bunsenges. Phys. Chem.* **87**, 1143 (1983).
78. F. Heisel and J. A. Miehé, *Chem. Phys. Lett.* **100**, 183 (1983).
79. F. Heisel and J. A. Miehé, *Chem. Phys.* **98**, 233 (1985).
80. Y. Wang, M. McAuliffe, F. Novak, and K. B. Eisenthal, *J. Phys. Chem.* **85**, 3736 (1981).
81. Y. Wang and K. B. Eisenthal, *J. Chem. Phys.* **77**, 6076 (1982).
82. W. S. Struve, P. M. Rentzepis, and J. Jortner, *J. Chem. Phys.* **59**, 5014 (1973); W. S. Struve and P. M. Rentzepis, *J. Chem. Phys.* **60**, 1533 (1974); *Chem. Phys. Lett.* **29**, 23 (1974).

83. D. Huppert, S. D. Rand, P. M. Rentzepis, P. F. Barbara, W. S. Struve, and Z. R. Grabowski, *J. Chem. Phys.* **75**, 5714 (1981).
84. S. R. Meech and D. Phillips, *Chem. Phys. Lett.* **116**, 262 (1985).
85. C. Reichardt and E. Harbusch-Görnert, *Liebigs Ann. Chem.* 721 (1983).
86. W. Rettig, M. Vogel, and A. Klock, *Europ. Photochem. Ass. Newsletter* **27**, 41 (1986).
87. W. Rettig, *J. Lumin.* **26**, 21 (1981).
88. F. Heisel, J. A. Miehé, and J. M. G. Martinho, *Chem. Phys.* **98**, 243 (1985).
89. F. Heisel and J. A. Miehé, *Chem. Phys. Lett.* **128**, 323 (1986).
90. Y. T. Mazurenko and N. G. Bakhshiev, *Opt. and Spectrosc.* **28**, 490 (1970).
91. W. Rapp, H. H. Klingenberg, and H. E. Lessing, *Ber. Bunsenges. Phys. Chem.* **75**, 883 (1971).
92. T. Kobayashi, H. Ohtani, and K. Kurokawa, *Chem. Phys. Lett.* **121**, 356 (1985).
93. N. G. van Kampen, *Int. J. Quant. Chem. Symp.* **16**, 101 (1982).
94. E. Lippert, in *2. Intern. Farbensymposium*, W. Foerst (Ed.). Verlag Chemie, Weinheim, 1965.
95. M. Haba, *J. Lumin.* (to be published in *J. Luminescence*).
96. R. J. Visser and C. A. G. O. Varma, *J. Chem. Soc. Faraday II*, **76**, 453 (1980); R. J. Visser, C. A. G. O. Varma, J. Konijenberg, and P. Bergwerf, *J. Chem. Soc., Faraday II* **79**, 347 (1983); R. J. Visser, C. A. G. O. Varma, J. Konijenberg, and P. C. M. Weisenborn, *J. Mol. Struct.* **114**, 105 (1984).
97. R. J. Visser, P. C. M. Weisenborn, and C. A. G. O. Varma, *Chem. Phys. Lett.* **113**, 330 (1985); R. J. Visser, P. C. M. Weisenbron, P. J. M. van Kan, B. H. Huizer, C. A. G. O. Varma, J. M. Warman, and M. P. De Haas, *J. Chem. Soc. Faraday II* **81**, 689 (1985).
98. Y. Wang, *Chem. Phys. Lett.* **116**, 286 (1985).
99. W. Hub, S. Schneider, F. Dörr, I. D. Oxman, and F. D. Lewis, *J. Am. Chem. Soc.* **106**, 701 (1984).
100. D. L. Huber, *Phys. Rev.* **158**, 843 (1967); **178**, 93 (1969).
101. L. Salem and C. Rowland, *Angew. Chem., Int. Ed. Engl.* **11**, 92 (1972).
102. J. Michl, *Mol. Photochem.* **4**, 257 (1972).
103. W. T. Borden (Ed.), *Biradicals*. Wiley, New York, 1982.
104. L. Salem, *Electrons in Chemical Reactions*. Wiley, New York, 1982.
105. C. E. Wulfman and S. Kumai, *Science* **172**, 1061 (1971).
106. R. J. Buenker, V. Bonačić-Koutecký, and L. Pogliani, *J. Chem. Phys.* **73**, 1836 (1980).
107. V. Bonačić-Koutecký, L. Pogliani, M. Persico, and J. Koutecký, *Tetrahedron* **38**, 741 (1982).
108. V. Bonačić-Koutecký, J. Köhler, and J. Michl, *Chem. Phys. Lett.* **104**, 440 (1984).
109. M. Persico and V. Bonačić-Koutecký, *J. Chem. Phys.* **76**, 6018 (1982).
110. L. Salem, *Acc. Chem. Res.* **12**, 87 (1979).
111. V. Bonačić-Koutecký, *J. Am. Chem. Soc.* **100**, 396 (1978).
112. C. M. Meerman-van Benthem, H. J. C. Jacobs, and J. J. C. Mulder, *Nouv. J. Chim.* **2**, 123 (1978).

113. H. Kollmar and V. Staemmler, *Theor. Chim. Acta* **48**, 223 (1978).
114. R. J. Buenker and S. D. Peyerimhoff, *Chem. Phys.* **9**, 75 (1975).
115. W. T. Borden and E. R. Davidson, *J. Am. Chem. Soc.* **99**, 1789 (1977).
116. A. Weller and H. Beens, *Chem. Phys. Lett.* **3**, 668 (1969).
117. W. Feller, *An Introduction to Probability Theory and its Applications*, 2nd ed. Wiley, New York, 1971, Vol. II.
118. R. L. Stratonovich, *Topics in the Theory of Random Noise*. Gordon and Breach, New York, 1963, Vols. I and II.
119. A. Papoulis, *Probability, Random Variables, and Stochastics Processes*. McGraw-Hill, New York, 1965.
120. A. Blanc-Lapierre and B. Picinbono, *Fonctions aléatoires*. Masson, Paris 1981.
121. A. T. Bharucha-Reid, *Elements of the Theory of Markov Processes and their Applications*. McGraw-Hill, New York, 1960.
122. E. Parzen, *Stochastic Processes*. Holden-Day, San Francisco, 1967.
123. E. W. Montroll and J. L. Lebowitz, *Fluctuation Phenomena*. North-Holland, Amsterdam, 1979.
124. E. W. Montroll and B. J. West, in *Fluctuation Phenomena*, E. W. Montroll and J. L. Lebowitz (eds.). North Holland, Amsterdam, 1979.
125. B. Mandelbrot, *Fractals, Form, Chance and Dimension*. W. H. Freeman, San Fransisco, 1977.
126. B. Mandelbrot, *Les objets fractals*. Flammarion, Paris, 1984.
127. P. Lévy, *Théorie de l'addition des variables aléatoires*. Gauthier-Villars, Paris, 1937.
128. B. V. Gdenko and A. N. Kolmogorov, *Limit Distributions for Sums of Independent Random Variables*. Addison-Wesley, Cambridge, MA, 1954.
129. G. Williams and D. C. Watts, *Trans. Faraday Soc.* **66**, 2503 (1970).
130. G. Williams, D. C. Watts, S. B. Dev, and A. M. North, *Trans. Faraday Soc.* **67**, 1323 (1970).
131. C. T. Moynihan, L. P. Boesch, and N. L. La Berge, *Phys. Chem. Glasses* **14**, 122 (1973).
132. C. P. Lindsey and G. D. Patterson, *J. Chem. Phys.* **73**, 3348 (1980).
133. E. W. Montroll and J. T. Bendler, *J. Stat. Phys.* **43**, 129 (1984).
134. D. Kenan, *An Introduction to Stochastic Processes*. North-Holland, Amsterdam, 1979.
135. N. G. van Kampen, *Stochastic Processes in Physics and Chemistry*. North-Holland, Amsterdam, 1981.
136. I. Oppenheim, K. E. Shuler, and G. H. Weiss, *Stochastic Processes in Chemical Physics—The Master Equation*. M.I.T. Press, Cambridge, 1977. I. Oppenheim, *Adv. Chem. Phys.* **15**, 1 (1969).
137. C. W. Gardiner, *Handbook of Stochastic Methods*. Springer Verlag, Berlin, 1983.
138. N. S. Goel and N. Richter-Dyn, *Stochastic Models in Biology*. Academic Press, New York, 1974.
139. I. Oppenheim, K. E. Shuller, and G. H. Weiss, *Adv. Mol. Relaxation Processes* **1**, 13 (1967).
140. D. A. McQuarrie, *Stochastic Approach to Chemical Kinetics*. Methuen, London, 1968.

141. D. A. McQuarrie, *J. Appl. Prob.* **4**, 413 (1967).
142. D. A. McQuarrie, *Adv. Chem. Phys.* **15**, 149 (1969).
143. E. Teramoto, N. Shiegsada, H. Nakajima, and K. Sato, *Adv. Biophys.* **2**, 1955 (1971).
144. D. T. Gillespie and M. Mangel, *J. Chem. Phys.* **75**, 704 (1981).
145. M. Kac, in *Selected Papers on Noise and Stochastic Processes*, N. Wax (ed.). Dover, New York, 1954.
146. N. Wax, *Selected Papers on Noise and Stochastic Processes*. Dover, New York, 1954.
147. S. Chandrasekhar, in *Selected Papers on Noise and Stochastic Processes*. N. Wax (ed.). Dover, New York, 1954.
148. J. E. Moyal, *J. R. Stat. Soc.* **11**, 151 (1949).
149. N. G. van Kampen, *Can. J. Phys.* **39**, 551 (1961); *Adv. Chem. Phys.* **34**, 245 (1976).
150. P. Hänggi, H. Grabert, P. Talkner, and H. Thomas, *Phys. Rev.* **A29**, 371 (1984).
151. B. J. Matkowsky, Z. Schuss, and C. Knessl, *Phys. Rev.* **A29**, 3559 (1984).
152. C. Knessl, M. Mangel, B. J. Matkowsky, Z. Schuss, and C. Tier, *J. Chem. Phys.* **81**, 1285 (1984).
153. G. Nicolis and I. Prigogine, *Self Organisation in Nonequilibrium Systems*, Wiley, New York, 1977.
154. G. H. Weiss, *Stat. Phys.* **6**, 179 (1972).
155. T. Hida, *Brownian Motion*. Springer Verlag, New York, 1980.
156. R. Kubo, N. Hashitsume, and M. Toda, *Statistical Physics II*. Springer Verlag, Berlin, 1985.
157. Z. Schuss, *Theory and Applications of Stochastic Differential Equations*. Wiley, New York, 1980.
158. R. F. Fox, *Phys. Rep.* **48**, 179 (1978).
159. D. A. McQuarrie, *Statistical Mechanics*. Harper & Row, New York, 1976.
160. R. S. Berry, S. A. Rice, and J. Ross, *Physical Chemistry*. Wiley, New York, 1980.
161. W. Horsthemke and R. Lefever, *Noise-Induced Transitions*. Springer Verlag, Berlin, 1984.
162. A. M. Il'in and R. Z. Khas'minskii, *Theory of Prob. Appl.* **9**, 421 (1964).
163. G. Wilemski, *J. Stat. Phys.* **14**, 153 (1976).
164. N. G. van Kampen, *Phys. Rep.* **124**, 69 (1985).
165. L. Arnold, *Stochastic Differential Equations: Theory and Applications*. Wiley, New York, 1974.
166. I. I. Gihman and A. V. Skorohod, *Stochastic Differential Equations*. Springer, Berlin, 1972.
167. S. Kim and I. Oppenheim, *Physica* **57**, 469 (1972).
168. H. C. Brinkmann, *Physica* **22**, 29 (1956).
169. P. B. Visscher, *Phys. Rev.* **B13**, 3272 (1976).
170. C. Blomberg, *Physica* **A86**, 49 (1977).
171. O. Edholm and O. Leimar, *Physica* **A98**, 313 (1979).
172. M. Lopez de Haro, *Physica* **A111**, 65 (1982).
173. P. B. Visscher, *Phys. Rev.* **B14**, 347 (1976).
174. M. Mangel, *J. Chem. Phys.* **72**, 6606 (1980).

175. C. W. Gardiner, *J. Stat. Phys.* **30**, 157 (1983).
176. R. Landauer and J. A. Swanson, *Phys. Rev.* **121**, 1668 (1961).
177. R. J. Donnelly and P. H. Roberts, *Proc. Roy. Soc.* **A312**, 519 (1969).
178. J. L. Skinner and P. G. Wolynes, *J. Chem. Phys.* **72**, 4913 (1980).
179. D. K. Garrity and J. L. Skinner, *Chem. Phys. Lett.* **95**, 46 (1980).
180. J. A. Montgomery Jr., D. Chandler, and B. J. Berne, *J. Chem. Phys.* **70**, 4056 (1979).
181. M. Buttiker, E. P. Harris, and R. Landauer, *Phys. Rev.* **B28**, 1268 (1983).
182. B. H. Lavenda, *Lett. Nuovo Cim.* **37**, 200 (1983).
183. B. Carmeli and A. Nitzan, *Phys. Rev. Lett.* **51**, 233 (1983).
184. N. G. van Kampen, *J. Stat. Phys.* **17**, 71 (1977).
185. M. Mörsch, H. Risken, and H. D. Vollmer, *Z. Physik* **B32**, 245 (1979).
186. A. Kaufmann, D. Grouchko, and R. Cruon, *Modèles mathématiques pour l'étude de la fiabilité des systèmes*. Masson, Paris, 1975.
187. W. Hess and R. Klein, *Adv. Phys.* **32**, 173 (1983).
188. R. F. Grote and J. T. Hynes, *J. Chem. Phys.* **74**, 4465 (1981); *J. Chem. Phys.* **77**, 3736 (1982).
189. P. Hänggi and F. Mojtabai, *Phys. Rev.* **A26**, 1168 (1982).
190. P. Hänggi, *Phys. Rev.* **A26**, 2996 (1982).
191. S. A. Adelman, *Adv. Chem. Phys.* **53**, 61 (1983); *J. Chem. Phys.* **64**, 124 (1976).
192. P. Hänggi, H. Thomas, H. Grabert, and P. Talkner, *J. Stat. Phys.* **18**, 155 (1978).
193. B. Carmeli and A. Nitzan, *J. Chem. Phys.* **79**, 393 (1983); *Isr. J. Chem.* **22**, 360 (1982); *Phys. Rev.* **A32**, 2439 (1985).
194. A. Szabo, K. Schulten, and Z. Schulten, *J. Chem. Phys.* **72**, 4350 (1980).
195. B. Bagchi, G. R. Fleming, and D. W. Oxtoby, *J. Chem. Phys.* **78**, 7375 (1983); B. Bagchi, S. Singer, and D. W. Oxtoby, *Chem. Phys. Lett.* **99**, 225 (1983).
196. A. J. F. Siegert, *Phys. Rev.* **81**, 617 (1951).
197. G. H. Weiss, *Adv. Chem. Phys.* **13**, 1 (1967).
198. G. H. Weiss, K. E. Shuler, and K. Lindenberg, *J. Stat. Phys.* **31**, 255 (1983).
199. Landolt-Börnstein, *Atomic and Molecular Physics, II, 5*, K. H. Hellwege (ed.). Springer Verlag, Berlin, 1967, p. 190.
200. K. M. Hong and J. Noolandi, *Surface Sci.* **75**, 561 (1978).
201. W. Rettig, E. Lippert, and H. Otto, *Berliner Elektronen-Speicherring für Synchrotronstrahlung (BESSY) Jahrbuch*, 1984, p. 171.
202. A. Siemiarczuk, Z. R. Grabowski, A. Krówczyński, M. Asher, and M. Ottolenghi, *Chem. Phys. Lett.* **51**, 315 (1977).
203. A. Siemiarczuk, J. Koput, and A. Pohorille, *Z. Naturforsch.* **A37**, 598 (1982).
204. W. Baumann, F. Petzke, and K.-D. Loosen, *Z. Naturforsch.* **A34**, 1070 (1979).
205. B. Schwager and W. Baumann, *12th International Conference on Photochemistry*. Tokyo, Aug. 1985.
206. W. Liptay, H.-J. Schumann, and F. Petzke, *Chem. Phys. Lett.* **39**, 427 (1975).
207. W. Rettig (unpublished results).
208. G. Jones II, W. R. Jackson, and A. M. Halpern, *Chem. Phys. Lett.* **72**, 391 (1980).
209. G. Jones II, W. R. Jackson, S. Kanoktanaporn, and A. M. Halpern, *Opt. Commun.* **33**, 315 (1980).

210. G. Jones II, W. R. Jackson, C. Choi, and W. R. Bergmark, *J. Phys. Chem.* **89**, 294 (1985).

211. M. Vogel and W. Rettig, *Ber. Bunsenges. Phys. Chem.* **89**, 962 (1985).

212. Th. Förster and G. Hoffmann, *Z. Phys. Chem. N. F.* **75**, 63 (1971).

213. G. Hoffmann, *Z. Physik. Chem. N. F.* **71**, 132 (1970).

214. G. Hoffmann, A. Schönbucher, and H. Steidl, *Z. Natf.* **28a**, 1136 (1973).

215. M. Vogel and W. Rettig, Contribution to the XI. IUPAC Symposium on Photochemistry, Lisbon, Portugal, 1986.

216. D. J. Erskine, A. J. Taylor, and C. L. Tang, *J. Chem. Phys.* **80**, 5338 (1984).

217. D. Ben-Amotz and C. B. Harris, *Chem. Phys. Lett.* **119**, 305 (1985).

218. V. Sundström and T. Gillbro, *Chem. Phys. Lett.* **110**, 303 (1984); *J. Chem. Phys.* **81**, 3463 (1984).

219. C.A. Cremers and M. W. Windsor, *Chem. Phys. Lett.* **71**, 27 (1980).

220. G.S. Beddard, T. Doust, and M. W. Windsor, *Springer Series in Chemical Physics* **14**, 167 (1980).

221. T. Doust, *Chem. Phys. Lett* **96**, 522 (1983).

222. R. Trebino and A. E. Siegman, *J. Chem. Phys.* **79**, 3621 (1983).

223. R. Menzel, C. W. Hoganson, and M. W. Windsor, *Chem. Phys. Lett.* **120**, 29 (1985).

224. R. Kołos and Z. R. Grabowski, *J. Mol. Struct.* **84**, 251 (1982).

225. a) W. Baumann, *Z. Natf.* **36A**, 868 (1981); b) W. Baumann, in proceedings of the "Meeting on Photoinduced Electron Transfer and Related Phenomena—Fundamental, Functional and Biological Aspects," N. Malaga, Ed., Kyoto, Japan, Aug. 12–13, 1985, page 28; c) ensuing discussion, page 35; d) W. Baumann, H. Bischof, J.-C. Fröhling, W. Rettig, and K. Rotkiewicz, to be published.

226. R. J. Visser, P. C. M. Weissenborn, C. A. G. O. Varma, M. P. De Haas, and J. M. Warman, *Chem. Phys. Lett.* **104**, 38 (1984).

227. K. Rotkiewicz and W. Rubaszewska, *J. Lumin.* **29**, 329 (1984).

228. E. M. Kosower and H. Dodiuk, *J. Amer. Chem. Soc.* **98**, 924 (1976).

229. H. Labhart, *Chem. Phys. Lett.* **1**, 263 (1967).

230. R. J. Buenker, S. D. Peyerimhoff, *Theor. Chim. Acta* **35**, 33 (1974); R. J. Buenker, S. D. Peyerimhoff, W. Butscher, *Mol. Phys.* **35**, 771 (1978); R. J. Buenker, in "Studies in Physical and Theoretical Chemistry", Vol. 21 (Current Aspects of Quantum Chemistry 1981), ed. R. Carbo, Elsevier Scientific Publ. Co., Amsterdam 1982, pp. 17–34; R. J. Buenker, in "Proceedings of Workshop on Quantum Chemistry and Molecular Physics" in Wollongong, Australia, February 1980; R. J. Buenker and R. A. Phillips, *J. Mol. Struct., Theochem* **123**, 291–300 (1985).

231. O. S. Wolfbeis, E. Lippert, and H. Schwarz, *Ber. Bunsenges. Phys. Chem.* **84**, 1115 (1980).

232. F. Marchesoni, *Adv. Chem. Phys.* **63**, 603 (1985).

233. T. Fonseca, J. A. N. F. Gomes, P. Grigolini, and F. Marchesoni, *Adv. Chem. Phys.* **62**, 425 (1985).

234. T. Okada, N. Mataga, and W. Baumann, private communication (to be published in *J. Chem. Phys.*)

KINETIC THEORY FOR CHEMICALLY REACTING GASES AND PARTIALLY IONIZED PLASMAS

YU. L. KLIMONTOVICH

Moscow State Lomonossov University
Faculty of Physics, Moscow, 117234, USSR

D. KREMP

Wilhelm-Pieck-Universität Rostock
Sektion Physik, Rostock, 2500, DDR

W. D. KRAEFT

Ernst-Moritz-Arndt-Universität Greifswald
Sektion Physik/Elektronik, Greifswald, 2200, DDR

CONTENTS

INTRODUCTION

The year 1872 was the year of the formulation of the famous Boltzmann equation (BE), which is one of the most important equations of statistical physics. One of the remarkable consequences of the BE is the *H*-theorem. Furthermore, the BE is the basic equation for transport processes in macroscopic systems.

The BE was found intuitively. Here, the hypothesis of molecular chaos, that is, the assumption that any pair of particles enters the collision process uncorrelated (statistical independence) is the most important one. Only this assumption allows the formulation of a closed equation for the single-particle distribution function.

Until 1935, the BE was the only kinetic equation. The success of kinetic theory was so great that the latter seemed to be perfect. However, since the formulation of BE, there exists the problem of its foundation from the equation of motion of the many-particle system. One early generalization of kinetic theory was given by Leontovich.[1] Further essential progress in this direction was achieved by the work of Landau, Vlasov, Bogolyubov, Born, Green, Kirkwood, Yvon, Balescu, and Lenard. Bogolyubov, especially, used the idea that the assumption of molecular chaos can be applied as an initial condition for the solution of the equations of motions for the distribution functions. With this idea, it was possible to derive the BE applying perturbation theory in first order with respect to the density. In spite of its importance, the BE is of approximate character only. There are phenomena, such as large-scale fluctuations, bound states, and higher density and nonideality effects, that cannot be described in terms of the BE. In order to include such phenomena in the kinetic description, the assumptions made so far much be dropped, at least partially. Mainly, the condition of total weakening of initial correlations must be replaced by the more general one of the particle weakening of correlations.

In this chapter generalizations of the BE in the directions mentioned are given.

Among such generalizations, an essential step is the inclusion of bound states, that is, a kinetic theory of gases and plasmas with chemical reactions. The kinetic theory taking into account the formation and the decay of bound states (i.e., inelastic collisions) is rather complicated. There exist different approaches corresponding to different levels of description. In the first part of our chapter, we will start from the quantum BBGKY hierarchy, and formulate the generalization of the Bogolyubov weakening condition for systems with two-particle bound states. Using this condition and a cluster expansion for the nonequilibrium density operator, we will derive a kinetic equation for atoms and free particles taking into account three-particle reactions. Moreover, nonideality effects will be considered, which, for example, in contrast to the usual Boltzmann equation lead to the conservation of the full energy.

In the second part we will consider the formulation of kinetic equations for partially ionized plasmas including ionization and recombination of the atoms.

Owing to the long-range character of Coulomb forces, the formulation of kinetic equations for plasmas is more complicated than that for neutral gases. Therefore, the Coulomb systems show a collective behavior, and we observe for example, the dynamical screening of the Coulomb potential.

Therefore, it is convenient to start with another concept. We will use the concept of compostite particles. A further generalization, which is essentially also of importance for chemically reacting systems, is the inclusion of fluctuations. In this connection the early work of Leontovich is an essential one.[1] In that part, there was given a generalized BE for the N-particle distribution function. While that equation reduces to the usual BE, if the condition of the total weakening of initial correlations is adopted, it includes, in general, large-scale fluctuations (see also Ref. 2). Only much later were equations of a similar type again derived in connection with the kinetic theory of fluctuations.[2] One of the main results of the kinetic theory of fluctuations is the fact that besides the usual collision terms in the kinetic equation corresponding to scattering events between the elementary particles or clusters of particles, respectively, an additional source term appears which is due to large-scale fluctuations. The effects of large-scale fluctuations are of special importance for systems far from equilibrium.

In the last section of our chapter we deal briefly with the problem of fluctuations. As an example, we will discuss the influence of fluctuations on chemical rate equations.

The authors did not set out to present all methods of the derivation of kinetic equations. They presented only certain methods that seem to work rather effectively.

I. QUANTUM STATISTICS OF MANY-PARTICLE SYSTEMS

A. The Density Operator

The properties and the behavior of systems composed of many elementary particles, atoms, and molecules, are described by quantum statistics.[3-5] Let $b_1 \cdots b_N$ be a complete set of observables of an N-particle system, where b_1 is a complete observable of the single-particle systems, for example, $b_1 = r_1 s_1$, with r_1 being the position and s_1 being the spin projection. Then the microscopic state is given by a vector $|b_1 \cdots b_N\rangle$ in the space of states $\mathcal{H}_N$.

The observable physical quantities A are presented by Hermitan operators acting on the space $\mathcal{H}_N$. If the state of the system is given by $|\psi\rangle$, the result of the measurement of the observable A is characterized by the expectation value given as

$$\langle A \rangle = \langle \psi | A | \psi \rangle \, . \tag{1.1}$$

Special attention must be paid in systems of identical particles, where we have to take into account the symmetry postulate of quantum mechanics. This means that the space of states for fermions is the antisymmetric subspace $\mathcal{H}_N^-$ of $\mathcal{H}_N$, while the symmetric subspace $\mathcal{H}_N^+$ refers to bosons.

A very useful and convenient method that fulfils this postulate is the method of second quantization. That means, we describe all quantities (relevant for our system under consideration) in terms of so-called annihilation and creation operators $a(b)$ and $a^+(b)$, respectively.

In order to construct symmetrical or antisymmetrical vectors of state $|b_1 \cdots b_N\rangle^{\pm}$, we introduce the vacuum state $|0\rangle$; then the a^+, are defined by

$$|b_1 \cdots b_N\rangle^{\pm} = \frac{n_1! \cdots n_N!}{\sqrt{N!}} \, a^+(b_1) \cdots a^+(b_N)|0\rangle \, .$$

The condition for the states to be antisymmetrized for fermions and symmetrized for bosons, respectively, may be expressed by the following commutation rules:

$$a(b)a(b') \mp a(b')a(b) = 0 \, ,$$

$$a^+(b)a^+(b') \mp a^+(b')a^+(b) = 0\,,$$
$$a(b)a^+(b') \mp a^+(b')a(b) = \delta(b - b')$$

(upper sign, boson; lower, fermion). In the special case of coordinate representation, we will write

$$a(b_1t_1) \equiv a(r_1s_1t_1) \equiv \psi(r_1s_1t_1) \equiv \psi(1)\,.$$

In this case, the ψ's are called field operators. In the formalism of second quantization an additive operator can be expressed in terms of $\psi(1)$, $\psi^+(1)$ in the following way:

$$A = \int d1\, d1'\, \langle 1|A|1'\rangle^{\pm}\psi^+(1)\psi(1')\,, \tag{1.2}$$

and for a binary operator we obtain

$$A = \frac{1}{2}\int d1\, d2\, d1'\, d2'\, \langle 12|A|2'1'\rangle^{\pm}\psi^+(1)\psi^+(2)\psi(2')\psi(1')\,. \tag{1.3}$$

In this way we get for the Hamiltonian in coordinate representation:

$$H = -\sum_a \left(\frac{\hbar^2}{2m_a}\int dr_1\, \psi_a^+(1)\nabla^2\psi_a(1)\right)$$
$$+\frac{1}{2}\sum_{a,b}\int dr_1\, dr_2\, \psi_a^+(1)\psi_b^+(2)V_{ab}(r_1 - r_2)\psi_b(2)\psi_a(1)\,. \tag{1.4}$$

Let us now consider the time development of the system. In the Heisenberg picture the time evolution is determined by the equation of motion for the field operators:

$$i\hbar\,\frac{\partial}{\partial t}\,\psi_a(1) = [\psi_a(1), H]\,.$$

Using the Hamiltonian (1.4) one can determine the commutator:

$$i\hbar\,\frac{\partial}{\partial t}\,\psi_a(1) = -\frac{\hbar^2}{2m_a}\,\Delta_1\psi_a(1) + \sum_b\int dr_2\, V_{ab}(r_1 - r_2)$$
$$\times\, \psi_a(r_1t_1s_1)\psi_b^+(r_2t_1s_2)\psi_b(r_2t_1s_2)\,, \tag{1.5}$$

which determines completely the dynamical behavior of the many particle system.

Until now we assumed that we have the maximum information on the many-particle system. Now we will consider a large many-body system in the so-called thermodynamic limit ($N \to \infty$, $V \to \infty$, $n = N/V$ finite); that means a macroscopic system. Because of the (unavoidable) interaction of the macroscopic many-particle system with the environment, the information of the microstate is not available, and the quantum-mechanical description is to be replaced by the quantum-statistical description. Thus, the state is characterized by the density operator ρ with the normalization

$$\operatorname*{Tr}_{\mathscr{H}} \rho = 1 . \tag{1.6}$$

The average value of any observable is determined by

$$\langle A \rangle = \operatorname*{Tr}_{\mathscr{H}} (A\rho) . \tag{1.7}$$

The averages of additive and binary operators, respectively, read

$$\langle A \rangle = \int d1\, d1'\, \langle 1|A_1|1'\rangle^{\pm} \langle \psi^{+}(1)\psi(1')\rangle|_{t_1 = t_1'} , \tag{1.8}$$

and

$$\langle A \rangle = \frac{1}{2} \int d1\, d2\, d1'\, d2'\, \langle 12|A_{12}|2'1'\rangle^{\pm} \times \langle \psi^{+}(1)\psi^{+}(2)\psi(2')\psi(1')\rangle|_{\substack{t_1 = t_2 = t \\ t_1' = t_2' = t}} . \tag{1.9}$$

Therefore the average values

$$\langle \psi^{+}(1)\psi(1')\rangle|_{t_1 = t_1' = t} = \operatorname{Tr}(\rho\psi^{+}(rst)\psi(r's't)) \equiv nF_1(rr't) \tag{1.10}$$

and

$$\langle \psi^{+}(1)\psi^{+}(2)\psi(2')\psi(1')\rangle|_{\substack{t_1 = t_2 = t \\ t_1' = t_2' = t}} = \operatorname{Tr}[\rho\psi^{+}(r_1 s_1 t)\psi^{+}(r_2 s_2 t)\psi(r_2' s_2' t)\psi(r_1' s_1' t)] \equiv n^2 F_{12}(r_1 s_1 r_2 s_2 r_1' s_1' r_2' s_2' t) , \tag{1.11}$$

which are called one- and two-particle density matrices, are quantities of essential information in quantum statistics.[6,7] From the normalization of ρ we obtain the normalization for the density matrices, that is,

$$\frac{1}{V}\int dr\, F_1(rrt) = 1 \tag{1.12}$$

and

$$\frac{1}{V^2}\int dr_1\, dr_2\, F_{12}(r_1 r_2 r_1 r_2 t) = 1\,. \tag{1.13}$$

Generally, the s-particle density matrix reads

$$\begin{aligned} n^s F_{\{s\}}(r_1 \cdots r_s r'_1 \cdots r_{s'} t) \\ = \langle \psi^+(r_1 s_1 t) \cdots \psi^+(r_s s_s t)\psi(r_{s'} s_{s'} t) \cdots \psi(r'_1 s'_1 t)\rangle\,, \end{aligned} \tag{1.14}$$

with $\{s\} = 12 \cdots s$ and

$$\frac{1}{V^s}\int dr_1 \cdots dr_s\, F_{\{s\}}(r_1 \cdots r_s r_1 \cdots r_s) = 1\,. \tag{1.15}$$

In many cases it is useful to write F_1 in terms of new variables

$$r \rightarrow r + \frac{\hbar\gamma}{2}\,, \qquad r' \rightarrow r - \frac{\hbar\gamma}{2}\,.$$

Thus we get the Wigner distribution by Fourier transformation with respect to γ

$$f(rpt) = \hbar^3 \frac{N}{V}\int d\gamma e^{-i\gamma p} F_1\left(r + \frac{\hbar\gamma}{2}, r - \frac{\hbar\gamma}{2}, t\right). \tag{1.16}$$

Here we have normalization

$$\int f(rpt)\, \frac{dr\, dp}{(2\pi\hbar)^3} = N\,. \tag{1.17}$$

The Wigner distribution function $f(rpt)$ has many properties in common with the classical distribution function. Unfortunately, however, this distribution function is not positive definite, and, therefore, it cannot be interpreted as a usual distribution function.

Finally let us introduce the density correlation function (second moment):

$$\langle \hat{N}(1)\hat{N}(2)\rangle = \langle \psi(1)\psi^+(1)\psi(2)\psi^+(2)\rangle\,.$$

We study the relation of this function to the density matrix. Using the commutation relation it follows easily that

$$\langle \hat{N}(1)\hat{N}(2)\rangle|_{t_1=t_2} = \frac{N(N-1)}{V^2} F_{12}(r_1r_1r_2r_2t) \pm \frac{N}{V} \delta(r_1 - r_2)F_1(r_1r_1t) .$$

Furthermore, we introduce the density fluctuation (central moments) by

$$\delta N = \hat{N}(1) - \langle N(1)\rangle .$$

Using the fact that $\langle \delta N\rangle = 0$, we find

$$\langle \delta N(1)\delta N(2)\rangle|_{t_1=t_2} = \langle \hat{N}(1)\hat{N}(2)\rangle - \langle \hat{N}(1)\rangle\langle \hat{N}(2)\rangle .$$

It is possible to express the density fluctuation correlation function by the correlation operator $g_{12} = F_{12} - F_1F_2$, and F_1 may be obtained easily:

$$\begin{aligned}\langle \delta N, \delta N\rangle = {} & \frac{N(N-1)}{V^2} g_{12}(r_1r_1r_2r_2t) \\ & + \frac{N}{V}\left\{\delta(r_1 - r_2)F_1(r_1t) - \frac{1}{V} F_1(r_1t)F_2(r_2t)\right\} .\end{aligned}$$

This formalism is the quantum-mechanical analog of the Klimontovich phase-space density formalism, and is particularly effective in the theory of long-range correlation and fluctuation in the kinetic theory of plasmas.

B. Equation of Motion for the Density Operator and the Wigner Distribution

In order to determine the reduced density matrix, let us write equations of motion for these quantities. Because the density matrices are expressed by the field operators $\psi(r)$, $\psi^+(r)$, such equations may be obtained from the equations of motion for ψ and ψ^+, given, for example, by (1.5).

Using the definition (1.10) of the single-particle density matrix and Eq. (1.11), it is easy to obtain an equation for the single-particle density matrix:

$$\begin{aligned}&\left\{i\hbar \frac{\partial}{\partial t} + \frac{\hbar^2}{2m_1} (\Delta_1 - \Delta_{1'})\right\}F_1(r_1r_1't) \\ &\qquad = n \int dr_2 \,\{V_{12}(r_1 - r_2) - V_{12}(r_1' - r_2)\}F_{12}(r_1r_2r_1'r_2t) . \qquad (1.18)\end{aligned}$$

Because of the interaction this equation is not closed; it is coupled with the two-particle density operator. Therefore, it is necessary to find an equation of motion for $F_{12}(r_1 r_2 r_1' r_2' t)$. Using the definition of F_{12} given by (1.11) and the equations of motion for ψ, ψ^+, we obtain

$$\left\{ i\hbar \frac{\partial}{\partial t} - H_{12}(r_1 r_2) + H_{12}(r_1' r_2') \right\} F_{12}(r_1 r_2 r_1' r_2' t)$$
$$= n \sum_{a=1}^{2} \int dr_3 \, \{V_{a3}(r_2 - r_3) - V_{a3}(r_2' - r_3)\} F_{123}(r_1 r_2 r_3 r_1' r_2' r_3 t) \,. \quad (1.19)$$

Here

$$H_{12}(r_1 r_2) = -\frac{\hbar^2}{2m_1} \Delta_1 - \frac{\hbar^2}{2m_2} \Delta_2 + V_{12}(r_1 - r_2)$$

is the Hamiltonian of an isolated pair a, b. Equation (1.19) is also not closed. Therefore, Eqs. (1.18) and (1.19) are members of a hierarchy of equations for F_1, F_{12}, $F_{123}, \ldots$.

The general form of this hierarchy can be obtained from the definition of $f_{\{s\}}$ (1.14), and the equations of motion for ψ, ψ^+. We find

$$\left\{ i\hbar \frac{\partial}{\partial t} - H_{\{s\}}(r_1 \cdots r_s) + H_{\{s\}}(r_1' \cdots r_s') \right\} F_{\{s\}}(r_1 \cdots r_s r_1' \cdots r_s' t)$$
$$= n \sum_{i=1}^{s} \int dr_{s+1} \, \{V(r_i - r_{s+1}) - V(r_i' - r_{s+1})\} F_{\{s+1\}}$$
$$\times (r_1 \cdots r_{s+1} r_1' \cdots r_s' r_{s+1} t) \,. \quad (1.20)$$

Let us consider the first equation in more detail. If we take into account the relation

$$F_{12}(r_1 r_2 r_1' r_2' t) = F_{12}^*(r_1' r_2' r_1 r_2 t) \,,$$

we can write Eq. (1.18) in the useful form

$$\left\{ i\hbar \frac{\partial}{\partial t} + \frac{\hbar^2}{2m} (\Delta_1 - \Delta_{1'}) \right\} F_1(r_1 r_1' t) = I \quad (1.21)$$

with the collision integral

$$I = n \int dr_2 \, V(r_1 - r_2) \, \mathrm{Im} \, [F_{12}(r_1 r_2 r_1' r_2 t)] \,.$$

Another useful form we get is the Wigner representation. Let us take into account additionally an external potential $U(r, t)$. In this case it is convenient to introduce the correlation density matrix g_{12} by

$$F_{12}(r_1 r_2 r'_1 r'_2 t) = F_1(r_1 r'_1 t) F_2(r_2 r'_2 t) \pm F_1(r_1 r'_2 t) F_2(r_2 r'_1 t) + g_{12}(r_1 r_2 r'_1 r'_2 t) . \tag{1.22}$$

Using (1.16) for the Wigner distribution, we obtain the equation

$$\frac{\partial f(rpt)}{\partial t} + \frac{p}{m} \frac{\partial f}{\partial r} - \frac{1}{\hbar} \int \left[\bar{U}\left(r - \frac{\hbar \gamma}{2}\right) - \bar{U}\left(r + \frac{\hbar \gamma}{2}\right) \right] \times \exp(i\gamma(p - p')) f(rp't) \frac{d\gamma \, dp'}{(2\pi)^3} = I(rpt) , \tag{1.23}$$

where $I(rpt)$ is the collision integral in the Wigner distribution:

$$I = \frac{i}{\hbar} \int \left[V_{12}\left(\left| r - r_2 - \frac{\hbar\gamma}{2} \right|\right) - V_{12}\left(\left| r - r_2 + \frac{\hbar\gamma}{2} \right|\right) \right] \times g_{12}\left(r + \frac{\hbar\gamma}{2}, r - \frac{\hbar\gamma}{2}, r_2, r_2 t\right) e^{-i\gamma p} \frac{\hbar^3}{V} d\gamma \frac{dr_2}{V} . \tag{1.24}$$

The effective external potential U_a includes the Hartree–Fock contribution and reads

$$\bar{U}(rt) = U(rt) + N \int V_{12} f(r'p't) \frac{dr' \, dp'}{(2\pi\hbar)^3} . \tag{1.25}$$

Equation (1.24) is very similar to that of the single-particle distribution function of classical statistical mechanics. In the limit $\hbar \to 0$ we get the first equation of the BBGKY hierarchy.

Previously we have considered the reduced density matrices and the equations of motion. These quantities are the representation of the reduced density operators (Bogolyubov,[6] Gurov[7]) defined by

$$F_{\{s\}} = V^s \operatorname*{Tr}_{s+1 \cdots N} \rho_{\{N\}} ; \qquad \frac{1}{V^s} \operatorname*{Tr}_{1 \cdots s} F_{\{s\}} = 1 . \tag{1.26}$$

Then the average of an observable of the s-particle type

$$A_N = \sum_{\{1 \cdots s\}}^{N} A_{1 \cdots s} \tag{1.27}$$

is given by

$$\langle A\rangle = \sum_{\{1\cdots s\}}^{N} \frac{n^s}{s!} \operatorname*{Tr}_{1\cdots s} F_{\{s\}} A_{1\cdots s} . \tag{1.28}$$

Comparing with Eqs. (1.8) and (1.9), the s-particle density matrix (1.14) turns out to be the coordinate representation

$$\langle x_1 \cdots x_s | F_{\{s\}} | x'_s \cdots x'_1 \rangle \equiv F_{\{s\}}(x_1 \cdots x_s x'_1 \cdots x'_s t) .$$

of F_s. Then from Eq. (1.20) it is easy to obtain the Bogolyubov hierarchy for the reduced density operator[6]

$$i\hbar \frac{\partial F_{\{s\}}}{\partial t} + [H_{\{s\}}, F_{\{s\}}] = n \operatorname*{Tr}_{s+1} \sum_{i=1}^{s} [V_{i,s+1}, F_{\{s+1\}}] . \tag{1.29}$$

For many formal considerations this compact operator formulation is more convenient. For example, it is possible to write a formal solution of this equation in the form

$$\begin{aligned} F_{\{s\}}(t) &= e^{-iH_{\{s\}}(t-t_0)/\hbar} F_{\{s\}}(t_0)\, e^{iH_{\{s\}}(t-t_0)/\hbar} \\ &+ \frac{n}{i\hbar} \int_0^{t-t_0} dt'\, e^{-iH_{\{s\}}t'/\hbar} \operatorname*{Tr}_{s+1} \sum_{i=1}^{s} [V_{i,s+1}, F_{\{s+1\}}(t-t')]\, e^{iH_{\{s\}}t'/\hbar} . \end{aligned} \tag{1.30}$$

So far we considered the density operators in coordinate representation. In many cases, especially for homogeneous systems, it is more convenient to use the momentum representation

$$\langle p_1 | F_1 | p'_1 \rangle \equiv F_1(p_1 p'_1 t) .$$

For homogeneous systems the density matrix is diagonal:

$$\langle p_1 | F_1 | p'_1 \rangle = \delta(p_1 - p'_1) F(p_1) .$$

$F(p_1)$ is the quantum momentum distribution function. The normalization follows from (1.26):

$$\int \frac{dp}{(2\pi\hbar)^3} F(pt) = 1 .$$

Sometimes it is more convenient to use other distribution functions given by

$$nF(pt) = f(pt)$$

with the normalization

$$\int \frac{dp}{(2\pi\hbar)^3} f(pt) = n \, .$$

II. KINETIC EQUATION FOR NONIDEAL GASES

A. Binary Density Matrix in Two-Particle Collision Approximation—Boltzmann Equation

The essential information about transport properties in many-particle systems is given by the single-particle density matrix or by the single-particle Wigner distribution. The equations of motion (1.18) and (1.23) for these important quantities are called kinetic equations. For the further consideration we write the latter equation in the momentum representation:

$$i\hbar \frac{\partial F_1(p_1 p_1' t)}{\partial t} + \left(\frac{p_1^2}{2m} - \frac{p_1'^2}{2m}\right) F_1(p_1 p_1' t) = I(p_1 p_1') \tag{2.1}$$

with the collision integral

$$I(p_1 p_1') = n \int dp_2 \, V(p_1 - p_2) \, \mathrm{Im}\, [F_{12}(p_1 p_2 p_1' p_2 t)]$$

and (2.2)

$$V(p) = \int dr \, V(r) e^{-ipr} \, .$$

In order to get a close equation we must determine the collision integral, that is, we must determine F_{12} as a function of F_1 (or f_1, respectively). For this purpose we can use the formal solution of the Bogolyubov hierarchy, especially for F_{12}:

$$\begin{aligned} F_{12}(t) &= e^{-iH_{12}(t-t_0)/\hbar} F_{12}(t_0) e^{iH_{12}(t-t_0)/\hbar} \\ &\quad + \frac{n}{i\hbar} \int_0^{t-t_0} dt' \, e^{-iH_{12}t'/\hbar} \, \mathrm{Tr}_3 \, [V_{13} + V_{23}, F_{123}(t-t')] e^{iH_{12}t'/\hbar} \, . \end{aligned} \tag{2.3}$$

To use this solution it is necessary to solve two problems. The first problem is the determination of $F_{12}(t_0)$. Usually, for deriving a closed equation for $F_1(t)$, Bogolyubov's condition of the complete weakening of the initial correlation is used. This is given by[8]

$$\lim_{t_0 \to -\infty} F_{\{s\}}(t_0) = \lim_{t_0 \to -\infty} F_1(t_0) \cdots F_s(t_0) . \tag{2.4}$$

With this condition, we neglect those initial correlations for which the correlation time τ_{corr} is much smaller than the relaxation time τ_{rel} of F_1. Thus, we may neglect the correlations for which

$$\tau_{\text{corr}} \ll \tau_{\text{rel}} .$$

Therefore, the condition of complete weakening of the initial correlations corresponds to the assumption that the long-living correlations (with $\tau_{\text{corr}} \gtrapprox \tau_{\text{rel}}$) do not play an essential role. We shall see later, especially in connection with the problem of bound states in kinetic theory, that this is not always ture.

The second problem is the decoupling of the chain of equations for the formal solutions. It is necessary to make approximations in order to truncate this chain. The simplest approximation is the binary collision approximation. That means, we neglect three-particle collisions. Then, we obtain

$$F_{12}(t) = e^{-iH_{12}(t-t_0)/\hbar} F_{12}(t_0) e^{iH_{12}(t-t_0)/\hbar} . \tag{2.5}$$

With (2.4) the Bogolyubov condition takes the form

$$\lim_{t \to -\infty} \| e^{-iH_{12}(t-t_0)/\hbar} F_{12}(t_0) e^{iH_{12}(t-t_0)/\hbar} - e^{-iH_{12}^{(0)}(t-t_0)/\hbar} F_1(t_0) F_2(t_0) e^{iH_{12}^{(0)}(t-t_0)/\hbar} \| = 0 . \tag{2.6}$$

From this we get for any t_0

$$F_{12}(t_0) = \Omega_{12} F_1(t_0) F_2(t_0) \Omega_{12}^{\dagger} , \tag{2.7}$$

where Ω_{12} are Møller operators of scattering theory (see e.g., Ref. 9) given by

$$\Omega_{12} = \lim_{t \to -\infty} e^{-iH_{12}t/\hbar} e^{iH_{12}^{(0)}t/\hbar} . \tag{2.8}$$

It follows that Ω_{12} is isometric,

$$\Omega_{12}^{\dagger}\Omega_{12} = 1 . \tag{2.8a}$$

If the potential does not support bound states, we have, furthermore,

$$\Omega_{12}\Omega_{12}^{\dagger} = 1 . \tag{2.8b}$$

Also very important is the intertwining relation

$$\Omega_{12}H_{12}^{(0)} = H_{12}\Omega_{12} . \tag{2.9}$$

Let us now again consider the solution (2.5). With (2.7) we may obtain for arbitrary t_0

$$F_{12}(t) = e^{-iH_{12}(t-t_0)/\hbar}\Omega_{12}F_1(t_0)F_2(t_0)\Omega_{12}^{\dagger}e^{iH_{12}(t-t_0)/\hbar} .$$

Thus the binary density operator is given only by the single-particle density operator F_1, which depends on the earlier time t_0. Therefore, retardation effects appear. In order to eliminate this time we use the formal solution (1.30) for F_1:

$$F_1(t) = e^{-iH_1(t-t_0)/\hbar}F_1(t_0)e^{iH_1(t-t_0)/\hbar} + O(n) . \tag{2.10}$$

Thus the difference between $F_1(t)$ and $F_1(t_0)$ is of higher order in the density. In lowest order of the density, therefore, the retardation may be neglected.

Applying Eq. (2.10) we can write

$$F_{12}(t) = \Omega_{12}F_1(t)F_2(t)\Omega_{12}^{\dagger} + O(n) . \tag{2.11}$$

Here the relation (2.9) was used.

For the further considerations it is convenient to express the relevant quantities in terms of eigenstates of the two-particle Hamiltonian, that is, with states obeying the following equation:

$$H_{12}|\alpha_{12}P_{12}\rangle = E_{\alpha_{12}P_{12}}|\alpha_{12}P_{12}\rangle . \tag{2.12}$$

The subscript α denotes the quantum numbers

$$\alpha_{12} = \begin{cases} p_{12} & \text{for scattering states ,} \\ n_{12} & \text{for bound states ,} \end{cases}$$

with

$$p_{12} = (m_1 p_1 - m_2 p_2)/M\,, \qquad P_{12} = p_1 + p_2\,,$$
$$M = m_1 + m_2\,.$$

In coordinate and momentum representation, we have, respectively,

$$\langle r_1 r_2 | \alpha_{12} P_{12} \rangle = \psi_{\alpha P}(r_1 r_2) \tag{2.13}$$

and

$$\langle p_1 p_2 | \alpha_{12} P_{12} \rangle = \psi_{\alpha P}(p_1 p_2)$$
$$= \frac{1}{(2\pi\hbar)^3} \int \psi_{\alpha P}(r_1 r_2) \exp\left\{ -\frac{i}{\hbar}(r_1 p_1 + r_2 p_2) \right\} dr_1\, dr_3\,. \tag{2.14}$$

Using the normalization of the eigenstates

$$\langle \alpha P | P' \alpha' \rangle = \begin{cases} \delta_{nn'} \delta(P - P') & \text{for } \alpha, \alpha' = n, n'\,, \\ \delta(p - p')\delta(P - P') & \text{for } \alpha, \alpha' = p, p'\,, \end{cases}$$

we obtain the following normalization of the eigenfunctions:

$$\int dr_1\, dr_2\, \psi_{\alpha P}(r_1 r_2) \psi^*_{\alpha' P'}(r_1 r_2) = \begin{cases} \delta_{nn'} \delta(P - P') \\ \delta(p - p')\delta(P - P')\,. \end{cases} \tag{2.15}$$

From scattering theory it is known that the two-particle scattering states, $|p_1 p_2\rangle^+$, may be constructed with the Møller operators from the two-particle noninteracting states $|p_1 p_2\rangle$ by the relation

$$\Omega_{12} | p_1 p_2 \rangle = | p_1 p_2 \rangle^+\,, \tag{2.16}$$

or, in momentum representation,

$$\Omega_{12} \psi^{(0)}_{\bar{p}_1 \bar{p}_2}(p_1 p_2) = \psi^{+}_{\bar{p}_1 \bar{p}_2}(p_1 p_2)\,. \tag{2.17}$$

Using the relation, the binary density matrix in momentum representation may be expressed in terms of scattering wave functions.

If the potential does not support bound states, we get

$$\langle p_1 p_2 | F_{12} | p_2' p_1' \rangle = F_{12}(p_1 p_2 p_1' p_2' t)$$
$$= \int d\bar{p}_1\, d\bar{p}_2\, \psi^{+}_{\bar{p}_1 \bar{p}_2}(p_1 p_2) \psi^{+*}_{\bar{p}_1 \bar{p}_2}(p_1' p_2') F_1(p_1 t) F_2(p_2 t)\,. \tag{2.18}$$

Inserting (2.18) into (2.1), it follows that

$$\frac{\partial F_1(p_1t)}{\partial t} = I^{(2)}(p_1t) \tag{2.19}$$

with the collision integral

$$I^{(2)}(p_1t) = n \int V\left(\frac{p_1 - p_1'}{\hbar}\right)\delta(p_1 + p_2 - p_1' - p_2')F_1(\bar{p}_1t)F_1(\bar{p}_2t)$$
$$\times \operatorname{Im}\left[\psi^*_{\bar{p}_1\bar{p}_2}(p_1'p_2')\psi_{\bar{p}_1\bar{p}_2}(p_1p_2)\right] dp_1'\, dp_2'\, d\bar{p}_1\, d\bar{p}_2\, dp_2 \,. \tag{2.20}$$

Now we will show that this expression can be transformed into the usual form of the Boltzmann collision integral with the quantum cross section. For this purpose we can use the following relations from the scattering theory:

1. The definition of the two particle T-matrix by $\psi^+_{p_1p_2(p_1'p_2')}$ according to

$$\langle p_1'p_2'|T|p_1p_2\rangle = \delta(p_1 + p_2 - p_1' - p_2')T(p_1'p_2' \rightarrow p_1p_2)$$
$$= \int V\left(\frac{p_1 - p_1'}{\hbar}\right)\delta(p_1 + p_2 - p_1' - p_2')$$
$$\times \psi^+_{p_1'p_2'}(\bar{p}_1\bar{p}_2)\, d\bar{p}_1\, d\bar{p}_2 \,. \tag{2.21}$$

2. The Lippmann–Schwinger equation for the scattering states

$$\psi^+_{p_1p_2}(p_1'p_2') = \delta(p_1' - p_1)\delta(p_2 - p_2') + \frac{\langle p_1p_2|T|p_1'p_2'\rangle}{E_{p_1p_2} - E_{p_1'p_2'} + i\hbar\Delta} \,. \tag{2.22}$$

3. The optical theorem:

$$\operatorname{Im}\langle p_1p_2|T|p_2p_1\rangle = -\pi \int d\bar{p}_1\, d\bar{p}_2\, \delta(E_{p_1p_2} - E_{\bar{p}_1\bar{p}_2})|\langle p_1p_2|T|\bar{p}_1\bar{p}_2\rangle|^2 \,. \tag{2.23}$$

With these relations it is easy to get the usual form of the quantum Boltzmann collision integral:

$$I(p_1t) = \frac{N(2\pi\hbar)^3}{\hbar} \int dp_2\, d\bar{p}_1\, d\bar{p}_2\, |\langle p_1p_2|T|\bar{p}_2\bar{p}_1\rangle|^2$$
$$\times 2\pi\delta(E_{p_1p_2} - E_{\bar{p}_1\bar{p}_2})\{F_1(\bar{p}_1)F_1(\bar{p}_2) - F_1(p_1)F_1(p_2) \tag{2.24}$$

To complete this equation we write the well-known equation for the T-matrix given by

$$\begin{aligned}\langle p_1 p_2 | T | p_2' p_1' \rangle = \langle p_1 p_2 | V | p_2' p_1' \rangle \\ + \int d\bar{p}_1 \, d\bar{p}_2 \, dp_3 \, dp_4 \langle p_1 p_2 | V | \bar{p}_2 \bar{p}_1 \rangle \\ \times G^0(\bar{p}_1 \bar{p}_2 p_3 p_4) \langle p_3 p_4 | T | p_2' p_1' \rangle \, . \end{aligned} \tag{2.25}$$

We can solve this equation in first Born approximation. Thus, it follows for the T-matrix that

$$\langle p_1 p_2 | T | p_2' p_1' \rangle = V\left(\frac{p_1 - p_1'}{\hbar}\right) \delta(p_1 + p_2 - p_1' - p_2') \, . \tag{2.26}$$

In this way the collision integral is completely determined.

We recall, that this result was obtained under three conditions:

1. Determination of F_{12} in binary collision approximation.
2. The condition of the complete weakening of initial correlation that means long-living correlations, especially bound states, are not taken into account.
3. The retardation in (2.5) was completely neglected.

The general properties of the Boltzmann collision integral are well known and will not be discussed here.

An essential feature of this equation is that the Boltzmann kinetic equation is only valid for ideal systems. Indeed, as can be shown, this kinetic equation does not take into account the contributions of the interaction to the thermodynamic functions. For example, the conservation of the energy following from the Boltzmann equation has the form

$$\frac{d}{dt}\left\langle \frac{p^2}{2m} \right\rangle = 0 \, , \qquad \left\langle \frac{p^2}{2m} \right\rangle = \int \frac{p^2}{2m} f_1(pt) \, dp \, . \tag{2.27}$$

This means from the Boltzmann equation follows only the conservation of the kinetic energy. But, in nonideal systems, the average of kinetic and potential energy as a sum should be covered.

B. Binary Density Operator in Three-Particle Collision Approximation—Boltzmann Equation for Nonideal Gases

Generalizations of the Boltzmann equation are possible in many directions. For example, we can take into account bound states, or we can construct kinetic equations, including triple- and higher-order collisions, by expanding the binary density operator into a cluster series. We want to obtain especially kinetic equations for nonideal systems, kinetic equations that describe the influence of the pair collisions to the conservation of energy and on the thermodynamic functions. It seems that the cluster expansion of the binary density operator is a general method for the construction of very general kinetic equations. But the investigation carried out by Bogolyubov,[8] Cohen and co-workers,[10] and Weinstock[11] have shown that this program is connected with serious difficulties, because time-divergent contributions appear. In simple cases these divergent terms can be eliminated by taking into account the retardation $(t - t_0)$ in a consistent manner.

Here we will first consider the cluster expansion of F_{12} in the triple-collision approximation. In the triple-collision approximation the first three equations of the hierarchy (1.29) have the form (in the spatially homogeneous gas)

$$\frac{\hbar}{i}\frac{\partial F_1}{\partial t} = n \operatorname*{Tr}_2 [V_{12}, F_{12}] , \tag{2.28}$$

$$\frac{\hbar}{i}\frac{\partial F_{12}}{\partial t} + [H_{12}, F_{12}] = n \operatorname*{Tr}_3 [V_{13} + V_{23}, F_{123}] , \tag{2.29}$$

$$\frac{\hbar}{i}\frac{\partial F_{123}}{\partial t} + [H_{123}, F_{123}] = 0 . \tag{2.30}$$

We solve these equations using the Bogolyubov condition:

$$\lim_{t\to-\infty} \| F_{1\cdots s}(t) - F_1(t)F_2(t) \times \cdots \times F_s(t) \| = 0 . \tag{2.31}$$

Furthermore, we obtain from (2.30) the solution

$$F_{123}(t) = e^{-iH_{123}(t-t_0)/\hbar} F_{123}(t_0) e^{iH_{123}(t-t_0)/\hbar} . \tag{2.32}$$

From (2.5) and (2.32) it follows in well-known manner that

$$F_{123}(t_0) = \Omega_{123} F_1(t_0) F_2(t_0) F_3(t_0) \Omega_{123}^{\dagger} . \tag{2.33}$$

Using (1.30) we may obtain for F_{12}

$$F_{12}(t) = e^{-iH_{12}(t-t_0)/\hbar} F_{12}(t_0) e^{iH_{12}(t-t_0)/\hbar}$$

$$+ \frac{n}{i\hbar} \int_0^{t-t_0} dt' \, e^{-iH_{12}t'/\hbar} \operatorname*{Tr}_3 [V_{13} + V_{23}, F_{123}(t-t')] e^{iH_{12}t'/\hbar} . \tag{2.34}$$

Let us discuss two approximations for the integral term. The first approximation is a more exact form of the binary collision approximation. Up to now this approximation has the form (2.11). However, this approximation is not quite consistent. Indeed we can show that the integral term of (2.34) contains terms that contribute to the binary collisions. Since V_{13} is different from zero only if the particles 1 and 3 are in finite distance, in binary collision approximations for the second term in (2.34) can be found:

$$\frac{n}{i\hbar}\int_0^{t-t_0} dt'\, e^{-iH_{12}t'/\hbar}\{\mathop{\mathrm{Tr}}_3 [V_{13}, F_2F_{13}(t-t')] + \mathop{\mathrm{Tr}}_3 [V_{23}, F_1F_{23}(t-t')]e^{iH_{12}t'/\hbar}\,, \quad (2.35)$$

and because the first equation of the hierarchy, for spatially homogeneous systems follows[12]

$$F_{12}(t) = e^{-iH_{12}(t-t_0)/\hbar}F_{12}(t_0)e^{iH_{12}(t-t_0)/\hbar} + \frac{n}{i\hbar}\int_0^{t-t_0} dt'\, e^{-iH_{12}t'/\hbar}\,\frac{\partial}{\partial t'}\,[F_1(t-t')F_2(t-t')]e^{iH_{12}t'/\hbar}\,. \quad (2.36)$$

This generalized binary collision approximation is discussed more in detail by Klimontovich.[12]

The second approximation that we will consider is the three-particle approximation. In the case that three-particle collisions are taken into account we must start with the more general expression (2.34). Using the identity

$$V_{13} + V_{23} = H_{123} - H_{12} - H_1\,,$$

the integration over t can be carried out. Thus we find the first term of the cluster expansion of the binary density operator

$$\begin{aligned} F_{12}(t) &= e^{-iH_{12}(t-t_0)/\hbar}F_{12}(t_0)e^{iH_{12}(t-t_0)/\hbar} \\ &\quad + n\mathop{\mathrm{Tr}}_3\{F_{123}(t) - e^{-iH_{12}(t-t_0)/\hbar}F_{123}(t_0)e^{iH_{12}(t-t_0)/\hbar}\} \\ &= e^{-iH_{12}(t-t_0)/\hbar}\Omega_{12}F_1(t_0)F_2(t_0)\Omega_{12}^{\dagger}e^{iH_{12}(t-t_0)/\hbar} + n\mathop{\mathrm{Tr}}_3\cdots\,. \end{aligned} \quad (2.37)$$

In connection with this expression, two problems occur: The first one is the problem of the secular divergencies. It is well known from the literature[10] that this equation contains terms that grow with τ/τ_{coll} by $\tau\to\infty$. The physical origin of the terms are succesive binary collisions.

The second problem is the problem of retardations that means in (2.5) $F_{12}(t)$ is given by $F_1(t_0)F_2(t_0)$. These problems are closely connected.

In order to analyze the retardation problem we consider the formal solution (1.30) for F_1. Using the binary collision approximation for F_{12}, as before, the t' integration can be carried out. The result is

$$F_1(t) = e^{-iH_1(t-t_0)/\hbar}F_1(t_0)e^{iH_1(t-t_0)/\hbar} + n \operatorname*{Tr}_2 \{e^{-iH_{12}(t-t_0)/\hbar}F_{12}(t_0)e^{iH_{12}(t-t_0)/\hbar} - e^{-iH_1(t-t_0)/\hbar}F_{12}(t_0)e^{iH_1(t-t_0)/\hbar}\} . \tag{2.38}$$

Neglecting the retardation in the second term, it follows

$$F_1(t) = e^{-iH_1(t-t_0)/\hbar}F_1(t_0)e^{iH_1(t-t_0)/\hbar} + n \operatorname*{Tr}_2 \{\Omega_{12}F_1(t)F_2(t)\Omega_{12}^\dagger - F_1(t)F_2(t)\} . \tag{2.39}$$

As can be seen the difference between $F_1(t)$ and $F_1(t_0)$ is of the order n. If we take into account three-particle collisions, it is therefore not possible to neglect the retardation. Using the expansion (2.39) it is easy to eliminate $F_1(t_0)$. Thus, we obtain the import density expansion[13–15]

$$\begin{aligned} F_{12}(t) = {} & \Omega_{12}F_1(t)F_2(t)\Omega_{12}^\dagger \\ & + n \operatorname*{Tr}_3 \{\Omega_{123}F_1(t)F_2(t)F_3(t)\Omega_{123}^\dagger - \Omega_{12}F_1(t)F_2(t)F_3(t)\Omega_{12}^\dagger\} \\ & - n \operatorname*{Tr}_3 \{\Omega_{12}(\Omega_{13}F_1(t)F_2(t)F_3(t)\Omega_{13}^\dagger - F_1(t)F_2(t)F_3(t))\Omega_{12}^\dagger \\ & + \Omega_{12}(\Omega_{23}F_1(t)F_2(t)F_3(t)\Omega_{23}^\dagger - F_1(t)F_2(t)F_3(t))\Omega_{12}^\dagger\} . \end{aligned} \tag{2.40}$$

To obtain this result we have used

$$\Omega H = H^0 \Omega .$$

The first term in this expansion describes the binary collision and leads, as we have shown, to the Boltzmann collision integral. Among the terms of the order n, the first describes the three-particle collision and has, as will be shown, the same structure as the Boltzmann collision term.

The second term arose from the elimination of the retardation. It describes two successive binary collisions of three particles. This term plays an essential role, since the three-particle collision term also contains such special three-particle collisions, the contribution of which is thus canceled. This is very important since processes of such types produce

secular divergencies. Clearly the contribution of the order n is in this way well defined.

We will show that this nondivergent term determines the interaction contribution and the thermodynamic functions. The substitution of the expansion (2.38) into the right-hand side of (2.1) yields an expression for the collision integral describing triple collision and the time-retardation in the form

$$\frac{\partial f_1(p_1)}{\partial t} = I(p_1 t) = I^{(2)}(p_1 t) + I^{(R)}(p_1 t) + T^{(3)}(p_1 t) . \qquad (2.41)$$

where $I^{(2)}(p_1 t)$ is the Boltzmann collision integral. $I^{(3)}(p_1 t)$ is the triple collision integral, which can be obtained by similar methods as in the case of $I^{(2)}(p_1 t)$. The result is

$$\begin{aligned} I^{(3)}_{(p_1 t)} = \frac{(2\pi\hbar)^3}{V\hbar} \frac{1}{2} \int & dp_2\, dp_3\, d\bar{p}_1\, d\bar{p}_2\, d\bar{p}_3 \\ & \times |\langle p_1 p_2 p_3 | T(E + i\epsilon) | \bar{p}_3 \bar{p}_2 \bar{p}_1 \rangle|^2 2\pi\delta(E_{123} - \bar{E}_{123}) \\ & \times \{ f_1(\bar{p}_1) f_2(\bar{p}_2) f_3(\bar{p}_3) - f_1(p_1) f_2(p_2) f_3(p_3) \} - \tfrac{1}{2} I^{(2)}_{(p_1 t)} . \end{aligned} \qquad (2.42)$$

The retardation term we will write in the form

$$I^{(R)}_{(p_1 t)} = -\frac{i}{\hbar} n^2 \operatorname*{Tr}_{2,3} [V_{12}, \{\Omega_{12}(\Omega_{13} F_1 F_2 F_3 \Omega_{13}^\dagger)\Omega_{12}^\dagger + \Omega_{12}(\Omega_{23} F_1 F_2 F_3 \Omega_{23}^\dagger)\Omega_{12}^\dagger . \qquad (2.43)$$

The equation (2.41) is the quantum-mechanical version of the well-known Choh–Uhlenbeck kinetic equation and was obtained in Pal'tsev,[13] Klimontovich and Kremp,[14] and McLennan.[15]

C. Energy Conservation

As mentioned in Section 2.1, the usual Boltzmann equation conserves the kinetic energy only. In this sense the Boltzmann equation is referred to as an equation for ideal systems. For nonideal systems we will show that the binary density operator, in the three-particle collision approximation, provides for an energy conservation up to the next-higher order in the density (second virial coefficient). For this reason we consider the time derivative of the mean value of the kinetic energy,[12,16,17]

$$\frac{\partial}{\partial t}\langle T\rangle = \frac{\partial}{\partial t}\, n \operatorname*{Tr}_{1}\left(F_1 \frac{p_1^2}{2m}\right) \equiv \dot{E}_{\text{kin}}\,, \tag{2.44}$$

which reads after application of the first hierarchy equation (2.28)

$$\dot{E}_{\text{kin}} = \frac{1}{i\hbar}\, n^2 \operatorname*{Tr}_{12}\left\{\frac{p_1^2}{2m}\,[V_{12}, F_{12}(t)]\right\}.$$

For the binary density operator we use Eq. (2.40), and $\dot{E}_{\text{kin}}$ reads

$$\begin{aligned}\dot{E}_{\text{kin}} &= \frac{1}{i\hbar}\, n^2 \operatorname*{Tr}_{12}\left\{\frac{p_1^2}{2m}\,[V_{12}, \Omega_{12}F_1(t)F_2(t)\Omega_{12}^{\dagger}]\right. \\ &\quad + \frac{1}{i\hbar}\, n^3 \operatorname*{Tr}_{123}\left\{\frac{p_1^2}{2m}\,[V_{12}, \Omega_{123}F_1(t)F_2(t)F_3(t)\Omega_{123}^{\dagger}\right. \\ &\quad + \Omega_{12}F_1(t)F_2(t)F_3(t)\Omega_{12}^{\dagger} - \Omega_{12}\Omega_{13}F_1(t)F_2(t)F_3(t)\Omega_{13}^{\dagger}\Omega_{12}^{\dagger} \\ &\quad \left. - \Omega_{12}\Omega_{23}F_1(t)F_2(t)F_3(t)\Omega_{23}^{\dagger}\Omega_{12}^{\dagger}]\right\} \equiv \sum_{i=1}^{5} \dot{E}_{\text{kin}}^{i}\,. \end{aligned} \tag{2.45}$$

Let us now consider the different contributions of the right-hand side of (2.45), which are denoted by $\dot{E}_{\text{kin}}^{1} \cdots \dot{E}_{\text{kin}}^{5}$. $\dot{E}_{\text{kin}}^{1}$ corresponds to the level of the Boltzmann equation and is equal to zero. In the third term, the trace Tr_3 may be carried out, and we have the Boltzmann case, giving $\dot{E}_{\text{kin}}^{3} = 0$. The second term gives

$$\begin{aligned}\dot{E}_{\text{kin}}^{2} &= \frac{1}{i\hbar}\, n^3 \operatorname*{Tr}_{123}\left\{\frac{p_1^2}{2m}\,[V_{12}, \Omega_{123}F_1(t)F_2(t)F_3(t)\Omega_{123}^{\dagger}]\right\} \\ &= \frac{1}{i\hbar}\,\frac{n^3}{2} \operatorname*{Tr}_{123}\left\{\frac{p_1^2+p_2^2+p_3^2}{2m}\,[V_{12}, \Omega_{123}F_1(t)F_2(t)F_3(t)\Omega_{123}^{\dagger}]\right\}.\end{aligned}$$

Here we took into account that the addition of p_3^2 yields zero (cyclic invariance), while the addition of p_2^2 produces the factor $\frac{1}{2}$. Furthermore, we use the fact that V_{12} produces the same result as V_{13} and V_{23}. Thus, we may write

$$\dot{E}^2_{\text{kin}} = \frac{1}{i\hbar}\frac{n^3}{6}\,\underset{123}{\text{Tr}}\,\{H^0_{123}[V_{123}, \Omega_{123}F_1(t)F_2(t)F_3(t)\Omega^\dagger_{123}]\}\,. \tag{2.46}$$

Here we used

$$H^0_{123} = \frac{p_1^2}{2m} + \frac{p_2^2}{2m} + \frac{p_3^2}{2m}\,,$$

$$V_{123} = V_{12} + V_{13} + V_{23}\,, \qquad H_{123} = H^0_{123} + V_{123}\,.$$

Replacing of V_{123} in (2.46) by H_{123} is possible because $\text{Tr}\,\{A[A, B]\} = 0$. Taking into account

$$H_{123}\Omega_{123} = \Omega_{123}H^0_{123} \tag{2.47}$$

and the adjoint formula

$$\Omega^\dagger_{123}H_{123} = H^0_{123}\Omega^\dagger_{123}\,, \tag{2.48}$$

we get, instead of (2.46),

$$\dot{E}^2_{\text{kin}} = \frac{1}{i\hbar}\frac{n^3}{6}\,\underset{123}{\text{Tr}}\,\{H^0_{123}\Omega_{123}[H^0_{123}, F_1(t)F_2(t)F_3(t)]\Omega^\dagger_{123}\}\,. \tag{2.49}$$

The commutator vanishes, so that

$$\dot{E}^2_{\text{kin}} = 0\,.$$

By similar considerations and taking into account the cyclic invariance of the trace we may also write instead of (2.46)

$$-\frac{1}{i\hbar}\frac{n^3}{6}\,\underset{123}{\text{Tr}}\,\{V_{123}[H_{123}, \Omega_{123}F_1(t)F_2(t)F_3(t)\Omega^\dagger_{123}]\} = \dot{E}^2_{\text{kin}} = 0\,. \tag{2.50}$$

For the remaining contributions of (2.45) it is necessary to have commutation rules for Hamiltonians and products of two particle Møller operators.

From

$$H_{12}\Omega_{12} = \Omega_{12}H^0_{12} \quad \text{and} \quad [H^0_3, \Omega_{12}] = 0$$

it follows that

$$\Omega_{12}H^0_{123} - H^0_{123}\Omega_{12} = V_{12}\Omega_{12} \tag{2.51}$$

and

$$\Omega_{12}\Omega_{13}H^0_{123} - H^0_{123}\Omega_{12}\Omega_{13} = \Omega_{13}V_{13}\Omega_{13} + V_{12}\Omega_{12}\Omega_{13} \,. \tag{2.52}$$

The adjoint equation reads

$$H^0_{123}\Omega^\dagger_{13}\Omega^\dagger_{12} - \Omega^\dagger_{13}\Omega^\dagger_{12}H^0_{123} = \Omega^\dagger_{13}V_{13}\Omega^\dagger_{12} + \Omega^\dagger_{13}\Omega^\dagger_{12}V_{12} \,. \tag{2.53}$$

In a similar way we get for the fourth and the fifth terms of (2.45)

$$\begin{aligned}\dot{E}^{4+5}_{\text{kin}} &= -\frac{1}{i\hbar}\, n^3 \operatorname*{Tr}_{123}\left\{\frac{p_1^2}{2m}\,[V_{12}, \Omega_{12}\Omega_{13}F_1(t)F_2(t)F_3(t)\Omega^\dagger_{13}\Omega^\dagger_{12}\right.\\ &\quad \left. + \Omega_{12}\Omega_{23}F_1(t)F_2(t)F_3(t)\Omega^\dagger_{23}\Omega^\dagger_{12}]\right\}\\ &= \frac{1}{i\hbar}\frac{n^3}{2}\operatorname*{Tr}_{123}\{V_{12}[H^0_{123}, \Omega_{12}\Omega_{13}F_1(t)F_2(t)F_3(t)\Omega^\dagger_{13}\Omega^\dagger_{12}\\ &\quad + \Omega_{12}\Omega_{23}F_1(t)F_2(t)F_3(t)\Omega^\dagger_{23}\Omega^\dagger_{12}]\} \,.\end{aligned} \tag{2.54}$$

The application of (2.52) and (2.53) to (2.54) gives the result

$$\begin{aligned}\dot{E}^{4+5}_{\text{kin}} &= -\frac{1}{i\hbar}\frac{n^3}{2}\operatorname*{Tr}_{123}(V_{12}\Omega_{12}\{[V_{13}, \Omega_{13}F_1(t)F_2(t)F_3(t)\Omega^\dagger_{13}]\\ &\quad + [V_{23}, \Omega_{23}F_1(t)F_2(t)F_3(t)\Omega^\dagger_{23}]\}\Omega^\dagger_{12}) \,.\end{aligned} \tag{2.55}$$

Here we applied $\operatorname{Tr}\{A[A, B]\} = 0$ and $[H^0_{123}, F_1(t)F_2(t)F_3(t)] = 0$. According to Eq. (2.11) we have

$$F_{13}(t) = \Omega_{13}F_1(t)F_3(t)\Omega^\dagger_{13} + O(n) \,. \tag{2.56}$$

Taking into account $[V_{13}, F_2] = 0$ and

$$\frac{\partial}{\partial t}F_1 = \frac{1}{i\hbar}\, n \operatorname*{Tr}_{3}[V_{13}, F_{13}] \,,$$

we have

$$\begin{aligned}\dot{E}^{4+5}_{\text{kin}} &= -\frac{n^2}{2}\operatorname*{Tr}_{12}\left\{V_{12}\Omega_{12}\,\frac{\partial}{\partial t}\,(F_1(t)F_2(t))\Omega^\dagger_{12}\right\}\\ &= -\frac{\partial}{\partial t}\frac{n^2}{2}\operatorname*{Tr}_{12}\{V_{12}F_{12}(t)\} + O(n^3) \,.\end{aligned} \tag{2.57}$$

Equation (2.57) represents the negative time derivative of the mean value of the potential energy in the approximation of the second virial coefficient (binary collision approximation). Therefore, we have from (2.44) and (2.57)

$$\frac{\partial}{\partial t}\langle T\rangle = -\frac{\partial}{\partial t}\langle V\rangle . \tag{2.58}$$

It is essential to note that the contribution to the potential energy comes from the retardation correction of the two-particle density operator, which describes three-particle interactions of the self-energy type.

III. BOUND STATES IN KINETIC THEORY

A. Problem of Bound States— Generalization of the Bogolyubov Condition

So far we have ignored bound states, or composite particles, which may form as a result of the interaction due to an attractive part of the potential. Of course, the behavior of macroscopic systems such as thermodynamic, transport, and optical properties, is essentially influenced by the existence of bound states. A particular problem of special interest in connection with these bound states is the ionization phenomenon, or more general, the problem of chemical reactions.

The kinetic theory of chemically reacting gases and partially ionized plasmas is even more complex. Very few works have been devoted to the problem of bound states in kinetic theory. The works of Klimontovich,[12,18] Peletminsky,[19] Kolesnichenko,[20] Vashukov and Maruzin,[21] Ivanov,[22] Klimontovich and Kremp,[14] and McLennan,[15] represent attempts at constructing such kinetic equations. In papers by Schlanges and Kremp the method of nonequilibrium Green's functions is also applied.[23]

In order to discuss the fundamental problems that are connected with the bound states in kinetic theory, we first restrict ourselves to systems with two-particle bound states only. The states of the two-particle system are determined by Eq. (2.12). Furthermore, we remark that to describe the formation of two-particle bound states by a collision, at least three particles are necessary in order to fulfill energy and momentum conservation. Thus, it is necessary to consider the quantum mechanics of three-particle systems.

Because of the existence of two-particle bound states the three-particle scattering states split into channels; that means, we have to take into account scattering processes between free particles and bound pairs. To study the effect of composite particle scattering processes, we pay attention to the fact that the scattering states are determined by the asymptotic initial states. Thus, we must classify the asymptotic states into channels. In the three-particle case without three-particle bound states there are four channels,[9] that is, 1,2 bound, 3 free; 2,3 bound, 1 free; 1,3 bound, 2 free, and 1,2,3 all free. Let us denote the channels and the asymptotic states by κ in following way:

$$\begin{aligned}
\kappa &= 0\,, \quad 1+2+3\,, \qquad |0p\rangle = |p_1\rangle|p_2\rangle|p_3\rangle\,; \\
\kappa &= 1\,, \quad 1+(2+3)\,, \quad |1p\rangle = |p_1\rangle|n_{23}P_{23}\rangle\,; \\
\kappa &= 2\,, \quad (1+3)+2\,, \quad |2p\rangle = |p_2\rangle|n_{13}P_{13}\rangle\,; \\
\kappa &= 3\,, \quad (1+2)+3\,, \quad |3p\rangle = |p_3\rangle|n_{12}P_{12}\rangle\,.
\end{aligned} \tag{3.1}$$

Then from the many-channel scattering theory it is well known that the scattering states $|\kappa p+\rangle$ can be constructed with the Møller operators $\Omega^{\kappa\dagger}$ by

$$|\kappa p+\rangle = \Omega^{\kappa\dagger}_{123}|\kappa p\rangle \tag{3.2}$$

from the asymptotic states, where $\Omega^{\kappa\dagger}$ is defined by

$$\Omega^{\kappa\dagger}_{123} = \lim_{t\to-\infty} e^{iH_{123}t/\hbar}e^{-iH^{\kappa}_{123}t/\hbar} \tag{3.3}$$

with

$$H^{\kappa}_{123} + V^{\kappa} = H_{123}\,. \tag{3.4}$$

V^{κ} is the interaction in the channel κ_1, which describes the interaction between the free particle and a bound pair. Thus we have

$$\begin{aligned}
V^1 &= V_{12} + V_{13}\,, \qquad V^2 = V_{21} + V_{23} \\
V^3 &= V_{31} + V_{32}\,, \qquad V^0 = V_{12} + V_{13} + V_{23}\,.
\end{aligned} \tag{3.5}$$

Finally let us remark that the states $|\kappa p+\rangle$ are eigenstates of H_{123}:

$$H_{123}|\kappa p+\rangle = E_{\kappa p}|\kappa p+\rangle\,. \tag{3.6}$$

In this manner the eigenstates of the three-particle problem are classified with respect to the asymptotic initial states (channels).

Let us now consider the problem of the condition of weakening of the initial correlation in systems with two-particle bound states. There are several reasons to generalize this condition (Peletminsky,[19] Klimontovich and Kremp,[14]; see also Refs. 23 and 24). Using the Bogolyubov condition in the form (2.4) bound states are excluded, because bound states have no free asymptotes. Thus we restricted the space of states to the space of scattering states. The condition of the total weakening of initial conditions is valid only under the condition

$$\tau_{\text{corr}} \ll \tau_{\text{coll}} ,$$

but bound states are long-living correlations.

Clearly, it is necessary to generalize the Bogolyubov condition (2.4). It is obvious in generalizing (2.4) to assume that in systems with two-particle bound states, the pair 1 and 2 moved independently of all other particles as $t \rightarrow -\infty$. Then it is possible to write

$$\begin{aligned} \lim_{t \rightarrow -\infty} F_{12} &= \sum_{\alpha P} F_{\alpha}(P) |\alpha P\rangle \langle P\alpha| \\ &= F_{12}^{\text{sc}} + F_{12}^{\text{b}} . \end{aligned} \tag{3.7}$$

F_{12}^{sc} is the contribution to $F_{12}(t)$ for pairs in scattering states, and F_{12}^{b} is the part for pairs in bound states. Now, in the sense of Bogolyubov, we use for the scattering states the condition of the total weakening of initial correlation:

$$\lim_{t \rightarrow -\infty} F_{12}^{\text{sc}} = F_1 F_2 . \tag{3.8}$$

Because a weakening of the initial correlation for F_{12}^{b} is absent, a system with two-particle bound states cannot be described only by F_1, but by F_{12}^{b}, and it is not possible to obtain a closed equation for F_1.

Let us now consider the Bogolyubov condition for the three-particle density operator. In generalizing (2.4), the Bogolyubov condition now takes the form (if three-particle bound states are absent)

$$\lim_{t \rightarrow -\infty} F_{123}(t) = F_{12}^{\text{b}} F_3 + F_{13}^{\text{b}} F_2 + F_{23}^{\text{b}} F_1 + F_1 F_2 F_3 , \tag{3.9}$$

which means the part of F_{123} acting in the channel subspace $\mathcal{H}_{\kappa}$ splits asymptotically into a product of the density operator of free particles and two-particle subclusters. If bound states do not appear in the system, we get again the usual Bogolyubov condition

$$\lim_{t \rightarrow -\infty} F_{1 \cdots s}(t) = F_1 \cdots F_s . \tag{3.10}$$

B. Binary Collision Approximation for the Two-Particle Density Operator—Kinetic Equation for Free Particles and Atoms

Let us first consider the effect of bound states in the binary collision approximation. In this approximation F_{12} satisfies the equation

$$\frac{\partial F_{12}}{\partial t} + \frac{i}{\hbar}[F_{12}, H_{12}] = 0 , \tag{3.11}$$

with the solution

$$F_{12}(t) = e^{-iH_{12}(t-t_0)/\hbar} F_{12}(t_0) e^{iH_{12}(t-t_0)/\hbar}$$

For the determination of $F_{12}(t_0)$ we have to use the conditions (2.7) and (2.8). In the same manner as in Section II we obtain

$$F_{12}(t) = \Omega_{12} F_1(t) F_2(t) \Omega_{12}^{\dagger} + F_{12}^{\mathrm{b}}(\mathrm{t}) . \tag{3.12}$$

But now the relations (2.8a) and (2.8b) are replaced by

$$\begin{aligned} \Omega_{12}^{\dagger}\Omega_{12} &= 1 , \\ \Omega_{12}\Omega_{12}^{\dagger} &= 1 - P , \end{aligned} \tag{3.13}$$

where p is the projection operator onto the subspace of bound states and is given by

$$P = \sum_{nP} |nP\rangle\langle Pn| . \tag{3.14}$$

Because $\Omega\Omega^{\dagger}\Omega\Omega^{\dagger} = \Omega\Omega^{\dagger}$, the operator $\Omega\Omega^{\dagger}$ is the projection onto the scattering states. Substituting the solution (3.12) into the first equation of the hierarchy, we obtain the modified Boltzmann equation

$$\frac{\partial F_1}{\partial t} + \frac{i}{\hbar}[F_1, H_1] = \frac{-i}{\hbar} n \operatorname*{Tr}_2 [V_{12}, \Omega_{12} F_1(t) F_2(t) \Omega_{12}^{\dagger} + F_{12}^{\mathrm{b}}(t)] \tag{3.15}$$

with the additional equation for $F_{12}^{\mathrm{b}}(t)$:

$$\frac{\partial}{\partial t} F_{12}^{\mathrm{b}} + \frac{i}{\hbar}[F_{12}^{\mathrm{b}}, H_{12}] = 0 . \tag{3.15a}$$

Now it is useful to transform this equation into a chemical picture. It is well known that there is no essential difference between a bound state and an elementary particle. For many physical properties, for example, the pressure, the difference between a bound state and an elementary particle does not play any role. For this reason, in statistical mechanics it is convenient, too, to use the following interpretation[25]:

Bound state = new composite particle.

Therefore, we have now a two-component system consisting of free particles and atoms. For the description of this system it is necessary to introduce distribution functions for the free particles and atoms. It seems to be obvious that the distribution function of the atoms is connected with the bound-state contribution of F^{b}_{12},

$$F^{\mathrm{b}}_{12} = PF_{12} , \tag{3.16}$$

and the density operator of the free particles is given by

$$F^{\mathrm{f}}_{1} = F_{1} - n \operatorname*{Tr}_{2} F^{\mathrm{b}}_{12} . \tag{3.17}$$

Because of the normalization of F_1, we find the following relation:

$$\frac{1}{V} \operatorname*{Tr}_{1} F_{1} = \frac{1}{V} \operatorname*{Tr}_{12} F^{\mathrm{f}}_{1} + \frac{1}{V} n \operatorname*{Tr}_{12} F^{\mathrm{b}}_{12} = 1 . \tag{3.18}$$

Now we interpret the first term as a contribution corresponding to the free particles

$$\frac{1}{V} \operatorname*{Tr}_{1} F^{\mathrm{f}}_{1} = \frac{N_1}{n} , \tag{3.19}$$

where N_1 is the number of free particles, and the second term as the contribution of the atoms

$$\frac{n}{V} \operatorname*{Tr}_{12} F^{\mathrm{b}}_{12} = \frac{2N_{12}}{N} . \tag{3.20}$$

Thus we obtain the necessary condition

$$N_1 + 2N_{12} = N . \tag{3.21}$$

Therefore, we can give the following definition of the distribution functions. For the free particles the distribution function is given by

$$n\langle p_1 | F^{\mathrm{f}}_{1} | p_1 \rangle = \frac{V}{(2\pi\hbar)^3} f(p_1) \tag{3.22}$$

and for atoms by

$$\frac{n^2}{2} \langle n_{12}P_{12}|F^{\mathrm{b}}_{12}|P_{12}n_{12}\rangle = \frac{V}{(2\pi\hbar)^3} f_{n_{12}}(P_{12}) . \tag{3.23}$$

The normalization of these functions is given by

$$\begin{aligned} \int f(p^1) \frac{dp_2}{(2\pi\hbar)^3} &= n_1 , \\ \sum_{n_{12}} \int \frac{dp_{12}}{(2\pi\hbar)^3} f_{n_{12}}(P_{12}) &= 2n_{12} . \end{aligned} \tag{3.24}$$

Now we return to Eq. (3.15). Using the definition of F^{f}_1 we can eliminate the bound-state part F^{b}_{12} and obtain the Boltzmann equation for the free particles in the form

$$\frac{\hbar}{i} \frac{\partial F^{\mathrm{f}}_1}{\partial t} + [F^{\mathrm{f}}_1, H_1] = n \operatorname*{Tr}_2 [V_{12}, \Omega_{12}F^{\mathrm{f}}_1 F^{\mathrm{f}}_2 \Omega^{\dagger}_{12}] . \tag{3.25}$$

Here we have used Eq. (3.17) and the fact that $F_1 - F^{\mathrm{f}}_1$ is of the order n.

As in Section II.1, we get from (3.25) the usual shape of the Boltzmann equation for free particles. Obviously, such equations are of no interest for systems with bound states. Equation (3.15a) for atoms is collisionless, and the equation for the free particles does not contain contributions that account for the formation and the decay of bound states. In order to derive such equations we must take into account three-particle and higher-order collisions. Subsequently, this problem will be dealt with.

C. Binary Density Operator for Higher Densities—Bound States

Now we want to generalize the kinetic equation for free (unbound) particles; that is, we want to derive a kinetic equation for free particles that takes into account collisions between free and bound particles as well. For this purpose it is necessary to determine the binary density operator, occurring in the collision integral of the single-particle kinetic equation, at least in the three-particle collision approximation. An approximation of such type was given in Section II.2 for systems without bound states. Thus we have to generalize, for example, the approximation for f_{12} given by Eq. (2.40), to systems with bound states.

It turns out that it is impossible now to get a closed equation for the single-particle density operator alone. We also have to consider a kinetic equation for the bound states (composite particles) and thus to derive a coupled set of two kinetic equations for the free particles and the bound states.

We will carry out our program in two steps. In this section we will derive the two-particle density operator F_{12} in a three-particle collision approximation for the application in the collision integral of F_1. As compared with Section II.2, the main difference will be the occurrence of bound states and, especially, the generalization of the asymptotic condition, which now has to account for bound states too. For the purpose of the application in the kinetic equation of the atoms (bound states) we need an approximation of the next-higher-order density matrix, that is, F_{123}. This quantity will be determined under inclusion of certain four-particle interaction.

The application of F_{12} and F_{123} in the relevant kinetic equations for free particles and bound states, respectively, will be given in the next section.

Our starting point is the expansion (2.37) for the binary density operator,

$$F_{12}(t) = e^{-iH_{12}(t-t_0)/\hbar} F_{12}(t_0) e^{iH_{12}(t-t_0)/\hbar} + n \operatorname*{Tr}_{3} \{F_{123}(t) - e^{-iH_{12}(t-t_0)/\hbar} F_{123}(t_0) e^{iH_{12}(t-t_0)/\hbar}\} . \quad (3.26)$$

In this connection, F_{123} is given in the three-particle collision approximation by

$$F_{123}(t) = e^{-iH_{123}(t-t_0)/\hbar} F_{123}(t_0) e^{iH_{123}(t-t_0)/\hbar} . \quad (3.27)$$

As in Section II, the initial values $F_{12}(t_0)$ and $F_{123}(t_0)$ for the binary and the three-particle density operators must be determined. For this purpose we have to generalize the Bogolyubov condition of the weakening of initial correlations given by (2.4) for systems that do not support the formation of bound states.

For systems with bound states we now write

$$F_{12}(t_0) = F_1(t_0)F_2(t_0) + F^{\mathrm{b}}_{12}(t_0) , \qquad t_0 \to -\infty , \quad (3.28)$$

and

$$F_{123}(t_0) = \sum_{\kappa=0,1,2,3} F^{\kappa}_{123}(t_0) , \qquad t_0 \to -\infty . \quad (3.29)$$

Thus we get from (3.26)–(3.29) for the two-particle density operator

$$F_{12}(t) = e^{-iH_{12}(t-t_0)/\hbar} F_1(t_0) F_2(t_0) e^{iH_{12}(t-t_0)/\hbar}$$
$$+ e^{-iH_{12}(t-t_0)/\hbar} F^{\mathrm{b}}_{12}(t_0) e^{iH_{12}(t-t_0)/\hbar}$$
$$+ n \operatorname*{Tr}_3 \sum_\kappa \{ e^{-iH_{123}(t-t_0)/\hbar} F^\kappa_{123}(t_0) e^{iH_{123}(t-t_0)/\hbar}$$
$$- e^{-iH_{12}(t-t_0)/\hbar} F^\kappa_{123}(t_0) e^{iH_{12}(t-t_0)/\hbar} \}|_{t_0 \to -\infty} . \tag{3.30}$$

It is to be seen that Eq. (3.30) fulfills the initial condition (3.28).

Our next problem is the elimination of the (arbitrary) initial time t_0. Closely connected with this problem is the compensation of the successive binary collisions. This was discussed in detail in Section II.2. For our purpose we use Eq. (2.38) after a generaliztion to bound states:

$$F_1(t) = e^{-iH_1(t-t_0)/\hbar} F_1(t_0) e^{iH_1(t-t_0)/\hbar} + n \operatorname*{Tr}_2 \{ \Omega_{12} F_1(t) F_2(t) \Omega^\dagger_{12} - F_1(t) F_2(t) \}$$
$$+ n \operatorname*{Tr}_2 F^{\mathrm{b}}_{12}(t) . \tag{3.31}$$

We introduce (3.31) into (3.30) and carry out the limiting procedure $t_0 \to -\infty$. Furthermore, we introduce the density operator of free particles, F^{f}_1, by

$$F^{\mathrm{f}}_1 = F_1 - n \operatorname*{Tr}_2 F^{\mathrm{b}}_{12} . \tag{3.32}$$

Then we get

$$F_{12}(t) = \Omega_{12} F^{\mathrm{f}}_1(t) F^{\mathrm{f}}_2(t) \Omega^\dagger_{12} + \lim_{t_0 \to -\infty} e^{-iH_{12}(t-t_0)/\hbar} F^{\mathrm{b}}_{12}(t_0) e^{iH_{12}(t-t_0)/\hbar}$$
$$+ n \operatorname*{Tr}_3 \sum_\kappa \{ \Omega^\kappa_{123} F^\kappa_{123}(t) \Omega^{\kappa\dagger}_{123} - \lim_{t_0 \to -\infty} e^{-iH_{12}(t-t_0)/\hbar} F^\kappa_{123}(t_0)$$
$$\times e^{iH_{12}(t-t_0)/\hbar} \} - n \operatorname*{Tr}_3 \{ \Omega_{12} [\Omega_{13} F_1(t) F_2(t) F_3(t) \Omega^\dagger_{13}$$
$$- F_1(t) F_2(t) F_3(t)] \Omega^\dagger_{12} + \Omega_{12} [\Omega_{23} F_1(t) F_2(t) F_3(t) \Omega^\dagger_{23}$$
$$- F_1(t) F_2(t) F_3(t)] \Omega^\dagger_{12} \} . \tag{3.33}$$

Here we used

$$\Omega^\kappa_{123} = \lim_{t \to -\infty} e^{-iH_{123}t/\hbar} e^{iH^\kappa_{123}t/\hbar} , \tag{3.34}$$

Ω^κ_{123} being the channel Møller operators.

The contribution F^{r}_{12} of (3.33),

$$F^{\mathrm{r}}_{12} \equiv -n \operatorname*{Tr}_3 \sum_\kappa \lim_{t_0 \to -\infty} e^{-iH_{12}(t-t_0)/\hbar} F^\kappa_{123}(t_0) e^{iH_{12}(t-t_0)/\hbar} ,$$

contains only renormalization contributions for the free three-particle scattering, and for the channel, in which the pair (12) is asymptotically bound and 3 is free; that is,

$$F^{\mathrm{r}}_{12}(t) = -n \operatorname*{Tr}_3 [F_3(t)F^{\mathrm{b}}_{12}(t) + \Omega_{12}F_1(t)F_2(t)F_3(t)\Omega^{\dagger}_{12}] . \tag{3.35}$$

In (3.33) we have still to replace $F^{\mathrm{b}}_{12}(t_0)$. This is possible by the projection onto the bound states according to

$$F^{\mathrm{b}}_{12}(t) = \tilde{P}_{12}F_{12}(t) , \tag{3.36}$$

and we get from (3.33) an equation for the density operator $F_{12}(t)$, in which all types of density operators F_1, F_{12} and F_{123} depend only on t:

$$\begin{aligned} F_{12}(t) = {} & \Omega_{12}F^{\mathrm{f}}_1(t)F^{\mathrm{f}}_2(t)\Omega^{\dagger}_{12} + F^{\mathrm{b}}_{12}(t) + (1 - \tilde{P}_{12}) \\ & \times \Big(n \operatorname*{Tr}_3 \sum_{\kappa} \Omega^{\kappa}_{123}F^{\kappa}_{123}\Omega^{\kappa +}_{123} - n \operatorname*{Tr}_3 (F^{\mathrm{b}}_{12}F_3 + \Omega_{12}F_1F_2F_3\Omega^{\dagger}_{12}) \\ & - n \operatorname*{Tr}_3 \{\Omega_{12}(\Omega_{13}F_1F_2F_3\Omega^{\dagger}_{13} - F_1F_2F_3)\Omega^{\dagger}_{12} \\ & + \Omega_{12}(\Omega_{23}F_1F_2F_3\Omega^{\dagger}_{23} - F_1F_2F_3)\Omega^{\dagger}_{12}\}\Big) . \end{aligned} \tag{3.37}$$

This expression will be used in Section III.4 for the formulation of a kinetic equation for free particles.

Let us now deal with the inclusion of four-particle collisions in order to formulate a bound-state kinetic equation.

In order to construct a collision integral for a bound-state kinetic equation (kinetic equation for atoms, consisting of elementary particles), which accounts for the scattering between atoms and between atoms and free particles, it is necessary to determine the three-particle density operator in four-particle approximation. Four-particle collision approximation means that in the formal solution, for example, (1.30), for F_{1234} the integral term is neglected. Then we obtain the expression

$$F_{1234}(t) = e^{-iH_{1234}(t-t_0)/\hbar}F_{1234}(t_0)e^{iH_{1234}(t-t_0)/\hbar} . \tag{3.38}$$

Even the approximation (3.38) for F_{1234} contains, in a system supporting two-particle bound states, a large variety of scattering processes between subclusters. Among these possibilities, we consider only the scattering between the two (isolated) bound states (12) and (34). This means that we have to choose the following asymptotic condition

$$F_{1234}(t_0) = F^{\mathrm{b}}_{12}(t_0)F^{\mathrm{b}}_{34}(t_0)\,, \qquad t_0 \to -\infty\,. \tag{3.39}$$

Then Eq. (3.38) takes the following form (for $t_0 \to -\infty$):

$$F_{1234}(t) = \Omega_{(12)(34)}F^{\mathrm{b}}_{12}(t)F^{\mathrm{b}}_{34}(t)\Omega^{\dagger}_{(12)(34)}\,. \tag{3.40}$$

Here the special Møller operator is defined by

$$\Omega_{(12)(34)} = \lim_{t\to-\infty} e^{iH_{1234}t/\hbar}e^{-i(H_{12}+H_{34})t/\hbar}\,.$$

Equation (3.40) has to be used in the formal solution for F_{123}, Eq. (1.30). The latter reads

$$F_{123}(t) = e^{-iH_{123}(t-t_0)/\hbar}F_{123}(t_0)e^{iH_{123}(t-t_0)/\hbar} + \frac{n}{i\hbar}\int_0^{t-t_0} dt'\, e^{-iH_{124}t'/\hbar}\,\mathop{\mathrm{Tr}}_4\,[V_{14}+V_{24}+V_{34},\,F_{1234}(t-t')]e^{iH_{123}t'/\hbar}\,. \tag{3.41}$$

Using the identity

$$V_{14}+V_{24}+V_{34} = H_{1234} - H_{123} - H_4\,,$$

the t' integration in (3.41) may be carried out. The result is the nonequilibrium cluster expansion of F_{123} up to the order n:

$$F_{123}(t) = e^{-iH_{123}(t-t_0)/\hbar}F_{123}(t_0)e^{iH_{123}(t-t_0)/\hbar} + n\mathop{\mathrm{Tr}}_4\,[F_{1234}(t) - e^{-iH_{123}(t-t_0)/\hbar}e^{-iH_4(t-t_0)/\hbar}F_{1234}(t_0)e^{iH_4(t-t_0)/\hbar}e^{iH_{123}(t-t_0)/\hbar}]\,. \tag{3.42}$$

Here the time evolution operator including H_4 was additionally introduced using the cyclic invariance of the trace with respect to the index 4.

The next problem is the determination of the initial value of the three-particle density operator. Using the generalized Bogolyubov asymptotic condition in the approximation (3.39), we obtain finally

$$\begin{aligned} F_{123}(t) = {} & \sum_{\kappa} e^{-iH_{123}(t-t_0)/\hbar}F^{\kappa}_{123}(t_0)e^{iH_{123}(t-t_0)\hbar}\big|_{t_0\to-\infty} \\ & + n\mathop{\mathrm{Tr}}_4\,\{\Omega_{(12)(34)}F^{\mathrm{b}}_{12}(t)F^{\mathrm{b}}_{34}(t)\Omega^{\dagger}_{(12)(34)} \\ & - \Omega^{\dagger}_{\{123\}4}\Omega_{(12)(34)}F^{\mathrm{b}}_{12}(t)F^{\mathrm{b}}_{34}(t)\Omega^{\dagger}_{(12)(34)}\Omega_{\{123\}4}\} \end{aligned} \tag{3.43}$$

[cf. the remark after (3.44)].

We now have to consider the question of retardation, that is, the elimination of t_0. In the trace term, we have already neglected the retardation, which corresponds to neglecting higher-order concentration terms. As a result we introduced immediately the relevant Møller operators in (3.43). However, the retardation terms coming from the first right-hand-side contribution of (3.43) would be of the same order as the trace term; however, here the four-particle terms represent an approximation only (only certain contributions are taken into account). Therefore, we will neglect the mentioned retardation coming from F_{123}, and we write

$$F_{123}(t) = \sum_{\kappa} \Omega^{\kappa}_{123} F^{\kappa}_{123}(t) \Omega^{\kappa\dagger}_{123} + n \operatorname*{Tr}_{4} \{\Omega_{(12)(34)} F^{\mathrm{b}}_{12}(t) F^{\mathrm{b}}_{34}(t) \Omega^{\dagger}_{(12)(34)} - \Omega^{\dagger}_{\{123\}4} \Omega_{(12)(34)} F^{\mathrm{b}}_{12}(t) F^{\mathrm{b}}_{34}(t) \Omega^{\dagger}_{(12)(34)} \Omega_{\{123\}4}\} . \tag{3.44}$$

The Møller operators of the last contribution, $\Omega_{\{123\}4}$, describe scattering processes, in which particle 4 is asymptotically free. Thus, this last contribution vanishes.

D. Kinetic Equations for Free Particles and for Composite Particles (Atoms)

With Eq. (3.37) for F_{12} it is possible to write a kinetic equation for F_1 that describes the formation and the decay of two-particle bound states in three-particle collisions. Introducing (3.37) into the first equation of the hierarchy (1.29), we obtain in a similar way as in Section III.2 a kinetic equation for the density operator of free particles. This equation may be written in the following form:

$$i\hbar \frac{\partial}{\partial t} F^{\mathrm{f}}_1 + [H_1, F^{\mathrm{f}}_1] = I^{\mathrm{B}}_2 + I^{\mathrm{B}}_3 + I^{\mathrm{R}}_3 , \tag{3.45}$$

where

$$I^{\mathrm{B}}_2 = n \operatorname*{Tr}_{2} [V_{12}, \Omega_{12} F^{\mathrm{f}}_1 F^{\mathrm{f}}_2 \Omega^{\dagger}_{12}] \tag{3.46}$$

is the Boltzmann collision terms of the two-particle collision, and

$$I^{\mathrm{B}}_3 = \frac{n^2}{2} \operatorname*{Tr}_{2,3} (1 - \tilde{P}_{12}) \sum_{\kappa} [V_{12} + V_{13} + V_{23}, \Omega^{\kappa}_{123} F^{\kappa}_{123} \Omega^{\kappa +}_{123}] - h^2 \operatorname*{Tr}_{2,3} (1 - \tilde{P}_{12}) [V_{12}, \Omega_{12} F_1 F_2 F_3 \Omega^{\dagger}_{12} + F^{\mathrm{b}}_{12} F_3] \tag{3.47}$$

is the three-particle Boltzmann collision term.

On behalf of the projection $(1 - \tilde{P}_{12})$, the bound-state contribution (12) of the last term drops out. Equation (3.47) contains for $\kappa = 0$ (scattering of three unbound particles) the collision integral (2.42), and for $\kappa = 1, 2, 3$ it describes the formation and the decay of bound states in the three-particle collision. Here we have

$$F^{\kappa}_{123} = F^{f}_{\kappa} F^{b}_{ij} \qquad \text{with } i, j \neq \kappa .$$

Finally I^{R}_{3} is the renormalization contribution that comes from the elimination of retardation effects. It reads

$$I^{R}_{3} = -n^2 \operatorname*{Tr}_{2,3} [V_{12}, \Omega_{12}(\Omega_{13} F^{f}_{1} F^{f}_{2} F^{f}_{3} \Omega^{\dagger}_{13} - F^{f}_{1} F^{f}_{2} F^{f}_{3}) \Omega^{\dagger}_{12}$$
$$+ \Omega_{12}(\Omega_{23} F^{f}_{1} F^{f}_{2} F^{f}_{3} \Omega^{\dagger}_{23} - F^{f}_{1} F^{f}_{2} F^{f}_{3}) \Omega^{\dagger}_{12}] . \qquad (3.48)$$

The two-particle Boltzmann collision term I^{B}_{2} and the three-particle contribution for $\kappa = 0$ were considered in Section II. It was possible to express those collision integrals in terms of the two- and three-particle scattering matrices. It is also possible to introduce the T matrix in I^{B}_{3} for the channels $\kappa = 1, 2, 3$, that is, in those cases where three are asymptotically bound states. Here we use the multichannel scattering theory, as outlined in Refs. 9 and 26.

First let us introduce the T operator. In the multichannel theory we have the two possibilities:

$$T^{\kappa\kappa'}_{(\omega + i\epsilon)} = \Omega^{\kappa}_{(\omega + i\epsilon)} V^{\kappa'} \quad \text{and} \quad \tilde{T}^{\kappa\kappa'}_{(\omega - i\epsilon)} = V^{\kappa} \Omega^{\kappa'}_{(\omega - i\epsilon)} . \qquad (3.49)$$

We mention that $T^{\kappa\kappa'}$ and $\tilde{T}^{\kappa\kappa'}$ are equivalent on the energy shell. The Møller operator may be determined from the Lippman–Schwinger equation

$$\Omega^{\kappa} = 1 + G V^{\kappa} = 1 + G^{\kappa} V^{\kappa} \Omega^{\kappa} , \qquad (3.50)$$

where the Green's operators are given by

$$G(z) = (z - H)^{-1} \quad \text{and} \quad G^{\kappa}(z) = (z - H^{\kappa})^{-1} . \qquad (3.51)$$

Using the Lippman–Schwinger equation for the T operator, it follows that

$$T^{\kappa\kappa'} = V^{\kappa'} + V^{\kappa} G V^{\kappa} = V^{\kappa'} + V^{\kappa} G^{\kappa} T^{\kappa\kappa'}$$
$$\tilde{T}^{\kappa\kappa'} = V^{\kappa} + V^{\kappa} G V^{\kappa'} = V^{\kappa} + V^{\kappa} G^{\kappa} \tilde{T}^{\kappa\kappa'} . \qquad (3.52)$$

Finally, let us write the generalization of the optical theorem

$$2\,\mathrm{Im}\,\langle \kappa p|T^{\kappa\kappa}|p\kappa\rangle = -2\pi \sum_{\bar\kappa} \int d\bar p \,\langle \kappa p|T^{\kappa\bar\kappa}|\bar p\bar\kappa\rangle \times \delta(E_{\bar p\bar\kappa} - E_{p\kappa})\langle \bar\kappa\bar p|T^{\bar\kappa\kappa^\dagger}|p\kappa\rangle\,. \tag{3.53}$$

In order to work with the on-shell (physical) scattering T matrix, we must consider the kinetic equation in the space of the asymptotic states $\mathcal{H}_3^{\mathrm{as}}$, which is the direct sum of the channel subspaces $\mathcal{H}_3^{\kappa}$. In this space we have the following completeness relation

$$1 = \sum_{\kappa} \int dp\, |p\kappa\rangle\langle \kappa p| \tag{3.54}$$

and the orthonormality

$$\langle \kappa p\,|\,\bar p\bar\kappa\rangle = \delta_{\kappa\bar\kappa}\delta(p-\bar p)\,. \tag{3.55}$$

Furthermore, we take into account that

$$\langle \bar\kappa p|T^{\kappa\kappa'}|p'\bar\kappa'\rangle = \delta_{\kappa\bar\kappa}\delta_{\kappa'\bar\kappa'}\cdot\langle \kappa p|T^{\kappa\kappa'}|p'\kappa'\rangle\,, \tag{3.56}$$

and that

$$\langle \kappa p|F^{\kappa}|\bar p\bar\kappa\rangle = \delta_{\kappa\bar\kappa}\delta(p-\bar p)\langle \kappa p|F^{\kappa}|p\kappa\rangle \tag{3.57}$$

is diagonal up to terms of higher order in the density. Starting from (3.48) and using (3.49)–(3.53) we get the following explicit expression for I_3^{B}:

$$\begin{aligned} T_3^{\mathrm{B}} = {} & \frac{(2\pi\hbar)^3}{V\hbar}\,\frac{1}{2}\sum_{\bar\kappa=0}^{3}(1-\tilde P_{12})\int dp_2\,dp_3\,d\bar p\,|\langle p_1p_2p_3|\tilde T^{0\bar\kappa}|\bar\kappa\bar p\rangle|^2 \\ & \times 2\pi\delta(E_{p_1p_2p_3} - E_{\bar\kappa\bar p})\{f(\bar\kappa\bar p) - f(p_1)f(p_2)f(p_3)\} \\ & + \sum_{n_{23}}\int dP_{23}\,d\bar p\,|\langle p_1P_{23}n_{23}|\tilde T^{1\kappa}|\bar\kappa\bar p\rangle|^2 2\pi\delta(E_{p_1}n_{23}P_{23} - E_{\bar\kappa\bar p}) \\ & \times\{f(\bar\kappa\bar p) - f(p_1)f_{n_{23}}(P_{23})\} - VI_2^{\mathrm{B}}(p_1)\,. \end{aligned} \tag{3.58}$$

Here

$$f(\kappa p) = \begin{cases} f(p_1)f(p_2)f(p_3)\,, & \kappa = 0\,,\\ f(p_1)f_{n_{23}}(P_{23})\,, & \kappa = 1\,,\\ f(p_2)f_{n_{13}}(P_{13})\,, & \kappa = 2\,,\\ f(p_3)f_{n_{12}}(P_{12})\,, & \kappa = 3\,. \end{cases}$$

For the other contribution to the kinetic equation it is not possible in a simple way to introduce a T matrix. In order to get the kinetic equation for bound states (atoms), we have to start again from Eq. (1.29) for $s = 2$. To this equation we have to apply the projection operator $\tilde{P}_{12}$ [cf. Eq. (3.36)] in order to get the density operator for two-particle bound states. The result is

$$\frac{\hbar}{i} \frac{\partial}{\partial t} F^{\mathrm{b}}_{12}(t) + [F^{\mathrm{b}}_{12}(t), H_{12}] = n \operatorname*{Tr}_{3} \tilde{P}_{12}[V_{13} + V_{23}, F_{123}(t)] , \tag{3.59}$$

where $F_{123}(t)$ has to be inserted in the approximation (3.44). The collision operator, that is, the right-hand side of (3.59), contains three- and four-particle contributions. Instead of (3.59) we may write

$$\frac{\hbar}{i} \frac{\partial}{\partial t} F^{\mathrm{b}}_{12}(t) + [F^{\mathrm{b}}_{12}(t), H_{12}] = \hat{I}^{(3)}_{12} + \hat{I}^{(4)}_{12} , \tag{3.60}$$

where the three-particle collision operator $\hat{I}^{(3)}_{12}$ describes the scattering between free (unbound) particles and between a free particle and a bound state (atom). The reactive processes (including free scattering and rearrangement) are the following:

$$(12) + 3 \rightarrow (1'2') + 3'$$

$$(12) + 3 \rightarrow (1'3') + 2'$$

$$(12) + 3 \rightarrow 1' + (2'3')$$

$$(12) + 3 \rightarrow 1' + 2' + 3' .$$

According to the occurrence of $\tilde{P}_{12}$ in (3.59), only those collision processes are of relevance in which the pair 1,2 occurs initially as a bound state (12).

As already outlined, it is useful to express the collision operators in terms of the scattering T matrix. For the three-particle collision operator we may use the formulas given by (3.49–3.58). Then we arrive at

$$\hat{I}^{(3)}_{12} = n \operatorname*{Tr}_{3} \tilde{P}_{12}[V_{12} + V_{13} + V_{23}, F_{123}(t)] , \tag{3.61}$$

which may be rewritten as [analogously to (3.58)]

$$\begin{aligned} \hat{I}^{(3)}_{12} = \frac{(2\pi\hbar)^3}{V\hbar} \sum_{\kappa=0}^{3} \tilde{P}_{12} \Big\{ \int dp_3\, d\bar{p}\, |\langle p_1 p_2 p_3 | \tilde{T}^{0\kappa} | \kappa\bar{p} \rangle|^2 2\pi\delta(E_{p_1p_2p_3} - E_{\kappa\bar{p}}) \\ \times [f(\kappa\bar{p}) - f(p_1) f(p_2) f(p_3)] + \int dp_3\, d\bar{p}\, |\langle P_{12} n_{12} p_3 | \tilde{T}^{3\kappa} | \kappa\bar{p} \rangle|^2 \\ \times 2\pi\delta(E_{P_{12}n_{12}p_3} - E_{\kappa\bar{p}}) \{ f(\kappa\bar{p}) - f(p_3) f_{n_{12}}(P_{12})] \Big\} . \end{aligned} \tag{3.62}$$

The meaning of $f(\kappa\bar{p})$ is that of Eq. (3.58).

In (3.59) only those states are of relevance that include, initially, a bound state of the pair (1, 2). Thus the first contribution of Eq. (3.62) including $\tilde{T}^{9\kappa}$ gives a zero contribution. For the same reason contributions including $\tilde{T}^{1\kappa}$ and $\tilde{T}^{2\kappa}$ do not occur.

The four-particle collision operator is assumed to correspond to one channel only, and reads with (3.40)

$$\hat{I}_{12}^{(4)} = n^2 \operatorname*{Tr}_{3,4} \tilde{P}_{12}[V_{13} + V_{23}, \Omega_{(12)(34)} F_{12}^{\mathrm{b}}(t) F_{34}^{\mathrm{b}}(t) \Omega_{(12)(34)}^{\dagger}] \,. \tag{3.63}$$

In (3.63) we may use the channel Hamiltonian

$$H_{(12)(34)} = H_{1234} - V_{13} - V_{24} - V_{14} - V_{23} \,. \tag{3.64}$$

Then Eq. (3.63) reads

$$\hat{I}_{12}^{(4)} = \frac{n^2}{2} \operatorname*{Tr}_{3,4} \tilde{P}_{12}[V_{13} + V_{23} + V_{14} + V_{24}, \Omega_{(12)(34)} F_{12}^{\mathrm{b}}(t) F_{34}^{\mathrm{b}}(t) \Omega_{(12)(34)}^{\dagger}] \,, \tag{3.65}$$

and the collision operator $\hat{I}_{12}^{(4)}$ describes the scattering of two atoms (having internal degrees of freedom) in the potential

$$V^{\mathrm{atom}} = V_{13} + V_{23} + V_{14} + V_{24} \,. \tag{3.66}$$

With this "channel potential" we may define a T matrix as in (3.52), T^{atom},

$$T^{\mathrm{atom}} = V^{\mathrm{atom}} + V^{\mathrm{atom}} G T^{\mathrm{atom}} \,,$$

and the collision integral reads

$$\begin{aligned} \tilde{I}_{12}^{(4)} = \frac{(2\pi\hbar)^3}{V\hbar} \frac{1}{2} \hat{P}_{12} \sum_{n_{34}} \int dP_{34} \sum_{n,\bar{n}} \int dP \, d\bar{P} \, |\langle n_{12} P_{12} n_{34} P_{34} | T^{\mathrm{atom}} | n P \bar{n} \bar{P} \rangle|^2 \\ \times 2\pi\delta(E_{n_{12}P_{12}n_{34}P_{34}} - E_{nP\bar{n}\bar{P}}) \{ f_n(P) f_{\bar{n}}(\bar{P}) - f_{n_{12}}(P_{12}) f_{n_{34}}(P_{34}) \} \,. \end{aligned} \tag{3.67}$$

Here all functions $f_n(P)$ are atomic distribution functions. We will discuss subsequently the Born approximation of (3.67).

E. Properties of the Kinetic Equations

The basic equations of the kinetic theory for systems with two-particle bound states are given by Eqs. (3.45) and (3.60). Let us consider the properties of these equations.

First let us give some generalization of the kinetic equations. So far we have considered only a one-component system; now we will deal with a multicomponent system. In this case it is necessary to indicate the species of a particle by an index, $a, b, c, \ldots$. Moreover, however, we will consider the kinetic equations in a more simple form taking into account only for "Boltzmannlike" contributions to the collision terms, which means, we neglect, for example, the terms coming from the retardation. Then we may obtain

$$\frac{\partial f_a(p_{a1})}{\partial t} = I_{2a}^{\mathrm{B}}(p_{a1}) + I_{3a}^{\mathrm{B}}(p_{a1}) \tag{3.68}$$

and

$$\frac{\partial f_{n_{a_1b_1}}(P_{a_1b_1})}{\partial t} = I_{3n_{a_1b_1}}(P_{a_1b_1}) + I_{4n_{a_1b_1}}(P_{a_1b_1})\,. \tag{3.69}$$

Now the two-particle Boltzmann contribution I_{2a}^{B} is given by

$$\begin{aligned} I_{2a}^{\mathrm{B}}(p_{a_1}) = \frac{(2\pi\hbar)^3}{\hbar V} \sum_b \int dp_{b_2}\, d\bar{p}_{a_1}\, d\bar{p}_{b_2}\, |\langle p_{a_1} p_{b_2}| T_{ab} |\bar{p}_{b_2}\bar{p}_{a_1}\rangle|^2 \\ \times 2\pi\delta(E-\bar{E})\{f_a(\bar{p}_{a_1})f_b(\bar{p}_{b_2}) - f_a(\bar{p}_{a_1})f_b(\bar{p}_{b_2})\} \end{aligned} \tag{3.70}$$

and the three-particle Boltzmann contribution I_{3a}^{B} including the reaction term has the form

$$\begin{aligned} I_{3a}^{\mathrm{B}}(p_{a_1}) = \frac{(2\pi\hbar)^3}{\hbar V} \sum_{bc}\sum_{\bar{\kappa}} \Big\{ \int dp_{b_2}\, dp_{c_3}\, d\bar{p}\, |\langle p_{a_1} p_{b_2} p_{c_3}| T_{abc}^{0\bar{\kappa}} |\bar{\kappa}\bar{p}\rangle|^2 \\ \times 2\pi\delta(E-\bar{E})[f_{abc}(\bar{\kappa}\bar{p}) - f_a(p_{a_1})f_b(p_{b_2})f_c(p_{c_3})] \\ + \sum_{n_{bc}} \int dP_{b_2c_3}\, d\bar{p}\, |\langle p_{a_1} P_{b_2c_3} n_{b_2c_3}| T^{1\bar{\kappa}} |\bar{p}\bar{\kappa}\rangle|^2 2\pi\delta(E-\bar{E}) \\ \times [f_{abc}(\bar{\kappa}\bar{p}) - f_a(p_{a_1})f_{n_{b_2c_3}}(P_{b_2c_3})]\Big\} - VI_{2a}^{\mathrm{B}}(p_{a_1})\,. \end{aligned} \tag{3.71}$$

Here,

$$f_{abc}(\kappa p) = \begin{cases} f_a(p_{a_1})f_b(p_{b_2})f_c(p_{c_3})\,, & \kappa = 0\,, \\ f_a(p_{a_1})f_{n_{bc}}(P_{b_2c_3})\,, & \kappa = 1\,, \\ f_b(p_{b_2})f_{n_{ac}}(P_{a_1c_3})\,, & \kappa = 2\,, \\ f_c(p_{c_3})f_{n_{ab}}(P_{a_1b_2})\,, & \kappa = 3\,. \end{cases} \tag{3.72}$$

In these formulas the sum over the possible channels κ must be carried out separately for each summand of the sum over the species. We will demonstrate this for the example of an electron–proton system, assuming that the index a_1 is the electron e_1. The sum over the species produces the following three-particle combinations:

$$e_1 e_2 p_3 \; ; \quad e_1 p_2 p_3 \; ; \quad e_1 e_2 e_3 \; ; \quad e_1 p_2 p_3 \, .$$

For each of these combinations, there are the channels $\kappa = 0, 1, 2, 3$, which are, for example, for $e_1 e_2 p_3$:

$$\begin{aligned} \kappa = 0: &\quad e_1 + e_2 + p_3 \, , \\ \kappa = 1: &\quad e_1 + (e_2 + p_3) \, , \\ \kappa = 2: &\quad e_2 + (e_1 + p_3) \, , \\ \kappa = 3: &\quad p_3 + (e_1 + e_2) \quad \text{(forbidden)}. \end{aligned} \tag{3.73}$$

The channel $\kappa = 3$ is forbidden because there are no bound states between electrons. In similar way, the channels $\kappa = 0, 1, 2, 3$ may be found for the other three-particle combinations. Let us now consider the properties of I^{B}_{3a} in more detail. In order to discuss this expression in an understandable way, we split up I^{B}_{3a} into two parts:

$$I^{\mathrm{B}}_{3a} = [I_a]_1 + [I_a]_2 \, . \tag{3.74}$$

The first part $[I_a]_1$ is given by the terms with $\kappa = 0, 1$, which means that this term describes collisions in which we have bound states only between the particles 2 and 3. The particle 1 is always free. The second part is given by the contributions with $\kappa = 2, 3$ and describes collisions in which the particle 1 is bound. First let us discuss the structure of $[I_a]_1$. For $\kappa = 0$ we have two contributions. The first one is given by

$$\begin{aligned} [I_a]_1^0 = \frac{(2\pi\hbar)^3}{V\hbar} \sum_{ab} \int & dp_{b_2}\, dp_{c_3}\, d\bar{p}_{a_1}\, d\bar{p}_{b_2}\, d\bar{p}_{c_3} \\ & \times |\langle p_{a_1} p_{b_2} p_{c_3} | T^{00}_{abc} | \bar{p}_{c_3} \bar{p}_{b_2} \bar{p}_{a_1} \rangle|^2 2\pi\delta(E - \bar{E}) \\ & \times \tfrac{1}{2} \{ f_a(p_{a_1}) f_b(p_{b_2}) f_c(p_{c_3}) - f_a(\bar{p}_{a_1}) f_b(\bar{p}_{b_2}) f_{c_3}) \} - V I^{\mathrm{B}}_{2a}(p_{a_1}) \, . \end{aligned} \tag{3.75}$$

This contribution describes the scattering between three particles,

$$a_1 + b_2 + c_3 \rightleftharpoons \bar{a}_1 + \bar{b}_2 + \bar{c}_3 \, ,$$

and is considered in Section II.2.

The second contribution for $\kappa = 0$ is given by

$$[I_a]_1^1 = \frac{(2\pi\hbar)^3}{V\hbar} \sum_{n_{bc}} \sum_{bc} \int dP_{b_2c_3}\, d\bar{p}_{a_1}\, d\bar{p}_{b_2}\, d\bar{p}_{c_3} \times 2\pi\delta(E - \bar{E}) |\langle p_{a_1} P_{b_2c_3} n_{b_2c_3} | T^{10}_{abc} | \bar{p}_{a_1} \bar{p}_{b_2} \bar{p}_{c_3} \rangle|^2 \times \tfrac{1}{2}\{ f_1(\bar{p}_{a_1}) f_b(\bar{p}_{b_2}) f_c(\bar{p}_{c_3}) - f_a(p_{a_1}) f_{n_{bc}}(P_{b_2c_3}) \} . \tag{3.76}$$

This collision integral describes the reaction

$$a_1 + (b_2c_3) \rightleftharpoons \bar{a}_1 + \bar{b}_2 + \bar{c}_3 ,$$

which means the ionization of the bound pair (b_2c_3) through the collision with a free particle a_1.

For $\kappa = 1$ we have the following contribution:

$$[I_a]_1^2 = \frac{(2\pi\hbar)^3}{V\hbar} \sum_{n_{bc}} \sum_{bc} \int d\bar{P}_{b_2c_3}\, d\bar{p}_{a_1}\, dp_{b_2}\, dp_{c_3} \times 2\pi\delta(E - \bar{E}) |\langle p_{a_1} p_{b_2} p_{c_3} | T^{01}_{abc} | \bar{n}_{b_2c_3} \bar{P}_{b_2c_3} \bar{p}_{a_1} \rangle|^2 \times \tfrac{1}{2}\{ f_a(\bar{p}_{a_1}) f_{\bar{n}_{b_2c_3}}(\bar{P}_{b_2c_3}) - f_a(p_{a_1}) f_b(p_{b_2}) f_c(p_{c_3}) \} , \tag{3.77}$$

which corresponds to the reaction

$$a_1 + b_2 + c_3 \rightleftharpoons \bar{a}_1 + (\bar{b}_2\bar{c}_3)$$

and represents the recombination in presence of a free particle a. The second contribution to $\kappa = 1$ is given by

$$[I_a]_1^3 = \frac{(2\pi\hbar)^3}{V\hbar} \sum_{n_{b_2c_3}} \sum_{\bar{n}_{b_2c_3}} \sum_{cb} \int dP_{b_2c_3}\, d\bar{P}_{b_2c_3}\, d\bar{p}_{a_1} \times 2\pi\delta(E - \bar{E}) |\langle p_{a_1} P_{b_2c_3} n_{b_2c_3} | T^{11}_{abc} | \bar{n}_{b_2c_3} \bar{P}_{b_2c_3} \bar{p}_{a_1} \rangle|^2 \times \tfrac{1}{2}\{ f_a(\bar{p}_{a_1}) f_{\bar{n}_{b_2c_3}}(\bar{P}_{b_2c_3}) - f_a(p_{a_1}) f_{n_{b_2c_3}}(P_{b_2c_3}) \} . \tag{3.78}$$

This term describes the inelastic scattering of a bound pair (b_2c_3) with a free particle a_1,

$$a_1 + (b_2c_3) \rightarrow \bar{a}_1 + (\bar{b}_2\bar{c}_3) .$$

In all these four processes, the average number n_a of particles of type a

interacting with free particles or with atoms does not change. Therefore, the collision integral has the important property

$$\int dp_{a_1}[I_a(p_{a_1}t)]_1 = 0\,, \tag{3.79}$$

meaning that n_a is conserved in these processes.

Let us now consider $[I_a]_2$. Again we have four contributions. Two parts we get for $\kappa = 2$, the first of which is

$$\begin{aligned}[I_a]_2^1 = \frac{(2\pi\hbar)^3}{V\hbar}\,\frac{1}{2}\sum_{\bar{n}_{a_1c_3}}\sum_{bc}\int dp_{b_2}\,dp_{c_3}\,d\bar{p}_{b_2}\,d\bar{P}_{a_1c_3}\\ \times 2\pi\delta(E-\bar{E})|\langle p_{a_1}p_{b_2}p_{c_3}|T^{02}_{abc}|\bar{n}_{a_1c_3}P_{a_1c_3}\bar{p}_{b_2}\rangle|^2\\ \times\{f_b(\bar{p}_{b_2})f_{\bar{n}_{a_1c_3}}(\bar{P}_{a_1c_3}) - f_a(p_{a_1})f_b(p_{b_2})f_c(p_{c_3})\}\,,\end{aligned} \tag{3.80}$$

with the reaction

$$a_1 + b_2 + c_3 \rightarrow \bar{b}_2 + (\bar{a}_1\bar{c}_3)\,, \tag{3.81}$$

which is the recombination of the particles a_1 and c_3. The second contribution reads

$$\begin{aligned}[I_a]_2^2 = \frac{(2\pi\hbar)^3}{V\hbar}\sum_{n_{b_2c_3}}\sum_{\bar{n}_{a_1c_3}}\int dP_{b_2c_3}\,d\bar{P}_{a_1c_3}\\ \times|\langle p_{a_1}n_{b_2b_3}P_{b_2c_3}|T^{12}_{abc}|\bar{n}_{a_1c_3}\bar{P}_{a_1c_3}\bar{p}_{b_2}\rangle|^2\\ \times 2\pi\delta(E-\bar{E})\{f_b(\bar{p}_{b_2})f_{\bar{n}_{a_1c_3}}(\bar{P}_{a_1c_3}) - f_a(p_{a_1})f_{n_{b_2c_3}}(P_{b_2c_3})\}\,,\end{aligned} \tag{3.82}$$

which describes the exchange reaction

$$a_1 + (b_2c_3) \rightarrow \bar{b}_2 + (\bar{a}_1\bar{c}_3)\,. \tag{3.83}$$

Furthermore, we get two parts for $\kappa = 3$, that is,

$$[I_a]_2^3 = [I_a]_2^1(\bar{b}_2 \leftrightarrow \bar{c}_3)\,,$$

corresponding to the reaction

$$a_1 + b_2 + c_3 \rightarrow \bar{c}_3 + (\bar{a}_1\bar{b}_2) \tag{3.84}$$

and

$$[I_a]_2^4 = [I_a]_2^2(\bar{b}_2 \leftrightarrow \bar{c}_3) ,$$

corresponding to the exchange reaction

$$a_1 + (b_2 c_3) \rightarrow \bar{c}_3 + (\bar{a}_1 \bar{b}_2) . \tag{3.85}$$

As can be seen, the four processes (3.81) and (3.83)–(3.85) change the number of free particles of species a. That means the property (3.79) does not hold for $[I_a]_2$. Instead, we have

$$\int dp_{a_1} [I_a(p_{a_1} t)]_2 \neq 0 , \tag{3.86}$$

that is, the particle number n_a is not conserved.

Similarly, we may discuss the properties of the kinetic equations for atoms. For example, the collision integral $I_{3n_{a_1 b_2}}$ of Eq. (3.68) contains the following contributions for $\kappa = 3$:

$$[I_{n_{a_1 b_2}}(P_{a_1 b_2})]^3 = \frac{(2\pi\hbar)^3}{\hbar V} \sum_c \int dp_{c_3} \sum_{\bar{n}_{a_1 b_2}} \int d\bar{P}_{a_1 b_2}\, d\bar{p}_{c_3}$$
$$\times |\langle P_{a_1 b_2} n_{a_1 b_2} p_{c_3} | T^{33}_{abc} | \bar{p}_{c_3} \bar{n}_{a_1 b_2} \bar{P}_{a_1 b_2} \rangle|^2$$
$$\times 2\pi\delta(E - \bar{E})\{f_c(\bar{p}_{c_3}) f_{\bar{n}_{a_1 b_2}}(\bar{P}_{a_1 b_2}) - f_c(p_{c_3}) f_{n_{a_1 b_2}}(P_{a_1 b_2})\} . \tag{3.87}$$

This collision integral has the shape of the usual Boltzmann collision integral for the scattering of a bound pair and a free particle and corresponds to the process

$$(a_1 b_2) + c_3 \rightarrow (\bar{a}_1 \bar{b}_2) + \bar{c}_3 . \tag{3.88}$$

The contributions for $\kappa = 0, 1, 2$ describe reactions. We get, for example, for $\kappa = 0$

$$[I_{n_{a_1 b_2}}(P_{a_1 b_2})]^0 = \frac{(2\pi\hbar)^3}{\hbar V} \sum_c \int dp_{c_3}\, d\bar{p}_{c_3}\, d\bar{p}_{b_2}\, d\bar{p}_{a_1}$$
$$\times |\langle n_{a_1 b_2} P_{a_1 b_2} p_{c_3} | T^{30}_{abc} | \bar{p}_{a_1} \bar{p}_{b_2} \bar{p}_{c_3} \rangle|^2 2\pi\delta(E - \bar{E})$$
$$\times \{f_a(\bar{p}_{a_1}) f_b(\bar{p}_{b_2}) f_c(\bar{p}_{c_3}) - f_c(p_{c_3}) f_{n_{a_1 b_2}}(P_{a_1 b_2})\} , \tag{3.89}$$

corresponding to the ionization and recombination reaction. The collision integral $I_{4n_{a_1b_2}}(P_{a_1b_2})$ of (3.69) again is of the shape of the usual Boltzmann collision integral, now for the atom–atom collision. Of course, there are other four-particle processes; but, according to our approximation, these were neglected.

F. Rate Equations for Reactions—Equilibrium Solutions

In order to describe the reactions mentioned previously, quantitatively, we need the densities of the different components as a function of the time. As we deal with nonequilibrium situations, the densities do not obey a mass-action law. Instead, we have to discuss rate equations that determine the densities as a function of the time depending on the cross sections and the distribution functions.

In order to get equations for the density, we have to integrate the kinetic equations (3.68) and (3.69) with respect to the variables p_a, or $n_{a_1b_2}$, $P_{a_1b_2}$, respectively. From (3.68) we get

$$\frac{dn_a}{dt} = \int \frac{dp_{a_1}}{(2\pi\hbar)^3} \{I^{\mathrm{B}}_{2a}(p_{a_1}) + [I_a]_1 + [I_a]_2\} . \tag{3.90}$$

Taking into account that the Boltzmann collision integrals I^{B}_{2a} and $[I_a]_1$ conserve the particle number, we get

$$\frac{dn_a}{dt} = \int \frac{dp_{a_1}}{(2\pi\hbar)^3} [I_a]_2 . \tag{3.91}$$

For purposes of simplification we assume that bound states are possible only between different species. Furthermore, we neglect exchange reactions of types (3.83) and (3.85). We obtain the rate equation

$$\frac{dn_a}{dt} = \alpha n_a n_{(12)} - \beta n_a^3 . \tag{3.92}$$

This equation describes the change of the density n_a as a result of ionization and recombination of bound pairs, the density of which is denoted by $n_{(12)}$. For this reason we call α the ionization coefficient, given by ($n_a = n_b = n_c$)

$$\alpha = \frac{1}{2} \frac{2\pi}{V\hbar} \sum_{\substack{\bar{n}_{ac} \\ bc}} \int dp_a\, dp_b\, dp_c\, d\bar{p}_b\, d\bar{P}_{ac} |\langle p_a p_b p_c | T^{02}_{abc} | \bar{n}_{ac} \bar{P}_{ac} \bar{p}_b \rangle|^2$$

$$\times\, \delta(E_{abc} - \bar{E}_{(ac)b}) \frac{f_{\bar{n}_{ac}}(\bar{P}_{ac})}{n_{(12)}} \frac{f_b(\bar{p}_b)}{n_b}$$

$$+\frac{1}{2}\frac{2\pi}{V\hbar}\sum_{\substack{\bar{n}_{ab}\\ bc}}\int dp_a\, dp_b\, dp_c\, d\bar{p}_c\, d\bar{P}_{ab}|\langle p_a p_b p_c|T^{03}_{abc}|\bar{n}_{ab}\bar{P}_{ab}\bar{p}_c\rangle|^2$$

$$\times\,\delta(E_{abc}-\bar{E}_{(ab)c})\,\frac{f_{\bar{n}_{ab}}(\bar{P}_{ab})}{n_{(12)}}\,\frac{f_c(\bar{p}_c)}{n_c}\,. \tag{3.93}$$

The coefficient β describes the recombination and reads

$$\beta=\frac{1}{2}\frac{2\pi}{\hbar V}\sum_{\substack{\bar{n}_{ac}\\ bc}}\int dp_a\, dp_b\, dp_c\, d\bar{p}_b\, d\bar{P}_{ac}|\langle p_a p_b p_c|T^{02}_{abc}|\bar{n}_{ac}\bar{P}_{ac}\bar{p}_b\rangle|^2$$

$$\times\,\delta(E_{abc}-\bar{E}_{(ac)b})\,\frac{f_a(p_a)}{n_a}\,\frac{f_b(p_b)}{n_b}\,\frac{f_c(p_c)}{n_c}$$

$$+\frac{1}{2}\frac{2\pi}{V\hbar}\sum_{\substack{\bar{n}_{ab}\\ bc}}\int dp_a\, dp_b\, dp_c\, d\bar{p}_c\, d\bar{P}_{ab}|\langle p_a p_b p_c|T^{03}_{abc}|\bar{n}_{ab}\bar{P}_{ab}\bar{p}_c\rangle|^2$$

$$\times\,\delta(E_{abc}-\bar{E}_{(ab)c})\,\frac{f_a(p_a)}{n_a}\,\frac{f_b(p_b)}{n_b}\,\frac{f_c(p_c)}{n_c}\,. \tag{3.94}$$

In this way, (3.93) and (3.94) define the macroscopic rate coefficients in terms of cross sections and distribution functions.

In the case of thermodynamic equilibrium, we get from (3.92) a mass-action law that reads

$$\frac{n_{(12)}}{n_a^2}=\frac{\beta}{\alpha}\,. \tag{3.95}$$

In the equilibrium case, the relevant kinetic equations without retardation terms have the solutions

$$f_a^0(p_a)=\exp\left\{-\beta_{\mathrm{B}}\left(\frac{p_a^2}{2m_a}-\mu_a\right)\right\}\qquad(\beta_{\mathrm{B}}=(k_{\mathrm{B}}T)^{-1}) \tag{3.96}$$

for free particles and

$$f^0_{n_{ab}}(P_{ab})=\exp\left\{-\beta_{\mathrm{B}}\left(E_{n_{ab}}+\frac{P_{ab}^2}{2M_{ab}}-\mu_{ab}\right)\right\} \tag{3.97}$$

for bound pairs; here the parameters μ_a, μ_{ab} have to fulfill the condition

$$\mu_{ab}=\mu_a+\mu_b\,. \tag{3.98}$$

Thus the quantities μ are chemical potentials, and (3.98) represents the condition of chemical equilibrium. The chemical potentials may be determined from the normalization conditions of the distribution functions, namely,

$$\int \frac{dp_a}{(2\pi\hbar)^3} f_a(p_a) = n_a$$

and

$$\sum_{n_{ab}} \int \frac{d\bar{P}_{ab}}{(2\pi\hbar)^3} f_{n_{ab}}(P_{ab}) = 2n_{(12)} . \qquad (3.99)$$

Taking into account retardation terms and thus nonideality scattering effects (sc),[14,25,27] we get from the normalization conditions (3.99)

$$\mu_a = k_B T \ln (n_a \Lambda_a^3) - 2k_B T \sum_c n_c b_{ac}^{sc} . \qquad (3.100)$$

The virial coefficients b_{ac}^{sc} are given by

$$b_{ac}^{sc} = \mathrm{Tr}\,(e^{-\beta H_{ab}} - e^{-\beta H_{ab}^0}) .$$

For the bound pairs we get

$$\mu_{ab} = k_B T \ln (n_{(12)} \Lambda_{ab}^3) - k_B T \ln \sum_{n_{ab}} e^{-\beta E_{n_{ab}}} . \qquad (3.101)$$

The thermal wavelengths Λ are defined as

$$\Lambda_\alpha = h/(2\pi m_\alpha k_B T)^{1/2} .$$

For Λ_{ab} we have $m_{(ab)} = m_a + m_b$. From (3.98) we get with (3.100) and (3.101)

$$\frac{n_{(12)}}{n_a n_b} = \frac{\Lambda_a^3 \Lambda_b^3}{\Lambda_{ab}^3} \sum_{n_{ab}} e^{-\beta E_{n_{ab}}} e^{-2 \Sigma_c n_c (b_{ac}^{sc} + b_{bc}^{sc})} . \qquad (3.102)$$

The right-hand side of (3.102) is a temperature- and density-dependent mass-action constant $K(nT)$, which reads [with $\lambda_{ab} = \hbar/(2\mu_{ab} k_B T)^{1/2}$, μ_{ab} being the reduced mass)

$$K(n, T) = 8\pi^{3/2} \lambda_{ab}^3 \sum_{n_{ab}} e^{-\beta E_{n_{ab}}} e^{-2 \Sigma_c n_c (b_{ac}^{sc} + b_{bc}^{sc})} . \qquad (3.103)$$

Thus we determined the ratio β/α of eq. (3.95), that is,

$$\beta/\alpha = K(n, T) . \tag{3.104}$$

The exponentials of (3.103) are sometimes referred to as the effective ionization potential, I^{eff}:

$$K(n, T) = 8\pi^{3/2}\lambda_{ab}^{3} e^{\beta_B I^{\text{eff}}} . \tag{3.105}$$

As a consequence of the interaction of the two particles forming a bound state with the surrounding plasma, the effective ionization energy may become zero at a certain density n_M for a given temperature T,

$$I^{\text{eff}}(n_M, T) = 0 . \tag{3.106}$$

The vanishing of the effective ionization energy at n_M is connected with a steep increase of the degree of dissociation

$$\alpha = n_a^{\text{free}}/n_a^{\text{total}}$$

and is referred to as a Mott transition.[25]

G. The Weak Coupling Approximation

The collision integrals of the kinetic equations (3.68) and (3.69) are essentially given by the scattering matrix $\langle|T|\rangle$. For this reason, the determination of such quantities from the Lippmann–Schwinger equations

$$T_{123}^{\kappa\kappa'} = V^{\kappa} + V^{\kappa} G_{123}^{\kappa'} T_{123}^{\kappa\kappa'} \tag{3.107}$$

and

$$T_{12} = V_{12} + V_{12} G_{12}^{0} T_{12} \tag{3.108}$$

turns out to be an essential task of the kinetic theory. Under the condition that the potentials remain small, that is, in the weak coupling situation, we may solve the Lippmann–Schwinger equations (3.107) and (3.80) by a perturbation procedure. In the first Born approximation we have

$$T_{123}^{\kappa\kappa'} = V^{\kappa} \quad \text{and} \quad T_{12} = V_{12} . \tag{3.109}$$

Let us first consider Eq. (3.68) for the distribution function of free particles. The contribution of the two-particle collision then gives the well-known Landau collision integral

$$I_2^{\mathrm{B}} = \frac{(2\pi\hbar)^3}{\hbar V} \sum_b \int dp_{b_2}\, d\bar{p}_{b_2}\, d\bar{p}_{a_1} \,|\langle p_{a_1} p_{b_2}| V_{ab} | \bar{p}_{b_2} \bar{p}_{a_1} \rangle|^2$$
$$\times 2\pi\delta(E - \bar{E})\{ f_a(\bar{p}_{a_1}) f_b(\bar{p}_{b_2}) - f_a(p_{a_1}) f_b(p_{b_2})\} . \quad (3.110)$$

As before, the three-particle contribution is split into two contributions, $[I_a]_1$ and $[I_a]_2$. Let us write $[I_a]_1$ in the first Born approximation; we get

$$[I_a]_1 = \frac{(2\pi\hbar)^3}{V\hbar} \frac{1}{2} \sum_{\alpha_{b_2c_3}\bar{\alpha}_{b_2c_3}} \sum_{bc} \int dP_{b_2c_3}\, d\bar{P}_{b_2c_3}$$
$$\times d\bar{p}_{a_1}\, 2\pi\delta(E - \bar{E}) |\langle p_{a_1} P_{b_2c_3} \alpha_{b_2c_3} | V^{\kappa_1 0}_{abc} | \bar{\alpha}_{b_2c_3} \bar{P}_{b_2c_3} \bar{p}_{a_1} \rangle|^2$$
$$\times \{ f_a(\bar{p}_{a_1}) f_{\bar{\alpha}_{b_2c_3}}(\bar{P}_{b_2c_3}) - f_a(p_{a_1}) f_{\alpha_{b_2c_3}}(P_{b_2c_3})\} . \quad (3.111)$$

Here we have, for example,

$$f_{\bar{\alpha}_{b_2c_3}}(\bar{P}_{b_2c_3}) = \begin{cases} f_{\bar{n}_{b_2c_3}}(\bar{P}_{b_2c_3}) , & \bar{\alpha}_{bc} = \bar{n}_{bc} , \\ f_b(\bar{p}_{b_2}) f_c(\bar{p}_{c_3}) , & \bar{\alpha}_{bc} = \bar{p}_{bc} , \end{cases}$$

where p_{bc} is the momentum of relative motion. It is convenient to express the matrix elements of the potentials $V^{\kappa_1 0}$ in terms of two particle eigenfunctions. We consider, for example, $\langle | T^{10}_{abc} | \rangle$. The Born approximation gives $\langle | V_{a_1b_2} + V_{a_1c_3} + V_{b_2c_3} | \rangle$. The last contribution gives in this case a vanishing contribution, and we have

$$\langle | T^{10}_{abc} | \rangle \approx \langle p_{a_1} P_{b_2c_3} n_{b_2c_3} | V_{a_1b_2} + V_{a_1c_3} | \bar{n}_{b_2c_3} P_{b_2c_3} p_{a_1} \rangle .$$

We introduce coordinates according to

$$R = (m_b r_2 + m_c r_3)/M ; \quad r = r_3 - r_2 ; \quad M \equiv m_{(bc)} = m_b + m_c .$$

Then we have

$$V_{ab}(r_1 - r_2) + V_{ac}(r_1 - r_3) = V_{ab}\left(r_1 - R + \frac{m_c}{M} r\right) + V_{ac}\left(r_1 - R - \frac{m_b}{M} r\right) .$$

In the coordinate representation we get

$$\langle p_{a_1} P_{b_2c_3} n_{b_2c_3} | \left\{ V_{ab}\left(r_1 - R + \frac{m_c}{M} r\right) + V_{ac}\left(r_1 - R - \frac{m_b}{M} r\right)\right\} | \bar{n}_{b_2c_3} \bar{P}_{b_2c_3} \bar{p}_{a_1} \rangle$$
$$= \int dr_1\, dr\, dR\, \psi_{\bar{n}_{b_2c_3}}(r) \psi^*_{n_{b_2c_3}}(r) \exp\{iR(P_{b_2c_3} - P_{b_2c_3})\}$$
$$\times \frac{1}{(2\pi\hbar)^6} \exp\{ir_1(\bar{p}_{a_1} - p_{a_1})\}$$
$$\times \left\{ V_{ab}\left(r_1 - R + \frac{m_c}{M} r\right) + V_{ac}\left(r_1 - R - \frac{m_b}{M} r\right)\right\} \tag{3.112}$$

Now we perform different substitutions; in the first right-hand side contribution we put $\hat{R} = r_1 - R + m_c r/M$; while for the second one we take $\hat{R} = r_1 - R - m_b r/M$. Then we arrive at the more simplified expression

$$\langle |T^{10}| \rangle \approx \frac{1}{(2\pi\hbar)^3} \int dt\, \psi_{\bar{n}_{b_2c_3}}(r) \psi^*_{n_{b_2c_3}}(r)$$
$$\times \{V_{ab}(k) e^{im_c rk/M} + V_{ac}(k) e^{-im_b rk/M}\} \delta(\bar{p}_{a_1} - p_{a_1} + k)\,, \tag{3.113}$$

where we applied $k = \bar{P}_{b_2c_3} - P_{b_2c_3}$ and

$$V_{ab}(k) = \int dr\, V_{ab}(r) e^{ikr}\,. \tag{3.114}$$

Expression (3.113) must be applied in the collision integral (3.71). An important example is the Coulomb problem. In this case the bound states must be interpreted as atoms. For systems with Coulomb interaction we get from (3.113) an expression that has a clear physical meaning:

$$\langle |T^{10}| \rangle \approx \frac{1}{\epsilon k^2}\, e_a P_{\bar{n}_{b_2c_3} n_{b_2c_3}}(k) \delta(\bar{p}_{a_1} - p_{a_1} + k)\,, \tag{3.115}$$

where we used

$$P_{\bar{n}_{b_2c_3} n_{b_2c_3}}(k) = \int dr\, \psi_{\bar{n}_{b_2c_3}}(r) \psi^*_{n_{b_2c_3}}(r) \{e_b e^{im_c kr/M} + e_c e^{-im_b kr/M}\}\,. \tag{3.116}$$

The quantity (3.115) may be interpreted as effective interaction between the free particle a and the bound pair (bc). The expression (3.116) characterizes the static charge distribution in the atom (bc).

For large distances between the free particles and the atoms, or, equivalently $kr < 1$, we may expand the exponentials in (3.116). The result is, with $e_b = -e_c = e$,

$$P_{\bar{n}_{b_2c_3}n_{b_2c_3}}(k) = \int dr\, \psi_{\bar{n}_{b_2c_3}}(r)\psi^*_{n_{b_2c_3}}(r) ik\left\{e_b \frac{m_c}{M} r - e_c \frac{m_b}{M} r\right\}.$$

Using the dipole moment of the atom

$$\mathbf{m}_{bc} = \int dr\, \psi_{\bar{n}_{b_2c_3}}(r)\psi_{n_{b_2c_3}}(r) e\mathbf{r}\,, \tag{3.117}$$

we may write

$$\langle|I^{10}|\rangle \approx i \frac{\mathbf{k}}{\epsilon_0 k^2} e_a \mathbf{m}_{bc} \delta(\bar{p}_{a_1} - p_{a_1} + k)\,, \tag{3.118}$$

the latter expression represents the charge–dipole interaction of the free particle and the bound pair.

For system with Coulomb interaction, we get thus the following expression for $[I_a]_1$ [Eq. (3.111)] for the electrons:

$$\begin{aligned}[I_e]_1 = \sum_{b,c=e,p} \sum_{\bar{\alpha}_{b_2c_3}\alpha_{b_2c_3}} \int \frac{1}{(2\pi\hbar)^3}\, dP_{b_2c_3}\, d\bar{P}_{b_2c_3}\, d\bar{p}_e \\ \times \left|\frac{e}{\epsilon_0 k^2} P_{\alpha_{b_2c_3}\bar{\alpha}_{b_2c_3}}(k)\right|^2 2\pi\delta(E_{e(bc)} - \bar{E}_{e(bc)})\delta(\bar{p}_e - p_e + k) \\ \times \tfrac{1}{2}\{f_e(\bar{p}_e)f_{\bar{\alpha}}(\bar{P}_{b_2c_3}) - f_e(p_e)f_{\alpha_{b_2c_3}}(P_{\alpha_{b_2c_3}})\}\,.\end{aligned} \tag{3.119}$$

In order to complete the kinetic equation for f_e, we still have to consider $[I_e]_2$. As we have seen already, this term is of special importance because it does not conserve the particle number and thus determines the rate coefficients α, β. In the case of Coulomb potentials, the latter correspond to the ionization and recombination processes. Neglecting exchange processes, we get for Coulomb systems

$$\begin{aligned}[I_e]_2 = \sum_{\bar{n}_{ep_2}} \int d\bar{P}_{ep_2}\, d\bar{p}_p\, dp_{p_1}\, dp_{p_2} \left|\frac{e}{\epsilon_0 k^2} P_{p_e p_2 \bar{n}_{ep_2}}(k)\right|^2 \\ \times \delta(\bar{p}_p - p_{p_2} + k) 2\pi\delta(E_{epp} - \bar{E}_{(ep)p}) \\ \times \{f_p(\bar{p}_p)f_{\bar{n}_{ep_2}}(\bar{P}_{ep_2}) - f_e(p_e)f_p(p_{p_1})f_p(p_{p_2})\}\,.\end{aligned} \tag{3.120}$$

In a similar way we have to deal with the kinetic equation for the atoms, that is, Eq. (3.69).

The summation over the species in (3.67) yields in the case that only two bound pairs (e, p) are taken into account for the relevant matrix element of the interaction:

$$T^{\text{atom}} \equiv \mathcal{V}_{(ep)(ep)} \equiv \langle n_{e_1p_1}P_{p_1p_1}n_{e_2p_2}P_{e_2p_2}|V_{ee}(r_{e_1} - r_{e_2}) + V_{ep}(r_{e_1} - r_{p_2}) + V_{pe}(r_{p_1} - r_{e_2}) + V_{pp}(r_{p_1} - r_{p_2})|\bar{n}_{e_1p_1}\bar{P}_{e_1p_1}\bar{n}_{e_2p_2}\bar{P}_{e_2p_2}\rangle \,. \tag{3.121}$$

We use abbreviations and introduce new variables:

$$\begin{aligned} r_{e_1} - r_{p_1} &= r_1 \,, \\ r_{e_2} - r_{p_2} &= r_2 \,, \\ (r_{e_1}m_e + r_{p_1}m_p)/M &= R_1 \,, \\ (r_{e_2}m_e + r_{p_2}m_p)/M &= R_2 \,, \\ m_e + m_p &= M \,. \end{aligned}$$

Instead of (3.121) we write

$$\begin{aligned} \langle n_1P_1n_2P_2|V_{ee}\Big(R_1 - R_2 + \frac{m_p}{M}(r_1 - r_2)\Big) &+ V_{ep}\Big(R_1 - R_2 + \frac{m_p}{M}r_1 + \frac{m_e}{M}r_2\Big) + V_{pe}\Big(R_1 - R_2 - \frac{m_e}{M}r_1 - \frac{m_p}{M}r_2\Big) \\ &+ V_{pp}\Big(R_1 - R_2 - \frac{m_e}{M}(r_1 - r_2)\Big)|\bar{n}_1\bar{P}_1\bar{n}_2\bar{P}_2\rangle \,. \end{aligned} \tag{3.122}$$

Now we apply the eigenfunction of the isolated two-particle problem, and we get, in the coordinate representation,

$$\begin{aligned} \mathcal{V}_{(ep)(ep)} = \int dr_1\, dr_2\, dR_1\, dR_2\, &\psi^*_{n_1}(r_1)\psi^*_{n_2}(r_2)\psi_{\bar{n}_1}(r_1)\psi_{\bar{n}_2}(r_2) \\ &\times \frac{1}{(2\pi\hbar)^6}\, e^{-i(P_2-\bar{P}_2)R_2}e^{-i(P_1-\bar{P}_1)R_1} \\ &\times \Big\{V_{ee}\Big(R_1 - R_2 + \frac{m_p}{M}(r_1 - r_2)\Big) + \cdots + V_{pp}(\cdots)\Big\} \,. \end{aligned} \tag{3.123}$$

In the different summands of (3.123) we perform different substitutions;

for example, in the first one we write

$$R_1 - R_2 + \frac{m_b}{M}(r_1 - r_2) = \bar{R}_1 .$$

Then we use the Fourier transform of the potential and introduce the polarizations according to (3.116):

$$P^{ep}_{\bar{n}_1 n_1}(P_1 - \bar{P}_1) = \int dr_1\, \psi_{\bar{n}_1}(r_1)\psi^*_{n_1}(r_1) \times \{e_e e^{i(P_1 - \bar{P}_1)m_p r_1/M} + e_p e^{-i(P_1-\bar{P}_1)m_e r_1/M}\} . \qquad (3.124)$$

With the abbreviation $P_1 - \bar{P}_1 = k$, we get then the Coulomb potential

$$T^{\text{atom}} \equiv \mathcal{V}_{(ep)(ep)} = \frac{1}{\epsilon_0 k^2} P^{ep}_{n_1 n_2}(k) P^{ep^*}_{\bar{n}_2 n_2}(k) \frac{1}{(2\pi\hbar)^3} \delta(P_2 - \bar{P}_2 - k) . \quad (3.125)$$

As before, using dipole moments we may write approximately for the diagonal elements of the polarization

$$p^{ep}_{n_1 n_1}(k) = i\mathbf{m}_{ep} \cdot \mathbf{k} .$$

Using (3.125) we may then write for the collision integral I_4 of the atom kinetic equation (3.69), (3.67) with (3.125)

$$\frac{\partial f_{n_{e_1 p_1}}(P_{e_1 p_1})}{\partial t} = I_{4, n_{e_1 p_1}}(P_{e_1 p_1}) ,$$

$$I_4 = \frac{1}{V\hbar} \frac{V}{(2\pi\hbar)^3} \sum_{\bar{n}_1 \bar{n}_2 n_2} \int d\bar{P}_2\, d\bar{P}_1\, dP_2 \left| \frac{1}{\epsilon_0 k^2} P^{ep}_{\bar{n}_1 n_1}(k) P^{ep^*}_{\bar{n}_e n_e}(k) \right|^2 \frac{1}{(2\pi\hbar)^3}$$
$$\times \delta(P_2 - \bar{P}_2 - k) 2\pi\delta(E_{n_1 n_2 P_1 P_2} - E_{\bar{n}_1 \bar{n}_2 \bar{P}_1 \bar{P}_2})$$
$$\times \{f_{\bar{n}_1}(\bar{P}) f_{\bar{n}_2}(\bar{P}_2) - f_{n_1}(P_1) f_{n_2}(P_2)\} . \qquad (3.126)$$

Approximations of type (3.126) will be considered in the next section.

In principle, such approximations may serve as a basis of the description of partially ionized plasmas, if we have to take into account ionization and recombination. However, because of the long range of Coulomb interaction, the Landau collision integral (3.110) and such integrals of type (3.119) are divergent. Such divergencies may be avoided by an appropriate screening. The simplest way to do this is to replace the

Coulomb potential by a dynamically screened potential,

$$V(k) \rightarrow \frac{V(k)}{\epsilon(k, \omega)}$$

where $\epsilon(k, \omega)$ is the dielectric function, which is in the simplest case

$$\epsilon(k, \omega) = \int \frac{dq}{(2\pi\hbar)^3} \frac{f_e(k-q) - f_e(k+g)}{\omega + i\epsilon + E(k-g) - E(k+g)} .$$

In the static approximation we have the Debye screening

$$\epsilon(k, 0) = \frac{k^2}{\kappa_D^2 + k^2} , \quad \kappa_D^2 = \sum_c \frac{n_c e_c^2}{\epsilon_0 k_B T} \quad \text{(Debye length).}$$

However, a justification of the screening can only be found by a consequent derivation of kinetic equations for plasmas. In the subsequent sections we will deal with kinetic equations of charged particles and with that of bound state between charged particles.

IV. KINETIC EQUATIONS FOR PLASMAS

Let us now consider the problem of bound states in plasmas. The interaction between the plasma particles is given by the Coulomb force. A characteristic feature of this interaction is its long range. Therefore, Coulomb systems show a collective behavior, so we can observe, for instance, the dynamical screening of the Coulomb potential and plasma oscillations.

Obviously we may expect that the simple two- and three-particle collision approximation discussed in the previous sections is not appropriate, because a large number of particles always interact simultaneously. Formally this approximation leads to divergencies. In the previous sections we used in a systematic way cluster expansions for the two- and three-particle density operator in order to include two-particle bound states and their relevant interaction in three- and four-particle clusters. In the framework of that consideration we started with the elementary particles (e, p) and their interactions. The bound states turned out to be special states, and, especially, scattering states were dealt with in a consistent manner.

In order to take into account bound states in the kinetic theory of plasmas, we will start now with another concept that is more convenient for plasma problems. That is the concept of "composite particles," which was developed by Klimontovich,[12,18,28] Girardeau,[29] and Brittin and Sakakura.[30]

In a partially ionized gas there are two limiting situations, the state with zero degree of ionization, that is, a neutral gas, and the state of a fully ionized plasma. As a starting point we take the first state in which all particles are bound. We wish to find a suitable description of such a system of interacting composite particles (atoms) starting with the basic properties of the interacting elementary particles (e, p).

A. Pair Creation and Annihilation Operators

Let us consider a system of H atoms the elementary particles of which are electrons (e) and protons (p). We assume that we have solved the two-particle problem

$$H_{ab}|\alpha_{ab}P_{ab}\rangle = E_{\alpha_{ab}P_{ab}}|\alpha_{ab}P_{ab}\rangle \,. \tag{4.1}$$

The eigenvectors given by this equation are complete,

$$\sum_{\alpha} \int dP|\alpha P\rangle\langle P\alpha| = 1 \,, \tag{4.2}$$

and orthonormal

$$\langle P\alpha \mid \bar{\alpha}\bar{P}\rangle = \delta_{\alpha\bar{\alpha}}\delta(P - \bar{P}) \,. \tag{4.3}$$

Here

$$P \equiv P_{ab} = p_a + p_b \,,$$

and

$$\alpha \equiv \alpha_{ab} = \begin{cases} p \,, & \text{for scattering states,} \\ n \,, & \text{for bound states.} \end{cases}$$

The eigenvalues are

$$E_{\alpha P} = E_{\alpha} + \frac{P^2}{2M} \,, \qquad M = m_a + m_b \,.$$

Let us consider especially the bound state $|nP\rangle$. Then we define the creation operator of a bound state pair by

$$A^{+}_{\alpha P}|0\rangle = |\alpha P\rangle \,, \tag{4.4}$$

where $|0\rangle$ is the vacuum state. First, we consider how $A^+_{\alpha P}$ and the elementary particles' creation and annihilation operators $\psi^+(x)$, $\psi(x)$ are related. For this purpose let us write

$$A^+_{\alpha P}|0\rangle = \int dx_1\, dx_2\, |x_1x_2\rangle\langle x_2x_1|\alpha P\rangle$$
$$= \int dx_1\, dx_2\, \psi^+(x_1)\psi^+(x_2)|0\rangle\langle x_2x_1|\alpha P\rangle\,. \tag{4.5}$$

That means the pair creation operator $A^+_{\alpha P}$ is expressed linearly in terms of a product of elementary particles' operators

$$A^+_{\alpha P} = \int dx_1\, dx_2\, \psi^+(x_1)\psi^+(x_2)\langle x_2x_1|\alpha P\rangle\,. \tag{4.6}$$

In this section, we apply the eigenfunctions with another normalization. First, we introduce the relative and center-of-mass coordinates

$$r = r_\alpha - r_b\,; \qquad R = \frac{m_a r_a + m_b r_b}{m_a + m_b}\,. \tag{4.7}$$

Then we define

$$\langle r_a r_b|\alpha P\rangle = (2\pi\hbar)^{-3/2}V^{-1/2}\psi_{\alpha P}(Rr)$$
$$= (2\pi\hbar)^{-3/2}V^{-1/2}\psi_\alpha(r)\psi_P(R) \tag{4.8}$$

with the normalization

$$\int \psi^*_P(R)\psi_{\bar P}(R)\,\frac{dR}{V} = \frac{(2\pi\hbar)^3}{V}\,\delta(P-\bar P)\,,$$
$$\int \psi^*_\alpha(r)\psi_{\bar\alpha}(r)\,\frac{dr}{V} = \begin{cases} \delta_{n\bar n}\,, & \alpha = n,\, \bar\alpha = \bar n\,; \\ \dfrac{(2\pi\hbar)^3}{V}\,\delta(p-\bar p)\,, & \alpha = p,\, \bar\alpha = \bar p\,. \end{cases} \tag{4.9}$$

Now the expansion (4.6) can be written as

$$A^+_{\alpha P} = \int \frac{dr\, dR}{V^{1/2}}\,\frac{1}{(2\pi\hbar)^{3/2}}\,\psi_{\alpha P}(rR)C^+(rR) \tag{4.10}$$

with the creation operator

$$C^+(r_1r_2) = \psi^+(r_1)\psi^+(r_2)\,. \tag{4.11}$$

The inverse relation is

$$C^+(r_1r_2) = V^{-1/2}(2\pi\hbar)^{-3} \sum_{\alpha P} \psi^*_{\alpha P}(r_1r_2)A^+_{\alpha P} . \tag{4.12}$$

The pair operators do not satisfy simple commutation relations. Using the commutation relation for the Fermi field operators $\psi^+(x)$, $\psi(x)$ we find the following relation for $C_{ab}(x_1x_2)$:

$$\begin{aligned}[C^+_{ab}(12), C_{cd}(1'2')] = {} & \psi^+_a(1)\psi_d(2')\delta_{bc}\delta(1'2) - \psi^+_b(2)\psi_d(2')\delta_{ac}\delta(11') \\ & + \psi_c(1')\psi^+_a(1)\delta_{bd}\delta(22') - \psi_c(1')\psi^+_b(2)\delta_{ad}\delta(12') .\end{aligned} \tag{4.13}$$

Using the operators $A_{\alpha P}$ and $C(x_1x_2)$ we may introduce the particle number operators

$$A^+_{nP}A_{nP} = N_{nP} \quad \text{(number of atoms)} \tag{4.14}$$

for the bound pairs. This definition is important for the interpretation of the formalism to describe composite particles. Furthermore, we define

$$P_nC^+(x_1x_2)C(x_1x_2) = N(x_1x_2) \tag{4.15}$$

where P_n is the projection operator onto the bound-state part of C^+C.

With the relations (4.15) and (4.12) it is easy to find the connection between $N_{\alpha P}$ and $N(x_1x_2)$ by

$$N(rRrR') = \frac{1}{V(2\pi\hbar)^3} \sum_{\alpha\beta} \int dP\, d\bar{P}\; A^+_{\alpha P}A_{\beta\bar{P}}\psi^*_{\alpha p}(rR)\psi_{\beta\bar{P}}(r'R') . \tag{4.16}$$

Taking the diagonal elements we may obtain the connection between the particles' number operators. For this purpose let us now derive some relations that are important for further consideration. First, we introduce the operator of the density of pairs $\hat{\rho}_{ab}$ by

$$C^+_{ab}(r_1r_2)C_{ab}(r'_1r'_2) = \frac{N_{ab}}{V^2}\, \hat{\rho}_{ab}(r_1r_2r'_1r'_2) , \tag{4.17}$$

where N_{ab} is the number of bound pairs. Using the commutation relations between the operators ψ^+ and ψ it is easy to show

$$N_a(N_b - \delta_{ab}) = \int N_{ab}(r_1r_2r_1r_2)\, dr_1\, dr_2 = \sum_{\alpha P} A^+_{\alpha P}A_{\alpha p} = N_{ab} . \tag{4.18}$$

Then we find the connection between $\hat{\rho}_{ab}$ and $\hat{\rho}_a$ to be

$$\int \hat{\rho}_{ab}(r_1 r_2 r_1' r_2) \frac{dr_2}{V} = \hat{\rho}_b(r_1 r_1') , \tag{4.19}$$

where $\hat{\rho}_b$ is given by $\psi_b^+ \psi_b = (N_e/V)\hat{\rho}_b$; $N_e = N_{ab}$ for the pure atomic system. Finally we introduce the operator of pair density by

$$A^+_{\alpha P} A_{\bar{\alpha}\bar{P}} = \frac{N_{ab} V}{(2\pi\hbar)^3} \hat{\rho}_{\alpha\bar{\alpha}}(P\bar{P}) . \tag{4.20}$$

This operator $\hat{\rho}_{\alpha\bar{\alpha}}$ is related to $\hat{\rho}_{ab}$ by

$$\hat{\rho}_{ab}(R'r'Rrt) = \frac{V^2}{(2\pi\hbar)^6} \sum_{\alpha\beta} \int dP\, d\bar{P}\, \hat{\rho}_{\alpha\beta}(P\bar{P}) \psi^*_{\alpha p}(r'R') \psi_{\beta\bar{p}}(rR) . \tag{4.21}$$

The inverse relation is

$$\hat{\rho}_{\alpha\beta}(P\bar{P}) = \int \hat{\rho}_{ab}(rRr'R') \psi^*_{\beta\bar{p}}(r'R') \psi_{\alpha p}(rR) \frac{dr\, dr'\, dR\, dR'}{V^4} . \tag{4.22}$$

The normalization follows from (4.18) and (4.20):

$$\frac{V}{(2\pi\hbar)^3} \sum_{\alpha} \int dP\, \rho_{\alpha\alpha}(PP) = 1 . \tag{4.23}$$

Now we will introduce the distribution function of the atoms. Obviously, this function is related to the average of $\hat{\rho}_{\alpha\beta}(P\bar{P})$. In the spatially homogeneous system this quantity is diagonal (in the binary collision approximation)

$$\langle \hat{\rho}_{\alpha\beta}(P\bar{P}) \rangle = \delta_{\alpha\beta} \frac{(2\pi\hbar)^3}{V} \delta(P - \bar{P}) f_\alpha(Pt) . \tag{4.24}$$

Thus we introduced $f_\alpha(Pt)$ as distribution function of the atoms. The normalization of f_α is given by

$$\frac{V}{(2\pi\hbar)^3} \sum_{\alpha} \int dP\, f_\alpha(Pt) = 1 . \tag{4.25}$$

B. The Equations of Motion

For further consideration it is necessary to know the equations of motion for the pair operators C^+_{ab} and C_{ab}. From the equation of motion for the field operators ψ^+ and ψ and the definition of C_{ab}, we find

$$i\hbar \frac{d}{dt} C_{ab}(r_{a_1} r_{b_1} t) = H_{ab}(r_{a_1} r_{b_1}) C_{ab}(r_{a_1} r_{b_1} t) - \sum_c \int dr_{c_2} \{V_{ac}(r_{a_1} - r_{c_2}) + V_{bc}(r_{b_1} - r_{c_2})\} \times \frac{N}{V} \hat{\rho}_c(r_{c_2}) C_{ab}(r_{a_1} r_{b_1} t) . \tag{4.26}$$

We consider especially $a = e$, $b = p$; furthermore, we carry out the sum over c:

$$i\hbar \frac{d}{dt} C_{ep}(r_{e_1} r_{p_2} t) = H_{ep}(r_{e_1} r_{b_2}) C_{ep}(r_{e_1} r_{p_2}) - \frac{N_e}{V} \int dr_{e_3} [V_{ee}(r_{e_1} - r_{e_3}) + V_{pe}(r_{p_1} - r_{e_3})] \hat{\rho}_e(r_{e_3}) + \int dr_{p_3} [V_{ep}(r_{e_1} - r_{p_3}) + V_{pp}(r_{p_1} - r_{p_3})] \hat{\rho}_p(r_{p_3})\} C_{ep}(r_{e_1} r_{p_2}) . \tag{4.27}$$

Because of the interaction between the particles on the right-hand side, the pair operators C and the operator of the single particle density $\hat{\rho}_a$ occur. Using relation (4.19) we find

$$\begin{aligned} \hat{\rho}_e(r_{e_3}) &= \int \hat{\rho}_{ep}(r_{e_3} r_{p_4} r_{e_3} r_{p_4}) \frac{dr_4}{V} , \\ \hat{\rho}_p(r_{p_3}) &= \int \hat{\rho}_{pe}(r_{p_3} r_{e_4} r_{p_3} r_{e_4}) \frac{dr_4}{V} . \end{aligned} \tag{4.28}$$

With these equations and Eq. (4.17) an equation for $\hat{\rho}_{ep}$ may be obtained:

$$\begin{aligned} i\hbar \frac{d}{dt} \hat{\rho}_{ep}(r_{e_1} r_{p_1} r'_{e_1} r'_{p_1}) &= (H_{ep} - H'_{ep}) \hat{\rho}_{ep}(r_{e_1} r_{p_1} r'_{e_1} r'_{p_1}) \\ &\quad - \frac{N_e}{V} \int dr_3\, dr_4 [U_{epep}(r_{e_1} r_{e_3} r_{p_1} r_{p_3}) \hat{\rho}_{ep}(r_{e_3} r_{p_3} r_{e_3} r_{p_3}) \hat{\rho}_{ep}(r_{e_1} r_{p_1} r'_{e_1} r'_{p_1}) \\ &\quad - \hat{\rho}_{ep}(r'_{e_1} r'_{p_1} r_{e_1} r_{p_1}) \hat{\rho}_{ep}(r_{e_3} r_{p_3} r_{e_3} r_{p_3}) U_{epep}(r'_{e_1} r_{e_3} r'_{p_1} r_{p_3})] \end{aligned} \tag{4.29}$$

with

$$U_{epep}(r_{e_1} r_{p_1} r_{e_3} r_{p_3}) = V_{ee}(r_{e_1} - r_{e_3}) + V_{pe}(r_{p_1} - r_{e_3}) + V_{ep}(r_{e_1} - r_{p_3}) + V_{pp}(r_{p_1} - r_{p_3}) . \tag{4.30}$$

Instead of r_e, r_p, it is more convenient to use the variables

$$r_e = R + \frac{m_p}{M} r\,; \qquad r_p = R - \frac{m_e}{M} r$$
$$r = r_e - r_p\,; \qquad R = \frac{m_e r_e + m_p r_p}{m_e + m_p}\,. \tag{4.31}$$

Furthermore, let us introduce the function

$$U_{ep}(rR) = \int \frac{d\bar{R}\, d\bar{r}}{V}\, U_{epep}(rR\bar{r}\bar{R})\rho_{ep}(\bar{r}\bar{R}\bar{r}\bar{R})\,. \tag{4.32}$$

This function is the potential energy of the particles in the pairs due to the field of the surrounding particles. Then the equation for the operator of the density takes the form

$$i\hbar \frac{d}{dt}\,\hat{\rho}_{ep} = \{H_{ep}(rR) - H_{ep}(r'R')\}\,\hat{\rho}_{ep}(rRr'R')$$
$$- \frac{N_e}{V}\,\{U_{ep}(rR)\hat{\rho}_{ep}(rRr'R') - \hat{\rho}_{ep}(r'R'rR)U_{ep}(r'R')\}\,. \tag{4.33}$$

In order to derive an equation for the distribution function of the atom, we use the representation of Eq. (4.33) in terms of eigenfunction $\psi_{\alpha p}$ of H_{ep}. Using the expansion (4.22), we find the following equation:

$$i\hbar \frac{\partial}{\partial t}\,\hat{\rho}_{\alpha\beta}(PP't) = (E_{\alpha P} - E_{\beta P'})\hat{\rho}_{\alpha\beta}(PP't)$$
$$+ \frac{N_e}{V}\,\frac{V^2}{(2\pi\hbar)^6} \sum_{\gamma} \int d\bar{P}\,[U_{\alpha j}(p\bar{p}t)\hat{\rho}_{\gamma\beta}(\bar{p}p') - \hat{\rho}_{\gamma\beta}(P\bar{P}t)U_{\alpha\gamma}(\bar{P}P't)]\,, \tag{4.34}$$

with

$$U_{\alpha\beta}(P\bar{P}t) = \int \frac{dr\, dR}{V^2}\, \psi_\alpha^*(r)\psi_P^*(R)U_{ei}(rR)\psi_\beta(r)\psi_{\bar{P}}(R)\,. \tag{4.35}$$

Using (4.32) and expansion (4.21), this expression may be written in the form

$$U_{\alpha\beta}(PP't) = \frac{V}{2\pi\hbar)^3} \int dk\, d\bar{P}\, d\bar{P}' \sum_{\bar{\alpha}\bar{\beta}} P_{\alpha\beta}(k)P_{\bar{\alpha}\bar{\beta}}(k)$$
$$\times \frac{4\pi}{k^2}\,\delta\!\left(k - \frac{P - P'}{\hbar}\right)\delta(P - P' - \bar{p} + \bar{p}')\hat{\rho}_{\bar{\alpha}\bar{\beta}}(\bar{P}\bar{P}'t)\,, \tag{4.36}$$

where

$$P_{\alpha\beta}(k) = \int \frac{dr}{V} [e_e e^{im_p r/M} + e_p e^{-im_e r/M}] \psi^*_\alpha(r) \psi_\beta(r) .$$

For the physical interpretation of Eqs. (4.34) and (4.36) let us write the following expression:

$$\frac{4\pi}{|P - P'|^2} P_{\alpha\beta}(P - P') P^*_{\bar{\alpha}\bar{\beta}}(P - p') \delta(P - P' - \bar{P} + \bar{P}') . \quad (4.37)$$

The quantity (4.37) has a clear meaning, it represents interaction between the charge distributions of the electron shells of the two atoms. In a large-distance approximation, we have

$$|P_{\alpha\beta}(k)|^2 \rightarrow |\mathbf{k} \cdot \mathbf{d}_{\alpha\beta}|^2 .$$

This means Eq. (4.38) is the interaction between the atoms in dipole approximation. Thus, (4.37) turns out to be the average potential of a single atom in the field of the other atoms. In this sense, Eqs. (4.34) and (4.36) represent a system for the description of composites (atoms with internal degrees of freedom) and their interaction.

Now we are ready to write a kinetic equation for the atom distribution function $f_\alpha(Pt)$. For this reason we introduce the deviations of the mean values (fluctuations) by

$$\delta\rho_{\alpha\beta} = \hat{\rho}_{\alpha\beta} - \langle \hat{\rho}_{\alpha\beta} \rangle ; \qquad \delta U_{\alpha\beta} \equiv U_{\alpha\beta} .$$

Then we may obtain the kinetic equation for the distribution function if in (4.34) we take $\alpha = \beta$, $P = P'$. Furthermore, we take the average of that equation and use the definition of $f_\alpha(Pt)$ [(4.24)] and take into account $\delta(P = 0) = V/(2\pi\hbar)^3$. Then we obtain the equation

$$\frac{\partial f_\alpha(Pt)}{\partial t} = -\frac{2N_e}{V} \frac{V}{(2\pi\hbar)^3 \hbar} \sum_\gamma \int \mathrm{Im} \langle \delta\rho_{\alpha\gamma}(PP't) \delta U_{\gamma\alpha}(p'pt) \rangle \, dP' . \quad (4.38)$$

This equation is essentially determined by $\langle \delta\rho_{\alpha\beta} \delta\rho_{\gamma\delta} \rangle$, which is related to $\langle \delta\hat{N}(12) \delta\hat{N}(1'2') \rangle$ according to (4.15) and (4.17). Therefore, it is useful to study the relation between the density fluctuations and the density matrix

$$\langle C^+(r_1r_2) C^+(r_3r_4) C(r_3'r_4') C(r_1'r_2') \rangle = \frac{N_{ab}(N_{ab} - \delta_{ab})}{V^4} \rho_{abab}(r_1r_2r_3r_4r_1'r_2'r_3'r_4't) . \quad (4.39)$$

Using the Bose commutation rules (which means neglecting the deviation from the Bose character of the pair operators), we may derive

$$\langle C^{+}_{ep}(12)C_{ep}(1'2')C^{+}_{ep}(34)C_{ep}(3'4')\rangle$$
$$= \frac{N_{ep}(N_{ep}-1)}{V^4}\,\rho_{epep}(12341'2'3'4') + \delta(1'3)\delta(2'4)\,\frac{N_{ep}}{V^2}\,\rho_{ep}(123'4')\,.$$

From this equation we get for $\langle \delta\rho\delta\rho \rangle$ the following equation:

$$\frac{N^2_{ep}}{V^4}\,\langle \delta\rho_{ep}(121'2')\delta\rho_{ep}(343'4')\rangle$$
$$= \frac{N_{ep}(N_{ep}-1)}{V^4}\big(\langle \hat{\rho}_{epep}(12341'2'3'4')\rangle - \langle \hat{\rho}_{ep}(121'2')\rangle\langle \hat{\rho}_{ep}(343'4')\rangle\big)$$
$$+ \frac{N^2_{ep}}{V^4}\,\langle \delta\rho_{ep}(121'2')\delta\rho_{ep}(343'4')\rangle^{\text{source}}\,,$$

where

$$\langle \delta\rho_{ep}(121'2')\delta\rho_{ep}(343'4')\rangle^{\text{source}}$$
$$= \frac{V^2}{N_{ep}}\,\delta(31')\delta(2'4)\langle \hat{\rho}_{ep}(123'4')\rangle - \frac{1}{N_{ep}}\,\langle \hat{\rho}_{ep}(121'2')\rangle\langle \hat{\rho}_{ep}(343'4')\rangle\,. \tag{4.40}$$

The quantities in (4.40) are single time quantities. According to Eq. (4.38) we need the special correlation function $\langle \delta\rho_{\alpha\gamma}(PP't)\delta U_{\gamma\alpha}(p'pt)\rangle$; this function is closely connected to the correlation function $\langle \delta\rho\delta\rho \rangle$. According to the discussions presented in Refs. 12 and 28 for the determination of such correlation functions, one has to start from a differential equation for the corresponding two-time correlation functions and use the relevant single-time correlation function as initial condition.

From the equation for the fluctuations $\delta\rho$, which reads

$$\left[i\hbar\,\frac{\partial}{\partial t} - (E_\alpha + E_P - E_\beta - E_{P'})\right](\delta\rho_{\alpha\beta}(PP't) - \delta\rho^{\text{source}}_{\alpha\beta}(PP't)) = I^{\text{B}}_{\alpha\beta}(PP't)$$

($I^{\text{B}}_{\alpha\beta}$ = Boltzmann-type collision integral), we get for the two-time correlation function

$$\left[i\hbar\,\frac{\partial}{\partial t} - (E_\alpha + E_P - E_\beta - E_{P'})\right]\langle \delta\rho_{\alpha\beta}(PP't)\delta\rho^{*}_{\mu\nu}(\bar{P}\bar{\bar{P}}t')\rangle^{\text{source}} = 0\,. \tag{4.41}$$

Here higher-order moments were neglected.[12,28] For this purpose we give a representation of Eq. (4.40) in terms of atomic eigenfunctions similar to (4.34). Moreover, we neglect the second term of (4.41) and write a symmetrical form, we arrive at

$$\langle \delta\rho_{\alpha\beta}(PP't)\delta\rho^*_{\mu\nu}(\bar{P}\bar{\bar{P}}t)\rangle^{\text{source}}$$

$$= \frac{1}{2N_{\text{ep}}} \frac{(2\pi\hbar)^3}{V} [\delta(\bar{P} - P')\delta_{\mu\beta}\langle \hat{\rho}_{\alpha\nu}(P\bar{\bar{p}}t)\rangle + \delta(\bar{\bar{p}} - p)\delta_{\alpha\nu}\langle \hat{\rho}_{\mu\beta}(p'\bar{p}t)\rangle] .$$

With this quantity as initial condition, the solution of (4.41) reads

$$(\delta\rho\delta\rho)^{\text{source}}_{t,t'} = e^{-(i/\hbar)(E_\alpha - E_\beta + E_P - E_{P'})(t-t')}(\delta\rho\delta\rho)^{\text{source}}_{t=t'} .$$

The Fourier transform of $(\delta\rho\delta\rho)^{\text{source}}_{t,t'}$ is given by

$$(\delta\rho\delta\rho)^{\text{source}}_{\omega} = 2\pi\hbar\delta(\hbar\omega - E_\alpha + E_\beta - E_P + E_{P'})(\delta\rho\delta\rho)^{\text{source}}_{t=t'} . \tag{4.42}$$

This function is also referred to as the spectral density.

C. Polarization Approximation for the Atomic Gas—Collision Integral

In order to derive a kinetic equation for the limiting situation of a system of bound states only, let us start from Eq. (4.34), which we write in the form of a kinetic equation:

$$i\hbar \frac{\partial}{\partial t} \hat{\rho}_{\alpha\beta}(PP't) - (E_\alpha + E_P - E_\beta - E_{P'})\hat{\rho}_{\alpha\beta}(PP't) = I_{\alpha\beta}(PP't) . \tag{4.43}$$

The collision integral is given by

$$I_{\alpha\beta}(PP't) = \frac{N_e}{V} \frac{V}{(2\pi\hbar)^3} \sum_\gamma \int [U_{\alpha\gamma}(P\bar{P}t)\hat{\rho}_{\gamma\beta}(\bar{P}P't) - U_{\gamma\beta}(\bar{P}P't)\bar{\rho}_{\alpha\gamma}(p\bar{p}t)]\, d\bar{P} . \tag{4.44}$$

In (4.44), there occurs the product $U\hat{\rho}$, which must be determined. We will do this in the so-called polarization approximation, which takes into account the polarization of the medium. This approximation is known from the theory of the fully ionized plasma and was discussed in detail in Refs. 12 and 28. A least approximately, the same method may be applied to an atomic gas with internal degrees of freedom.

The polarization approximation (including Born approximation) corresponds to the replacement in (4.44)

$$U_{\alpha\gamma}(PP't)\rho_{\gamma\beta}(P'P''t) \Rightarrow U_{\alpha\gamma}(PP't)\langle \hat{\rho}_{\gamma\beta}(P'P''t)\rangle \tag{4.45}$$

and condensing certain higher-order terms of $I_{\alpha\beta}(PP't)$ into a source term $\delta\rho^{\text{source}}$. For the deviation of the mean value of $\hat{\rho}$, $\delta\rho = \hat{\rho} - \langle \hat{\rho} \rangle$, we get from (4.43), using the diagonality of $\langle \hat{\rho} \rangle$ [Eq. (4.24)]

$$\left[i\hbar \frac{\partial}{\partial t} - (E_\alpha + E_P - E_\beta - E_{P'})\right](\delta\rho_{\alpha\beta}(PP't) - \delta\rho^{\text{source}}_{\alpha\beta}(PP't)) = \frac{N_e}{V} U_{\alpha\beta}(PP't)[f_\beta(P't) - f_\alpha(Pt)] \,. \tag{4.46}$$

The Fourier transformation of (4.46), which is, however, not extended to the distribution functions f_α, f_β, gives

$$[\hbar\omega - (E_\alpha + E_P - E_\beta - E_{P'})][\delta\rho_{\alpha\beta}(PP'\omega) - \delta\rho^{\text{source}}_{\alpha\beta}(PP'\omega)] = \frac{N_e}{V}[f_\beta(P't) - f_\alpha(Pt)]U_{\alpha\beta}(PP'\omega) \,. \tag{4.47}$$

The mean value of the potential energy $U_{\alpha\beta}$ is connected with potential fluctuations $\delta\varphi$ by

$$N_e U_{\alpha\beta}(pp'\omega) = \int P^*_{\alpha\beta}(k)\delta(\hbar k - P + P')\, \delta\varphi(k\omega)\, d\hbar k \,, \tag{4.48}$$

where the fluctuation $\delta\varphi$ reads

$$\delta\varphi(k\omega) = N_e V(k) \sum_{\mu\nu} P_{\mu\nu}(k)\delta\rho_{\mu\nu}(P\bar{P}\omega) \times \delta(\hbar k - P + \bar{P}) \frac{V}{(2\pi\hbar)^3}\, dP\, d\bar{P} \,; \qquad V(k) = \frac{4\pi}{k^2} \,. \tag{4.49}$$

We may then determine $\delta\rho_{\alpha\beta}$ from (4.47)F:

$$\delta\rho_{\alpha\beta}(PP'\omega) = \delta\rho^{\text{source}}_{\alpha\beta}(PP'\omega) + \frac{1}{V}\int \delta\varphi\,(k'\omega) \times \frac{P^*_{\alpha\beta}(k')\delta(\hbar k' - P + P')(P't) - f_\alpha(Pt)]}{\hbar(\omega + i\Delta) - E_\alpha - E_P + E_\beta + E_{P'}}\, d\hbar k' \,. \tag{4.50}$$

Here we added an imaginary part $i\Delta$ to the frequency ω with $\Delta \rightarrow +0$. Without this quantity $i\Delta$, the integrand in (4.50) has a pole at the real axis. In order to give the integral a well-defined meaning, we must add a rule on how to circumvent this pole; this was done by $i\Delta$. Physically, the quantity $i\Delta$ has its origin in the adiabatic switching on of the external perturbation at $t \to -\infty$.

The next problem is the determination of the fluctuation of the potential $\delta\varphi$. The latter is connected with that of the electric field by

$$\delta \mathbf{E}(k\omega) = i\mathbf{k}\delta\varphi(k\omega) . \tag{4.51}$$

If we use (4.50) in (4.49), we get

$$\delta \mathbf{E}(k\omega) = N_e V(k) \sum_{\mu\nu} |P_{\mu\nu}(k)|^2 \int \frac{dP\, d\bar{P}\, d\hbar k}{(2\pi\hbar)^3} \delta \mathbf{E}(k'\omega)\delta(\hbar k - P + \bar{P}) \times \frac{d(\hbar k' - p + \bar{p})[f_\nu(\bar{p}t) - f_\mu(Pt)]}{\hbar(\omega + i\Delta) - E_\mu - E_P + E_\nu + E_{\bar{P}}} + \delta \mathbf{E}^{\text{source}}_{(k\omega)} . \tag{4.52}$$

Here we used the abbreviation

$$\delta \mathbf{E}^{\text{source}}(k\omega) = i\mathbf{k}N_e V(k) \sum_{\mu\nu} P_{\mu\nu}(k) \int \delta\rho^{\text{source}}_{\mu\nu}(P\bar{P}\omega) \times \delta(\hbar k + \bar{P} - P) \frac{V}{(2\pi\hbar)^3}\, dP\, d\bar{P} . \tag{4.53}$$

The electrical field fluctuation is related to that which is due to the source according to the equation

$$\delta \mathbf{E}^{\text{source}}(k\omega) = \epsilon(k\omega)\delta \mathbf{E}(k\omega) , \tag{4.54}$$

where $\epsilon(k\omega)$ has the meaning of a dielectric constant. The latter is then given with (4.52) by[12,18,31]

$$\epsilon(k\omega) = 1 - \frac{N_e}{V} V(k) \sum_{\mu\nu} \int \frac{V}{(2\pi\hbar)^3}\, dP\, d\bar{P} \times \frac{|P_{\mu\nu}(k)|^2 \delta(\hbar k - P + \bar{P})[f_\nu(\bar{P}t) - f_\mu(Pt)]}{\hbar(\omega + i\Delta) - E_\mu - E_P + E_\nu + E_{\bar{P}}} . \tag{4.55}$$

This is a very important result. By this expression we give a statistical derivation of the dielectric function of the atomic gas, which is determined by the properties of the atoms.

Let us now give two expressions that are of relevance for the collision integral of the kinetic equation of atoms. The first one is $(\delta\mathbf{E}\delta\mathbf{E})_{\omega k}$, the spectral density of field fluctuations. According to (4.54) we have

$$(\delta\mathbf{E}\delta\mathbf{E})_{\omega k} = \frac{1}{|\epsilon(k\omega)|^2}(\delta\mathbf{E}\delta\mathbf{E})^{\text{source}}_{\omega k}\,. \tag{4.56}$$

For this reason it is sufficient to determine $(\delta\mathbf{E}\delta\mathbf{E})^{\text{source}}_{\omega k}$. The latter quantity is obtained from (4.53). The result is

$$\begin{aligned}(\delta\mathbf{E}\delta\mathbf{E})^{\text{source}}_{\omega k} &= N_e^2\frac{(4\pi)^2}{k^2}\sum_{\mu\nu}P_{\mu\nu}(k)\sum_{\alpha\beta}P_{\alpha\beta}(k)\\ &\quad\times(\delta\rho_{\mu\nu}(\bar{P}\bar{\bar{P}})\delta\rho_{\alpha\beta}(P'P''))^{\text{source}}_{\omega}\,\delta(\hbar k-\bar{P}+\bar{\bar{P}})\\ &\quad\times\delta(\hbar k-P'+P'')\frac{V^2}{(2\pi\hbar)^3}\,dP'\,dP''\,d\bar{P}\,d\bar{\bar{P}}\,.\end{aligned} \tag{4.57}$$

We use the expression (4.42) for $(\delta\rho\delta\rho)^{\text{source}}$, which reads in a symmetrized form

$$\begin{aligned}\langle\delta\rho_{\alpha,\beta}(PP't)\delta\rho^*_{\mu\nu}(\bar{P}\bar{\bar{P}}t)\rangle^{\text{source}} &= \frac{1}{2N_{ei}}\frac{(2\pi\hbar)^3}{V}[\delta(\bar{P}-P')\delta_{\mu\beta}\langle\hat{\rho}_{\alpha\nu}(P\bar{\bar{P}}t)\rangle\\ &\quad+\delta(\bar{\bar{P}}-P)\delta_{\alpha\nu}\langle\hat{\rho}_{\mu\beta}(P'\bar{P}t)\rangle]\,.\end{aligned} \tag{4.58}$$

The corresponding Fourier transform reads, according to (4.41),

$$\begin{aligned}[\delta\rho_{\alpha\beta}(PP')\delta\rho^*_{\mu\nu}(\bar{P}\bar{\bar{P}})]^{\text{source}}_{\omega} &= 2\pi\hbar\delta(\hbar\omega-E_\alpha-E_P+E_\beta+E_{P'})\\ &\quad\times\langle\delta\rho_{\alpha\beta}(PP't)\delta\rho^*_{\mu\nu}(\bar{P}\bar{\bar{P}}t)\rangle^{\text{source}}\,.\end{aligned} \tag{4.59}$$

Using, further, the diagonality of $\langle\hat{\rho}\rangle$, we get instead of (4.57)

$$\begin{aligned}(\delta\mathbf{E}\delta\mathbf{E})^{\text{source}}_{\omega k} &= \frac{N_e^2}{N_{e_i}}\frac{(4\pi)^2\pi\hbar}{k^2}\frac{V}{(2\pi\hbar)^3}\int dP'\,dP''\\ &\quad\times\sum_{\alpha\beta}|P_{\alpha\beta}(k)|^2\delta(\hbar k-P'+P'')\\ &\quad\times\delta(\hbar\omega-E_\alpha-E_{P'}+E_\beta+E_{P''})[f_\alpha(P't)-f_\beta(P''t)]\,.\end{aligned} \tag{4.60}$$

The second quantity we need is the imaginary part of the dielectric function. The latter we get directly from (4.55) using the identity

$$\operatorname{Im}\left(\frac{1}{x + i\Delta}\right) = -\pi\delta(x) .$$

With (4.48)–(4.50) we may determine the essential quantity of the collision integral $[\delta\rho_{\alpha\beta}(P'P'')\delta U_{\alpha\beta}(P'P'')]_{\omega}$. Moreover, we have to take the imaginary part, and the collision integral reads

$$\begin{aligned} I_{\alpha}(P't) = {} & \frac{1}{(2\pi\hbar)^3} \sum_{\beta} \int d\omega\, dk\, dP'' \,\frac{|P_{\alpha\beta}(k)|^2}{k^2} \\ & \times \delta(\hbar k - P' + P'')\delta(\hbar\omega - E_{\alpha} - E_{P'} + E_{\beta} + E_{P''}) \\ & \times \left\{ \frac{1}{N_e} (\delta\mathbf{E}\delta\mathbf{E})_{\omega k}[f_{\beta}(P''t) - f_{\alpha}(P't)] \right. \\ & \left. - \frac{4\pi \operatorname{Im} \epsilon(k\omega)}{|\epsilon(k\omega)|^2} \frac{N_e}{N_{ei}} [f_{\beta}(P''t) + f_{\alpha}(P't)] \right\} . \end{aligned} \tag{4.61}$$

Using (4.55), (4.56) and (4.60), both terms of the right-hand side of (4.61) may be combined to give

$$\begin{aligned} I_{\alpha}(P't) = {} & \frac{4N_e^2}{N_{ei}} \sum_{\alpha_1\beta_1\beta} \int \frac{dP_1'\, dP_1''\, dP''}{(2\pi\hbar)^3}\, d\omega\, dk\, \frac{|p_{\alpha\beta}(k)^2| p_{\alpha_1\beta_1}(k)|^2}{k^4|\epsilon(k\omega)|^2} \\ & \times \delta(\hbar k - p' + p'')\delta(p' - p'' + p_1'' - p_1') \\ & \times \delta(\hbar\omega - E_{\alpha} - E_{p'} + E_{\beta} + E_{p''}) \\ & \times \delta(E_{\alpha} + E_{P'} + E_{\beta_1} + E_{p_1''} - E_{\beta} - E_{P''} - E_{\alpha_1} - E_{P_1'}) \\ & \times [f_{\alpha_1}(P_1')f_{\beta}(p'') - f_{\beta_1}(p_1'')f_{\alpha}(p')] . \end{aligned} \tag{4.62}$$

D. Kinetic Equations for Reacting Systems

Until now we considered the limiting situation in which the plasma consists only of atoms. The kinetic equation was given by Eq. (4.62). However, we are also interested in describing the partially ionized plasma, and especially the ionization and recombination reactions. Formally, this corresponds to taking into account scattering states in Eq. (4.44). This means that the quantum numbers α may be discrete numbers and may also run over the continuous spectrum. Taking into account scattering states, we are faced with difficulties. These are connected essentially with the application of Eq. (4.28) in the equation of motion.

By that procedure, an additional factor V^{-1} appears in the equation of motion of ρ_{ep} [Eq. (4.29)]. This factor leads to the fact that the four-particle processes accounted for in this manner are not "real" and may vanish in the thermodynamic limit. At least this is true for four-particle scattering states. However, in the limiting case that we have only two-particle bound states, that is, the neutral gas, we can obtain a kinetic equation for the atoms if we use the special definition of the distribution function of the atoms (4.17) and (4.24). Using the ideas just outlined, the kinetic equation (4.62) was obtained.

If we have, in contrast to such a situation, bound and scattering states as well, we can overcome the difficulties outlined in a systematic way only if we use a cluster expansion in order to introduce four-particle processes into the kinetic equation of atoms as outlined in the previous sections. If we want to deal with scattering processes in Eq. (4.62), we must make certain assumptions.

1. For scattering states there appears a distribution function

$$f_\alpha(P) \to f_p(P) = f(p_e p_p)\,, \tag{4.63}$$

which should obey certain properties.

In the normalization condition, the sum has to be taken over discrete and continuous quantum numbers, that means, we have

$$\sum_\alpha \to \sum_n + \int \frac{V}{(2\pi\hbar)^3}\, dp$$

and

$$\frac{V}{(2\pi\hbar)^3} \int dp\; f(p) + \sum_n \int \frac{V}{(2\pi\hbar)^3}\, dP\; f_n(P) = 1\,, \tag{4.64}$$

where the connection with the single-particle (electron, proton) distribution function is given by the reduction formula

$$\int f(p_e p_p) \frac{V}{(2\pi\hbar)^3}\, dp_b = f(p_a)\,, \qquad b \neq a = e,\, p\,, \tag{4.65}$$

and the normalization reads

$$\frac{V}{(2\pi\hbar)^3} \int f(p)\, dp = \frac{N_1}{N}\,, \qquad \frac{V}{(2\pi\hbar)^3} \sum_n \int f_n(P)\, dP = \frac{N_{12}}{N}\,.$$

Therefore, we obtain the balance equation

$$N_1 + N_{12} = N .$$

The weakening of the initial condition should read in this case

$$f(p_e p_p) = N_e f(p_e) f(p_p) . \tag{4.66}$$

Here there appears an additional factor N, because in the relation (4.17) for scattering states, the factor $(N/V)^2$ must be taken instead of (n/V). Such replacements lead to results that follow immediately from the cluster expansion technique, if screening is neglected.

2. In order to determine the polarization quantities occurring in (4.36), we use the eigenfunctions of free states. Taking into account

$$\psi_p(r) = \exp\left(\frac{ipr}{\hbar}\right)$$

and

$$\delta(p=0) = \frac{V}{(2\pi\hbar)^3} ,$$

we may write

$$|P_{p'p''}(k)|^2 = \frac{(2\pi\hbar)^3}{V}\left[e_e^2 \delta\left(p' - p'' - \frac{m_p}{M}\hbar k\right) + e_p^2 \delta\left(p' - p'' + \frac{m_e}{M}\hbar k\right)\right]. \tag{4.67}$$

Using

$$P = p_e + p_p , \qquad p = \frac{1}{M}(m_p p_e - m_e p_p) ,$$

we get, with $\hbar k = P_1' - P_1''$,

$$p' - p'' + \frac{m_e}{M}\hbar k = p'_{e_1} - p''_{e_1} , \qquad p' - p'' - \frac{m_p}{M}\hbar k = p''_{p_1} - p'_{p_1} .$$

We may write then

$$|P_{p'p''}(k)|^2 = \frac{(2\pi\hbar)^3}{V}\{e_e^2 \delta(p''_{p_1} - p'_{p_1}) + e_p^2 \delta(p'_{e_1} - p''_{e_1})\} . \tag{4.68}$$

Our aim is now to consider kinetic equations for free particles and atoms as well, which include different elementary processes.

From Eqs. (4.62), (4.63), and (4.65) we get for the single-particle kinetic equation

$$\frac{\partial f_a(p_a t)}{\partial t} = \frac{V}{(2\pi\hbar)^3} \int I(p_a p_b t)\, dp_b \equiv I_a(p_a t)\,, \qquad b \neq a = e, p\,. \tag{4.69}$$

The corresponding equation for atoms follows directly from (4.62) with $\alpha \to n$,

$$\frac{\partial f_n(Pt)}{\partial t} = I_n(Pt)\,. \tag{4.70}$$

We first discuss Eq. (4.69) in more detail.

Equation (4.69) reads with (4.65)

$$\frac{\partial f_a(p_a t)}{\partial t} = I_a(p_a t)\,,$$

$$\begin{aligned} I_a(p_a t) = {} & \frac{4N}{V} \frac{V}{(2\pi\hbar)^3} \sum_{\alpha_1\beta_1\beta} \int dP_1'\, dP_1''\, dP'' \frac{V}{(2\pi\hbar)^3}\, dp_b\, d\omega\, dk \\ & \times \frac{|p_{\alpha\beta}(k)|^2 |p_{\alpha_1\beta_1}(k)|^2}{k^4 |\epsilon(\omega, k)|^2} \delta(\hbar k - p_a - p_b + P'') \\ & \times \delta(\hbar\omega - E_{p_a} - E_{p_b} + E_\beta + E_{P''}) \delta(p_a + p_b + P_1'' - P_1' - P'') \\ & \times \delta(E_{p_a} + E_{p_b} + E_{\beta_1} + E_{P_1''} - E_\beta - E_{P''} - E_{\alpha_1} - E_{P_1'}) \\ & \times [f_{\alpha_1}(P_1') f_\beta(P'') - f_{\beta_1}(P_1'') f(p_a p_b)]\,, \qquad b \neq a = e, p\,. \end{aligned} \tag{4.71}$$

This collision integral may be divided into two parts, if we take into account that the sum over β is replaced by

$$\beta \to p_a'\,, \qquad P'' \to p_a' + p_b'$$

$$\sum_\beta \to \frac{V}{(2\pi\hbar)^3} \int dp_a' + \sum_n\,. \tag{4.72}$$

This corresponds to

$$I_a(p_a t) = I_a^{(1)}(p_a t) + I_a^{(2)}(p_a t)\,; \qquad a = e, p\,. \tag{4.73}$$

The first term leaves the incident particle a unchanged, while the second one leads to processes, after which the incident particle is a constituent of a bound state. We may then list the process belonging to $I_a^{(1)}(p_a t)$ corresponding to the different possibilities of the choice of α_1 and β_1 (we indicate only the relevant quantum numbers)

$$\begin{aligned} p'_a + p''_{b_1} &\leftrightarrow p''_a + p'_{b_1} && \text{(free scattering)}\,, \\ p'_a + m_1 P''_1 &\leftrightarrow p''_a + p'_{a_1} + p'_{b_1} && \text{(ionization)}\,, \\ p'_a + p''_{a_1} + p''_{b_1} &\leftrightarrow p''_a + n_1 P'_1 && \text{(recombination)}\,, \\ p'_a + m_1 P''_1 &\leftrightarrow p''_a + n_1 P'_1 && \text{(inelastic scattering)}\,. \end{aligned} \tag{4.74}$$

As the number of particles of species a is conserved, we have for the first contribution $I_a^{(1)}(p'_a t)$

$$\frac{V}{(2\pi\hbar)^3} \int I_a^{(1)}(p'_a t)\, dp'_a = 0\,. \tag{4.75}$$

It is interesting to note that the processes (4.74) describe approximately only three-particle scattering processes. This follows from the fact that the integration over p_b may be carried out in (4.71) if we take into account the approximation (4.68) and use only the first contribution of (4.72). The first contribution of Eq. (4.71) then reads

$$\begin{aligned} I_a^{(1)}(p_a t) = \sum_b \frac{4N_b}{V} \frac{V}{(2\pi\hbar)^3} \sum_{\alpha_1\beta_1} \int dP'_1\, dP''_1\, dp'_b\, d\omega\, dk\, \frac{|P_{\alpha_1\beta_1}(k)|^2 e_b^2}{k^4 |\epsilon(k,\omega)|^2} \\ \times \delta(\hbar k - p_a + p'_b)\delta(\hbar\omega - E_{p_a} + E_{p'_b})\delta(p_a - p'_b + P''_1 - P'_1) \\ \times \delta(E_{pa} - E_{p'_b} + E_{\beta_1} + E_{P''_1} - E_{\alpha_1} - E_{P'_1}) \\ \times [f_{\alpha_1}(P'_1 t) f_b(p'_b t) - f_{\beta_1}(P''_1 t) f_a(p_a t)]\,. \end{aligned} \tag{4.76}$$

If we replace the quantities α_1, β_1 by discrete numbers n_1, n'_1, we get from (4.76) a kinetic equation that describes the inelastic scattering of (4.74).

As a further example, we write explicitly the kinetic equation that corresponds to the first process of (4.74). For this reason we apply (4.65) and (4.68). The result is

$$\begin{aligned} I_a(p_1 t) = \frac{1}{\hbar} \frac{V}{(2\pi\hbar)^3} \sum_b \int \frac{4e_a^2 e_b^2 n_a}{k^4 |\epsilon(k,\omega)|^2}\, dp'_1\, dp'_2\, dp_2\, d\omega\, dk \\ \times \delta(\hbar k - p_1 + p'_1)\delta\left(\hbar\omega - \frac{p_1^2}{2m_a} + \frac{p_1'^2}{2m_a}\right) \end{aligned}$$

$$\times \delta(p_1 + p_2 - p_1' - p_2')\delta\left(\frac{p_1^2}{2m_a} + \frac{p_2^2}{2m_b} - \frac{p_1'^2}{2m_a} - \frac{p_2'^2}{2m_b}\right)$$

$$\times [f_a(p_1't)f_b(p_2't) - f_a(p_1t)f_b(p_2t)] . \tag{4.77}$$

Of course, this is nothing else than the Lenard–Balescu quantum kinetic equation for charged particles (Silin,[32] Balescu[33]).

The contributions of $I_a^{(2)}(p_at)$ do not conserve the particle number of species a; thus, we have

$$\frac{V}{(2\pi\hbar)^3}\int I_a^{(2)}(p_a't)\, dp_a' \neq 0 . \tag{4.78}$$

The processes are

$$\begin{aligned}
p_1' + p_b' + p_{b_1}'' &\leftrightarrow mP'' + p_{a_1}' && \text{(recombination)} ,\\
p_a' + p_b' + m_1P_1'' &\leftrightarrow mP'' + p_{a_1}' + p_{b_1}' && \text{(inelastic exchange scattering)} ,\\
p_a' + p_b' + p_{a_1}'' + p_{b_1}'' &\leftrightarrow mP'' + n_1P_1'' && \text{(recombination)} ,\\
p_a' + p_b' + m_1P_1'' &\leftrightarrow mP'' + n_1P_1'' && \text{(recombination)}.
\end{aligned} \tag{4.79}$$

Let us now consider the atomic kinetic equation (4.70). There are also two contributions to the collision integral $I_n(Pt)$, the first one leaves the incident atom unchanged. The relevant processes are

$$\begin{aligned}
nP' + p_{b_1}'' &\leftrightarrow mP'' + p_{b_1}' && \text{(inelastic scattering)} ,\\
nP' + m_1P_1'' &\leftrightarrow mP'' + p_{a_1}' + p_{b_1}' && \text{(ionization)} ,\\
nP' + p_{a_1}'' + p_{b_1}'' &\leftrightarrow mP'' + n_1P_1' && \text{(recombination)} ,\\
nP' + m_1P_1'' &\leftrightarrow mP'' + n_1P_1' && \text{(inelastic scattering)} .
\end{aligned} \tag{4.80}$$

The four processes (4.80) correspond to the first part of $I_n(Pt)$, $I_n^{(1)}(Pt)$,

$$I_n(Pt) = I_n^{(1)}(Pt) + I_n^{(2)}(Pt) , \tag{4.81}$$

and we have the conservation law

$$\frac{V}{(2\pi\hbar)^3}\sum_n \int dP\, I_n^{(1)}(Pt) = 0 . \tag{4.82}$$

The second contribution does not conserve the number of ingoing atoms, thus

$$\frac{V}{(2\pi\hbar)^3} \sum_n \int dP\, I_n^{(2)}(Pt) = 0 \, . \tag{4.83}$$

The relevant processes are

$$\begin{aligned} nP' + p''_{a_1} &\leftrightarrow p''_a + p''_b + p'_{b_1} \, , \\ nP' + p''_{a_1} + p''_{b_1} &\leftrightarrow p''_a + p''_b + p'_{a_1} + p'_{b_1} \, , \\ nP' + m_1 P''_1 &\leftrightarrow p''_a + p''_b + p'_{a_1} + p'_{b_1} \, , \\ nP' + m_1 P''_1 &\leftrightarrow p''_a + p''_b + n_1 P'_1 \, . \end{aligned} \tag{4.84}$$

Again using Eqs. (4.65) and (4.67) we may write the collision integral for the first process of (4.80), corresponding to an inelastic scattering between an atom and free particles. We then have

$$\begin{aligned} I_n(P't) = {} & \frac{4N}{\hbar V} \frac{V}{2\pi\hbar)^3} \sum_{n'} \int dP''\, dk\, d\omega \, \frac{|P_{nn'}(k)|^2}{k^4 |\epsilon(k, \omega)|^2} \\ & \times \delta(\hbar k - P' + P'')\delta(\hbar\omega - E_n + E_{n'} - E_{P'} + E_{P''}) \\ & \times \sum_{a+e,p} e_a^2 \int dp'_a\, dp''_a\, \delta(P' - P'' + p''_a - p'_a) \\ & \times \delta(E_n - E_{n'} + E_{P'} - E_{P''} + E_{p'_a} - E_{p''_a}) \\ & \times \{ f(p'_a) f_{n'}(P'') - f_n(P') f(p''_a) \} \, . \end{aligned} \tag{4.85}$$

V. KINETIC THEORY OF FLUCTUATIONS IN PARTIALLY IONIZED PLASMAS

In this section we want to deal with the kinetic theory of fluctuations in partially ionized plasmas.

The theory of fluctuations in partially ionized plasmas may be developed in full analogy to the fluctuation theory of gases or fully ionized plasmas. The latter is developed in detail in Refs, 5, 12, and 28.

Following the ideas outlined in Ref. 28, let us consider the operator of the pair density

$$\hat{\rho}_{\alpha\bar{\alpha}}(P\bar{P}t) = \frac{N_{ab} V}{(2\pi\hbar)^3} A^+_{\alpha p} A_{\bar{\alpha}\bar{p}} \, . \tag{5.1}$$

The operator (5.1) describes the pair density in the microscopic state.

In statistical mechanics the pair density must be characterized by the average of the pair operator corresponding to the mixed state

$$\langle \hat{\rho}_{\alpha\bar{\alpha}}(P\bar{P}t)\rangle \,. \tag{5.2}$$

Then the deviation of the microstate (5.1) and the mixed state (5.2), that is, the fluctuation of the pair density, is given by the operator

$$\delta\rho_{\alpha\bar{\alpha}}(p\bar{p}t) = \hat{\rho}_{\alpha\bar{\alpha}}(p\bar{p}t) - \langle \hat{\rho}_{\alpha\bar{\alpha}}(p\bar{p}t)\rangle \,. \tag{5.3}$$

The fluctuations are essentially characterized by their correlation function

$$\langle \delta\rho_{\alpha\bar{\alpha}}(p\bar{p}t)\delta\rho_{\beta\bar{\beta}}(p'p''t)\rangle \,. \tag{5.4}$$

According to (4.38), the correlation function (5.4) determines the collision integral of the relevant kinetic equation for the distribution of pairs.

In Section IV, the correlation function of the fluctuations of the pair density was determined under the assumption of the weakening of the initial correlation between the pairs. This assumption is different from the total weakening of any initial correlation, because, in the bound-state case, we have correlations between the elementary particles of one bound pair.

Under the conditions just discussed, large time-scale fluctuations of the density of pairs were neglected. In order to take into account such large-scale fluctuations we introduce an average of the pair operator over a "small volume," $\tilde{\rho}$. This "small volume" is much smaller than the total volume of the system, however, it is large enough to smear out small-scale fluctuations. We call the quantity $\tilde{\rho}$ the "smoothed pair operator."

Then it is useful to define

$$\delta\tilde{\rho}_{\alpha\beta} = \tilde{\rho}_{\alpha\beta} - \langle \tilde{\rho}_{\alpha\beta}\rangle \,. \tag{5.5}$$

If we have only small-scale fluctuations, obviously $\delta\tilde{\rho}_{\alpha\beta}$ is zero. Therefore, $\delta\tilde{\rho}_{\alpha\beta}$ describes the large-scale fluctuations. Thus the entire fluctuation (5.3) can be divided into two parts:

$$\delta\rho_{\alpha\beta} = \delta\tilde{\rho}_{\alpha\beta} + \Delta\rho_{\alpha\beta} \,, \tag{5.6}$$

where we have

$$\Delta\rho_{\alpha\beta} = \hat{\rho}_{\alpha\beta} - \tilde{\rho}_{\alpha\beta} \,. \tag{5.7}$$

Consequently, we have the additional correlation function of fluctuations

$$\langle \delta\tilde{\rho}_{\alpha\bar{\alpha}}\,\delta\tilde{\rho}_{\beta\bar{\beta}} \rangle \, . \tag{5.8}$$

The correlator (5.8) does not have the property of weakening of initial correlations.

As a consequence of such additional correlations, we have an additional contribution to the collision integral, I_α, for the kinetic equation for f_α. Now, the kinetic equation has the structure

$$\frac{d}{dt} f_\alpha(pt) = I_\alpha(pt) + \tilde{I}_\alpha(pt) \, . \tag{5.9}$$

The usual collision integral, $I_\alpha(pt)$, is defined by (4.62). The additional term, which is produced by the large-scale fluctuations, is determined by the following expression:

$$\tilde{I}_\alpha(pt) = -\frac{2}{\hbar}\sum_\beta \int \operatorname{Im}\langle \delta\tilde{\rho}_{\alpha\beta}(pp't)\delta\tilde{U}_{\alpha\beta}(p'pt)\rangle \frac{V\,dP'}{(2\pi\hbar)^3} \, . \tag{5.10}$$

The contribution due to the collision integral $\tilde{I}_\alpha$ is important, if the state of the system is far from equilibrium. Here we shall discuss only the case of $\delta\tilde{\rho}_{\alpha\beta}(pp't)$ diagonal, that is,

$$\delta\tilde{\rho}_{\alpha\beta}(pp't) = \delta_{\alpha\beta}\,\frac{(2\pi\hbar)^3}{V}\,\delta(p'-p)\delta f_\alpha(p't) \, . \tag{5.11}$$

Under this assumption, the kinetic equation for the distribution function can be written as a Langevin equation:

$$\frac{\partial f_\alpha(pt)}{\partial t} = I_\alpha(pt) + y_\alpha(pt) \, , \tag{5.12}$$

where $y_\alpha(pt)$ is the random source. The moments of the random source are defined by

$$\begin{aligned} &\langle y_\alpha(pt)\rangle = 0 \, , \\ &\langle y_\alpha(pt)y_\beta(p't')\rangle = \delta(t-t')A_{\alpha\beta}(pp't) \, . \end{aligned} \tag{5.13}$$

In general, the intensity of the random source, $A_{\alpha\beta}$, consists of two parts

$$A_{\alpha\beta} = A^{\mathrm{B}}_{\alpha\beta} + \tilde{A}_{\alpha\beta} \, . \tag{5.14}$$

The first contribution, $A^{\mathrm{B}}_{\alpha\beta}$, is given by

$$A^{\mathrm{B}}_{\alpha\beta}(p'p''t) = \{(\delta\hat{I}_\alpha(p't) + \delta\hat{I}_\beta(p''t)) - [\delta\hat{I}_\alpha(p't) + \delta\hat{I}_\beta(p''t)]_0\} \times \frac{(2\pi\hbar)^3}{V}\,\delta_{\alpha\beta}\delta(p' - p'')f_\alpha(p't)\,. \tag{5.15}$$

The quantity δI_α is the linearized collision operator corresponding to (4.62). The subscript "0" in the second term means that the operator acts only on the function f_α and not on the delta distribution. The contribution $\tilde{A}_{\alpha\beta}$ in (5.14) is determined by $\tilde{I}_{\alpha\beta}$, that is by the large-scale fluctuations. We consider only the approximation that $\tilde{A}_{\alpha\beta}$ is equal to zero.

As discussed in Section IV, we may derive from (5.9) a kinetic equation for the distribution function of the electrons, which takes into account reactions and large-scale fluctuations. From such an equation we get a Langevin-type rate equation for the concentrations of electrons, ions, and atoms, which reads in the case of local equilibrium[35]

$$\frac{dn_e}{dt} = \alpha n_e n_{ei} - \beta n_e^2 n_i + \alpha_1 n_{ei}^2 - \beta_1 n_e n_i n_{ei} + \alpha_2 n_{ei}^2 - \beta_2 n_e^2 n_i^2 + \frac{N}{V}\, y_e(t)\,. \tag{5.16}$$

Here we have

$$n_e = n_i\,, \quad n_e + n_{ei} = N/V\,. \tag{5.17}$$

Equation (5.16) is a generalization of (3.92), α and β are the rate coefficients. An additional right-hand side contribution to (5.16) of the structure

$$\alpha_3 n_{ei} n_e - \beta_3 n_{ei} n_e$$

was omitted, since it vanishes in the local equilibrium case because $\alpha_3 = \beta_3$.

The moments of the Langevin source are now given by

$$\begin{aligned} \langle y_e(t)\rangle &= 0\,, \\ \langle y_e(t) y_e(t')\rangle &= \frac{N^2}{V^2}\, A_{ee}(t)\delta(t - t')\,, \end{aligned} \tag{5.18}$$

where the expression A_{ee} is given in the case that the Maxwell–

Boltzmann distribution is achieved, but not the chemical equilibrium[35]

$$A_{ee}(t) = \frac{V}{N^2} \{(\alpha n_e n_{ei} + \beta n_e^2 n_i) + (\alpha_1 n_{ei}^2 + \beta_1 n_{ei} n_e n_i) + 2(\alpha_2 n_{ei}^2 + \beta_2 n_e^2 n_i^2)\} . \tag{5.19}$$

Thus, the intensity of the random source involves three terms, each corresponding to one of the pair of processes (direct and inverse), defining the change in the concentrations. In the state of complete equilibrium (i.e., if a Saha equation is valid), the first and second terms in each bracket of (5.19) become equal.

For small deviations from the equilibrium, Eq. (5.16) reduces in linear approximation with respect to deviations δn_e from the equilibrium value n_e^0 to the equation

$$\frac{d}{dt} \delta n_e + \lambda_e \delta n_e = \frac{N}{V} y_e . \tag{5.20}$$

Here we used

$$\delta n_i = \delta n_e = -\delta n_{ei} , \tag{5.21}$$

and λ_e reads

$$\lambda_e = \left(1 + 2\frac{n_{ei}^0}{n_e^0}\right)(\alpha n_e^0 + \alpha_1 n_{ei}^0 + 2\alpha_2 n_{ei}^0) , \tag{5.22}$$

where the superscript "0" means equilibrium. From (5.20) it follows that the correlator $\langle (\delta n_e)^2 \rangle$ (density fluctuation) is connected with the quantity A_{ee}:

$$\langle (\delta n_e)^2 \rangle = \left(\frac{N}{V}\right)^2 A_{ee} \frac{1}{2\lambda_e} . \tag{5.23}$$

In the equilibrium case we have

$$A_{ee} = \frac{2V}{N^2} n_{ei}^0 (\alpha n_e^0 + \alpha_1 n_{ei}^0 + 2\alpha_2 n_{ei}^0) . \tag{5.24}$$

From (5.20)–(5.24) we get for the density fluctuations

$$\langle (\delta n_e)^2 \rangle = \langle (\delta n_i)^2 \rangle = \langle (\delta n_{ei})^2 \rangle = \frac{1}{V} \frac{n_{ei}^0 n_e^0}{2n_{ei}^0 + n_e^0} ; \qquad n_e^0 = n_i^0 . \tag{5.25}$$

The corresponding fluctuations of the total particle numbers read

$$\langle(\delta N_e)^2\rangle = \langle(\delta N_{ei})^2\rangle = N_e \frac{N - N_e}{2N - N_i} \qquad (N = N_e + N_{ei}) . \quad (5.26)$$

The second moments that refer to different species, however, read

$$\langle(\delta N_e)^2\rangle = -\langle(\delta N_e \delta N_{ei})\rangle . \quad (5.27)$$

The concentration fluctuations may, of course, be determined also by other methods, see, for example, Refs. 36 and 37.

The methods outlined in this section can be used even in states far from equilibrium when the contributions of the long-range fluctuations become important.

References

1. M. A. Leontovich, "Basic equations of the kinetic theory of gases including random processes (in Russian)," *Zh. Exp. Teor. Fiz.* **5**, 211 (1935).
2. Yu. L. Klimontovich, "Dissipative equations for many particle distribution functions" (in Russian), *Ukrain. Fiz. Zh.* **139**, 689 (1983).
3. L. D. Landau and E. M. Lifshits, *Theoretical Physics*, Vol. V—L. D. Landau and E. M. Lifshits, Statistical Physics, Part 1, Nauka, Moscow 1976; Vol. IX—E. M. Lifshits and L. P. Pitaevskii, Statistical Physics, Part 2, Nauka, Moscow 1978; Vol. X—E. M. Lifshits and L. P. Pitaevskii, Physical Kinetics, Nauka, Moscow 1979.
4. J. de Boer and G. E. Uhlenbeck (eds.), *Studies in Statistical Mechanics*. North Holland, Amsterdam, 1965, Vol. III.
5. Yu. L. Klimontovich, *Statistical Physics* (in Russian), Nauka, Moscow, 1982.
6. N.N. Bogolyubov, *Selected Works*. Naukova dumka, Kiev, 1971 (in Russian), Vol. 2; N. N. Bogolyubov and K. P. Gurov: *Zh. Exp. Teor. Fiz.* **17**, 614 (1947).
7. K. P. Gurov, *The Foundation of Kinetic Theory*. Nauka, Moscow, 1966.
8. N. N. Bogolyubov, *Problems of the Dynamical Theory in Statistical Mechanics*. Gostechisdat, Moscow, 1946. See also Ref. 6, Vol. 2.
9. J. R. Taylor, *Scattering Theory*. Wiley, New York, 1972.
10. E. G. D. Cohen, "The kinetic theory of dense gases," *Fundamental Problems in Statistical Mechanics*, North Holland, Amsterdam, 1968, Vol. II; J. R. Dorfman and E. G. D. Cohen, *Phys. Rev.* **A6**, 776 (1972); **12**, 292 (1975).
11. J. Weinstock, Phys. Rev. **132**, 454 (1963).
12. Yu. L. Klimontovich, *Kinetic Theory of Nonideal Gases and Nonideal Plasmas*. Nauka, Moscow, 1975 (Pergamon, Oxford, 1982).
13. L. A. Pal'tsev, *Teor. Mat. Fiz.* **30**, 114, 382 (1977) (in Russian).
14. Yu. L. Klimontovich and D. Kremp, *Physica* **109A**, 517 (1981).
15. J. A. McLennan, *J. Statist. Phys.* **28**, 521 (1982); S. Lagan and J. A. McLennan, *Physica* **128A**, 178 (1984).

16. Yu. L. Klimontovich and W. Ebeling, *Zh. Exp. Teor. Fiz.* **63**, 905 (1972).
17. W. D. Kraeft, M. Schlanges, and D. Kremp, *J. Phys.* **A19**, 3251 (1986).
18. Yu. L. Klimontovich, *Zh. Exp. Teor. Fiz.* **52**, 1233 (1967); **54**, 138 (1968).
19. S. V. Peletminsky, *Teor. Mat. Fiz.* **6**, 123 (1971).
20. E. G. Kolesnichenko, *Teor. Mat. Fiz.* **30** 114, 382 (1977).
21. S. I. Vashukov and V. V. Maruzin, *Teor. Mat. Fiz.* **29**, 255 (1976).
22. C. Ivanov, Ph.D. thesis, Rostock University, 1977.
23. D. Kremp, M. Schlanges, and T. Bornath, *J. Stat. Phys.* **41**, 661 (1985).
24. Yu. L. Klimontovich, D. Kremp and M. Schlanges, In W. Ebeling et al., *Transport Properties of Dense Plasmas*, Akademie-Verlag, Berlin, 1983 (Birkhäuser, Basel, 1984). (Ref. 27).
25. W. Ebeling, W. D. Kraeft, and D. Kremp, *Theor of Bound States and Ionization Equilibrium in Plasma and Solids*. Akademie-Verlag, Berlin, 1976 (Russian Translation: Mir, Moscow, 1979).
26. G. Newton, *Scattering Theory of Waves and Particles*, McGraw-Hill, New York, 1966.
27. W. Ebeling et al., *Transport Properties of Dense Plasmas*. Akademie-Verlag, Berlin, 1983 (Birkhäuser, Basel, 1984).
28. Yu. L. Klimontovich, "*Kinetic Theory of Electromagnetic Processes*." Nauka, Moscow, 1980 (in Russian).
29. M. D. Girardeau: *J. Math. Phys.* **4**, 1096 (1963); **19**, 2605 (1978); M. D. Girardeau and J. D. Gilbert, Physica **97A**, 42 (1979).
30. W. E. Brittin and A. Y. Sakakura, *Phys. Rev.* **A21**, 2050 (1980).
31. G. Röpke and R. Der, *Phys. Stat. Sol.* **B92**, 501 (1979).
32. V. P. Silin, *Zh. Exp. Teor. Fiz.* **40**, 1768 (1961).
33. B. Balescu, *Phys. Fluids* **4**, 94 (1961).
34. Yu. L. Klimontovich, "Kinetic theory of fluctuations in gases and plasmas," in *Fundamental Problems in Statistical Mechanics IV*, E. G. D. Cohen and W. Fizdon, (eds). Warsaw, 1978.
35. V. V. Belyi and Yu L. Klimontovich, *Zh. Exp. Teor. Fiz.* **74**, 1660 (1978).
36. G. Nicolis and J. Prigogine, *Self-Organization in Nonequilibirum Systems*. Moscow, Mir, 1978 (Russian translation).
37. N. G. Van Kampen, "The expansion of the master equation," *Adv. Chem. Phys.* **34**, 245 (1976).

INTERACTION OF CHARGED PARTICLES WITH MOLECULAR MEDIUM AND TRACK EFFECTS IN RADIATION CHEMISTRY

I. G. KAPLAN AND A. M. MITEREV

L. Ya. Karpov Institute of Physical Chemistry
ul. Obukha, 10 107120
Moscow B-120, USSR

CONTENTS

I. INTRODUCTION

As a charged particle makes its way through a substance, its interaction with surrounding atoms and molecules leads to the transfer of its energy to the medium. This energy is mainly spent on excitation and ionization of the molecules. The ejected electrons produce further ionization and excitation. Thus, the irradiation of a medium results in formation of molecules in electron-excitation states, of positive ions, and of cascades of ejected electrons with energy varying from the maximum possible value (determined by the type and energy of the initial flux of charged particles) to thermal level. The slow electrons may excite the molecules into optically forbidden states, such as the triplet ones,[1] and may also form negative ions when captured by molecules in a resonance process.

The energy transferred to molecules by fast charged particles may be much greater than their first ionization potential. This may lead to ejection of core electrons followed by a cascade of ionizations due to the shake-off and the Auger effect, as well as to formation of discrete states in the region of a continuous spectrum (the so-called *superexcitation* states,[2] Section III.B), in which predissociation competes with autoionization. In condensed media there are formed collective states of plasmon type[3,4] (see also Section III.C). Thus, the spectrum of the states produced by fast charged particles in a molecular medium in much more complicated than it is in the case of irradiation by light.

Another type of ionizing radiation is the high-energy electromagnetic emission (gamma quanta and X rays). The principal processes leading to losses of energy by gamma quanta are the Compton scattering, the photoeffect, and the formation of an electron–positron pair.[5] For the most common ^{60}Co gamma source, the average energy of Compton electrons is about 600 keV. On the other hand, the average energy lost by a fast electron in a single ionization event is about 30 eV. Thus, before degrading, such a Compton electron manages to produce some $6 \times 10^5/3 \times 10 = 2 \times 10^4$ secondary electrons. This means that the effect produced by gamma quanta in a medium is actually the effect produced by photo- and Compton electrons (in the case of X rays, by photoelectrons only).

The irradiation of a medium by neutrons also leads to formation of charged particles via the secondary processes. The most important among the latter is the elastic scattering of a neutron with formation of a recoil nucleus. This process is most efficient in media consisting of light elements. Besides this, the slow and thermal neutrons are efficiently captured by certain types of nuclei with ejection of either a proton (for example, by ^{14}N) or an alpha particle (for example, by ^{10}B). The newly formed nucleus has a smaller charge than the initial one and, thus, corresponds to a different chemical element. It is the recoil nuclei and the particles ejected in nuclear reactions following the capture of a neutron that are the principal physical causes of chemical effects produced by neutron irradiation in a medium.

Thus, the effect the different types of ionizing radiation produce in a medium is the effect produced by charged particles. It is to the different physical and physicochemical processes induced in a molecular medium by charged particles our present review is devoted.

When passing through a medium, a fast charged particle loses its energy at separate portions rather than in a continuous manner. The ions and excited molecules it produces are localized in the vicinity of the particle's trajectory. In solids the atoms may be knocked out of their equilibrium position, provided the transferred kinetic energy is sufficiently high. All this leads to formation of (reversible or irreversible) disturbance areas containing excited molecules, positive and negative ions, secondary electrons, and ejected atoms and radicals. Such areas of local disturbance make up the particle's track.

The tracks of alpha particles and of electrons ejected by X rays were first observed in Wilson chambers. Later the more advanced bubble and spark chambers were designed. Another type of detector, which is widely used for recording particle tracks, is one that fixes the changes in the structure of a medium when treated by certain chemical reagents. These are the photoemulsions and the different types of solid detectors.[6]

First of all, the experimental observation of tracks provides information about the charged particles themselves (their charge and mass) and plays an important role in discovering new elementary particles. As for the characteristics of the track, such studies allow one to measure only the density of ionization along the track in gaseous media and to determine the specific energy losses.

However, this information is absolutely insufficient for explaining the changes that occur in a molecular medium exposed to ionizing radiation. The final chemical transformations are determined by the microstructure of the short-lived primary excitation and ionization regions of the track and by the spatial distribution of the radicals produced. For instance, in order to make the theoretical model of water radiolysis agree with experimental data, it was necessary to give the nonhomogeneous distribution of radicals in electron tracks as initial conditions.[7]

With development of picosecond pulsed-radiolysis sets it became possible to study the processes occurring within the time intervals ~ 10 ps.[8,9] Processes that are of shorter duration are still inaccesible to direct experimental observation. Mozumder and Magee[10] have speculated on the existence of a "picosecond barrier" that cannot be surpassed by further improvement of experimental technique. The magnitude of this barrier can be estimated in the following way. If we take the size of the smallest reaction vessel to be, say 3 mm, the time taken for the detecting light to pass through the vessel will be $3 \times 10^{-1}/3 \times 10^{10} = 10$ ps.

Thus, the only way to study the early radiolysis stages is to study them theoretically, especially with mathematical simulation. The correctness of our theoretical conceptions is determined by how well all the consequences of the theory agree with the available experimental data concerning the subsequent, already measurable, radiolysis stages. In this respect a numerical experiment in modern high-speed computers proved to be very efficient. The processes that occur in a substance exposed to ionizing radiation are simulated by the Monte Carlo method.[11] At a sufficiently large number of simulated "histories," the reliability of the results obtained depends on the exactness of the interaction cross sections used. The latter are presented both in analytical form and an experimental curves.

The computer simulation makes it possible to calculate the spatial and energy distribution of ejected electrons and the distribution of ions and excited molecules at different distances from the axis of the track.[12–14]. Knowing the spatial and energy structure of the track, one can determine the features of primary radiation-chemical reactions in tracks of particles of different nature,[15] as well as to describe the evolution of the track and to calculate the yield of radiolysis products.[16]

The phase state of the medium in which the studied processes occur is very important in such calculations. For quite a long time all authors studying the primary processes in the condensed phase have used the traditional model of a medium as an ensemble of independent molecules with correction only for the greater density.[17] In the recent years there has been considerable process toward a more realistic account of the specific features of condensed media.[14,18,19]

We are fully aware of the extensive review literature concerning the mechanism of primary processes in interaction of the ionizing radiation with matter.[5,17,20–25] However, the most recent of the cited reviews has been written more than a decade ago.* The last ten years are marked by intensive development of experimental studies and by the appearance of new theoretical conceptions that change some of the traditional views on primary processes. In this review we discuss the modern ideas concerning the primary radiolysis stage that take into account the latest developments in this direction.

II. PRIMARY RADIOLYSIS PROCESSES (A QUALITATIVE PICTURE)

A. The Time Scale of Elementary Events

It is natural to begin our discussion of the succession of processes occurring in a medium irradiated by charged particles by considering the duration of the interaction itself. The interaction time depends both on

* Some of the aspects of the primary radiolysis stage have been recently discussed in Proceedings of NATO Advance Study Institute.[26]

the velocity of the bombarding particles and on the distance that separates the latter from the molecules with which they interact. If b denotes the distance of closest approach, the interaction time approximately equals

$$\tau \sim 2b/v \, . \tag{2.1}$$

For atomic distances ($\sim 10^{-8}$ cm) and relativistic particles we find the minimum τ_{int} to be about 10^{-18} s.

The amount of transferred energy ΔE can be regarded as a measure of the uncertainty in the energy of the charged particle and is related to the time of transfer by the uncertainty relation[27,28]

$$\Delta E \, \Delta t \geq \hbar \, . \tag{2.2}$$

The latter relation means that an event of energy transfer cannot be strictly localized in time. The moment it occurs can be determined more precisely the greater the transferred energy is. For example, the energy necessary to excite the valence electrons is transferred in about 10^{-16} s, whereas the transfer of the energy on excitation or ionization of the core electrons takes 10^{-19}–10^{-18} s.

It is conventional to relate the average energy losses of a charged particle to the unit path length it travels. In radiation physics this quantity ($-dE/dx$) is called the *stopping power*; in radiation chemistry, it is called the *linear energy transfer* (LET). The value of LET depends on the energy and charge of the particle, as well as on the properties of the surrounding medium. Before slowing down to energies below the lowest electron excitation level of the medium ($E_1 \lesssim 5$ eV), a fast electron mainly spends its energy on ionization and electron excitation of molecules. Knowing the LET in a sufficiently wide energy range, one is able to calculate the linear path length of a particle. For example, the linear ionization path length of an electron is

$$R_i(E_0) = \int_{I_1}^{E_0} \left(\frac{-dE}{dx} \right)^{-1} dE \, , \tag{2.3}$$

where I_1 is the first ionization potential. The total path length of an electron is actually greater than the value given by formula (2.3), since after its energy falls below I_1, the electron is still able to penetrate a large distance as the retardation becomes less efficient (see subsequently).

To be definite, let us take the medium to be liquid water. In this case, for electrons with energy from I to 10 MeV, the LET is about 2 MeV/cm. Since the velocity of an electron with $E = 1$ MeV and higher is very close to the speed of light (3×10^{10} cm/s), we can roughly estimate the time during which a relativistic electron loses 2 MeV of energy in water to be $\sim 3 \times 10^{-11}$ s. An electron with the initial energy $E_0 = 10$ MeV slows down in about 10^{-10} s.

The average energy W_i necessary to form a pair of ions in water is close to 30 eV. Thus, the absorption of 2 MeV of energy by the medium results in formation of $2 \times 10^6/3 \times 10^1 \approx 7 \times 10^4$ electrons of the second and subsequent generations. The majority of these electrons have small energies, below 100 eV. The time taken for a secondary electron to slow down from $E_0 = 100$ eV to $E = I_1$ can be roughly estimated as the ratio $R_i(100 \text{ eV})/\bar{v}$. The average velocity of an electron with energy between 10 and 100 eV is about 4×10^8 cm/s; for electrons with $E_0 = 100$ eV, the ionization path length in water is ~1.5 nm.[23] This gives $\tau \approx 4 \times 10^{-16}$ s, which means that ejected secondary electrons slow down to $E < I_1$ in about 10^{-16}–4×10^{-16} s.

When an electron slows down to energies below the lowest electron excitation level of molecules of the medium, the retardation becomes much less rapid. Such electrons, called the *subexcitation* electrons, can no longer excite the electron degrees of freedom by direct interaction with electrons of the medium. The remaining energy-transfer channels are less efficient. These are the excitation of optically active intra- and intermolecular vibrations (the resonance losses in the IR region), the dipole relaxation of the medium (the losses in the SHF region; $\lambda \approx 1$ cm), and elastic collisions. In water the rate of energy losses for subexcitation electrons is calculated to be about 4×10^{13} eV/s.[29] Thus, an electron with $E_0 = 4$ eV slows down to thermal velocities in about 10^{-13} s, its path length becoming accordingly longer. According to the measurements,[30] the path length of a 3-eV electron is ~4.5 nm in H_2O and ~9 nm in D_2O. (For comparison, let us recall that the ionization path length of a 100-eV electron in water is 1.5 nm.) The dependence on the isotopic composition of the medium indicates that the losses on excitation of vibrations are of principal importance. A more detailed discussion of the mechanism of retardation of slow electrons is presented in Section VI.

After thermalization, the electron may recombine with a positive ion or be captured by a molecule forming a negative ion, or it may be locked in a trap the role of which may be played by fluctuation cavities or structural disturbances in the medium, or by polarization pits that the electron "digs" when it interacts with surrounding molecules. Such captured electrons are called solvated electrons (in water they are sometimes called *hydrated* electrons).[31,32] According to the data obtained in picosecond pulsed-radiolysis sets,[33,34] the solvation time of an electron is $\sim 2 \times 10^{-12}$ s in water and $\sim 10^{-11}$ s in methanol.

Let us now consider the interaction between a charged particle and a medium from the point of view of the absorbing system. The ionization of a core electron needs large portions of transferred energy. According to relation (2.2), such energies are transferred during 10^{-19}–10^{-18} s. Apparently, the ionization time cannot be shorter than the time taken for the electron to leave the atom (the molecule), so, in any case, $\tau_i > 10^{-8}/v_e$. If the energy transferred to the electron is 10^3 eV, $\tau_i > 5 \times 10^{-18}$ s. In the next 10^{-15}–10^{-14} s the ion with a core hole formed in such a rapid manner experiences additional ionization via the shake-off and the Auger effect[35] (see Section III.A). Owing to these two effects the cascade of vacancies formed may lead to formation of multiple-charge ions. For example, the most probable charge of the ion formed after ejection of a K electron in a Kr atom is +5.

The characteristic frequencies of valence-electron excitations are about 10^{15} s^{-1}, which corresponds to a period of electron vibrations $\sim 10^{-15}$ s. Since the period of nuclear vibrations is 10^{-14}–10^{-13} s, before the nuclei get moving the electron excitation can be transferred to the neighboring molecules according to the resonance mechanism.[36] In long polymer molecules the electron excitation energy efficiently migrates along the polymer chain.[37]

If the redistribution of electron excitation energy results in the transfer of the energy to a repulsive molecular term, there occurs a *predissociation* from higher electron excitation states, leading to dissociation of the molecule into fragments within a period of nuclear vibrations ($\sim 10^{-13}$ s). If the predissociation does not occur, the electron excitation energy is transformed into vibrational energy in a nonradiative manner. Such an *internal conversion* results in a multiatomic molecule makes a nonradiative transition into the lowest electron excitation state within 10^{-13}–10^{-12} s. If the latter state is repulsive and its energy exceeds the energy of the chemical bond, the molecule experiences predissociation. Otherwise, the electron excitation energy degrades either in a radiative or a nonradiative manner within 10^{-8}–10^{-7} s, which is quite a long time on the atomic time scale. A triplet molecular state may live even longer: for molecules made of light atoms, its lifetime reaches 10^{-3} s, increasing up to several seconds at helium temperatures.[38]

The long lifetime of lower excited states favors the efficient transfer of electron excitation energy from molecules of the solvent to molecules of the solute, provided the emission spectrum of the donor overlaps the absorption spectrum of the acceptor. The mechanism of the nonradiative transfer of electron excitation energy was found by Förster and Dexter[39,40]. The distance over which the energy is transferred may be as large as 20–30 Å. The migration of electron excitation also occurs in crystals. For triplet excitons the range of migration may reach several hundreds of angstroms.[41]

The frequencies of rotational transitions are much smaller than vibrational frequencies, which means that the rotational motion is slower than the vibrational one. For a free molecule, the period of rotational motion is within 10^{-12}–10^{-9} s. In condensed media the rotational motion is even slower, its period being respectively greater. At this stage it is more correct to speak of the relaxation time of the molecules. The latter essentially depends on the phase state of the medium. For example, in liquid water the relaxation time of molecular dipoles in an external electric field is about 10^{-11} s, whereas in ice (at 0°C) it is $\sim 10^{-5}$ s.

The time between collisions in a liquid is about 10^{-13} s. In liquids with viscosity of water or hexane, the time during which the instantaneous configuration of molecules changes due to a diffusive jump amounts to 10^{-12}–10^{-11} s. Being surrounded by other molecules, the decay products cannot part at once (the so-called *cage effect*), owing to which their geminal recombination within the time of the diffusive jump (i.e., within 10^{-12}–10^{-11} s) becomes very probable in condensed media. The time of geminal recombination of electrons with parent positive ions has been estimated by Freeman.[42] According to his results, this time depends on the path length travelled by a subexcitation electron before it is thermalized, and is within 10^{-12}–10^{-10} s for *n*-hexane.

The most rapid bimolecular reactions must be the ion–molecular ones. Their duration can be limited only by the time of collision, thus being 10^{-13}–10^{-12} s. The recombination time of radicals that have escaped from the "cage" depends on their concentration in the track. For close pairs of radicals the recombination may already begin in 10^{-11} s. From this moment on we can consider the chemical stage of radiolysis to have begun.

B. The Stages of Radiolysis

The succession of radiolysis processes we have described previously is conventionally divided into three stages: the *physical*, the *physicochemical*, and the *chemical* stages.[20,21]* The physical stage includes all the processes of degradiation of the fast charged particle's energy leading to formation of excited and ionized molecules in the medium. It is also natural to attribute to this stage all the processes of redistribution of excitation and deactivation energy, which do not lead to any chemical changes. Thus, the physical stage of radiolysis lasts from 10^{-19} to 10^{-18} s

* The subsequently presented division of radiolysis processes into stages is different from the one originally proposed by Platzman.[21]

(the time taken for the energy sufficient to ionize the K electrons to be transferred) to 10^{-5}–10^{-3} s (the time of dipole relaxation in the solid phase and, also, the time of phosphorescence). Therefore, the physical stage is still in progress when the processes of the two subsequent stages are already taking place.

The physicochemical stage includes the chemical processes in electron excitation states, as well as the chemical transformations of the active intermediates under nonequilibrium conditions. These are the predissociation and the ion–molecular reactions that take about 10^{-13} s; the recombination of positive ions with thermalized electrons (10^{-12}–10^{-10} s); and the electron-solvation reactions (10^{-12}–10^{-10} s). Thus, the physicochemical stage lasts from 10^{-13} to 10^{-10} s.

TABLE I
Time Scale of Primary and Secondary Processes at Different Radiolysis Stages

Radiolysis Processes	Time (s)
Physical Stage	
Ionization of K electrons	5×10^{-18}–10^{-17}
Excitation of valence electrons	10^{-16}–10^{-15}
Additional ionization due to the Auger effect	10^{-15}–10^{-14}
Period of "electron oscillations"	10^{-15}
Period of molecular vibrations	10^{-14}–10^{-13}
Internal conversion	10^{-13}–10^{-12}
Period of molecular rotations	10^{-12}–10^{-9}
Time of dipole relaxation: in water	10^{-11}
in ice	10^{-5}
Fluorescence	10^{-8}–10^{-7}
Phosphorescence	10^{-4}–10^{-3}
Time during which an electron loses	
1 eV of energy in water: relativistic electron	10^{-17}
subexcitation electron	3×10^{-14}
Physicochemical Stage	
Predissociation	10^{-13}
Ion–molecular reactions	10^{-13}–10^{-12}
Geminal recombination of decay products	10^{-12}–10^{-11}
Recombination of positive ions with electrons	10^{-12}–10^{-10}
Solvation time	10^{-12}–10^{-10}
Chemical Stage	
Radical–"solute" reactions	10^{-11}
Radical–radical reactions: track period	10^{-11}–10^{-7}
homogeneous kinetics	$>10^{-7}$

The chemical stage of radiolysis includes the chemical reactions involving molecules and the active intermediates in a state of thermal equilibrium with surrounding medium. However, in contrast to traditional thermochemistry, one can usually distinguish two periods in the chemical stage in condensed media: the *track* period with nonhomogeneous kinetics, and the following period with a homogeneous distribution of reagents. The track period is characterized by a nonuniform distribution of radicals and chemically active particles in the track's volume. The diffusion of reagents competes with chemical reactions, and gradually the track becomes diffused. The numerical simulation of diffusion kinetics for tracks of fast electrons in water[16] has shown that by $t = 10^{-7}$ s the structure of the track becomes homogeneous. Thus, the track period lasts from 10^{-11} to 10^{-7} s. At $t > 10^{-7}$ s we have the usual homogeneous kinetics typical of thermochemistry.

The previously described classification of elementary processes at different stages of radiolysis is presented in Table I.

III. NATURE OF HIGHLY EXCITED MOLECULAR STATES PRODUCED BY IONIZING RADIATION

In contrast to photochemistry, the energy of different types of radiation we deal with in radiation chemistry is so high that the energy transferred to a molecule may considerably exceed its first ionization potential. This may lead to formation of discrete high-energy states in the autoionization region of the spectrum (the so-called *superexcitation* states, Section (III.B), as well as of the states in the region of continuous spectrum that were formed by ionization of a core electron and have a vacancy (a hole) in the corresponding inner shell of the molecule. Let us first discuss the latter type of states.

A. The Evolution of States with a Core Hole

We should consider an ion with a core hole to be in a highly excited state the energy of which is equal to the energy required to rearrange the corresponding shell of the molecule with formation of a hole. This energy may be transferred to one or several electrons of the molecule leading to their ionization or transition into vacant orbitals (the *Auger effect*[43,44]), or it may be emitted as an X-ray quantum (Fig. 1).

The Roentgen luminescence (i.e., the emission of an X-ray quantum) is significant only for elements with $Z > 20$. The ratio of the Auger effect probability to that of Roentgen luminescence is $W_A/W_X \simeq 10^6/Z^4$.[43] In Table II we present the values of the probabilities W_A and W_X together with the fluorescence yield for the *KLL* Auger transition for a number of

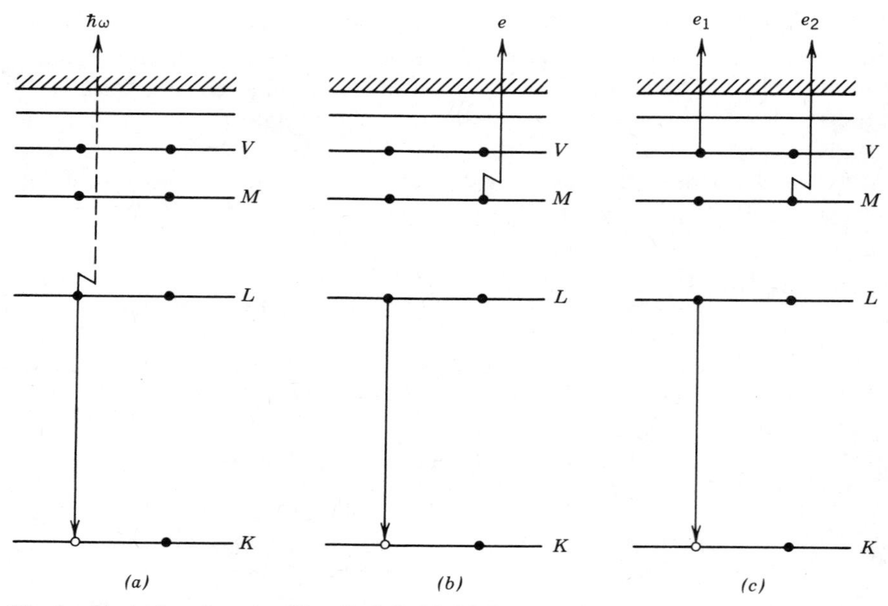

Fig. 1. Evolution of states with a "hole": (*a*) *KL* Roentgen luminescence; (*b*) *KLM* Auger effect; (*c*) *KLMV* double Auger effect.

atoms. For light elements (and their compounds) the emission probability is very small, whereas for heavy elements the situation is reversed. For example, for Tl ($Z = 81$) the K luminescence yield is 0.96.[43]

A more probable process for compounds of light elements is the transfer of the energy released with occupation of a core vacancy to one or several more weakly bound electrons. This process was named after its discoverer—the Auger effect. In contrast to luminescence tansitions, involving electrons of only two shells, an Auger transition involves electrons of three or more shells, and it is this the complicated notation of

TABLE II
Probabilities of X-Ray Emission and of the *KLL* Auger Transition for Different Atoms

Atom	Z	$W_A(10^{14}\ s^{-1})$	$W_X(10^{14}\ s^{-1})$	$W_X/(W_X + W_A)$
Ne	10	7.40	0.63	0.0085
Mg	12	8.07	0.130	0.016
Ar	18	11.29	0.83	0.068
Ni	28	12.23	4.85	0.284
Kr	36	13.46	14.39	0.517

the transition reflects. Figure 1*b* shows the *KLM* Auger transition; Fig. 1*c* features a double Auger transition with ionization of two external electrons (the corresponding notation is *KLMV*). The different types of Auger transitions are described in review.[45]

If the hole is in an inner shell, especially in the *K* shell, the energy of Auger electrons is quite high. For compounds containing oxygen it is about 500 eV. Figure 2 shows the Auger spectrum of an H_2O molecule after the ejection of a *K* electron by X rays.[46] The shape of the Auger spectrum does not depend on the source of radiation, since the Auger transitions occur within 10^{-15}–10^{-11} s (depending on the "allowedness" of the transition) after the formation of a hole. For example, the Auger spectra obtained in electron accelerators and with X-ray sources are practically identical. On the other hand, the shape of the Auger spectrum is very sensitive to the chemical surrounding of the atom from which a core electron is ejected. This is clearly seen in Fig. 3, which features the C(*KVV*) Auger spectra of three hydrocarbons differing in the type of hybridization of valence electrons of the carbon atom.[47]

The energy of an Auger electron is determined by the difference in energy between the corresponding electron shells with correction for the

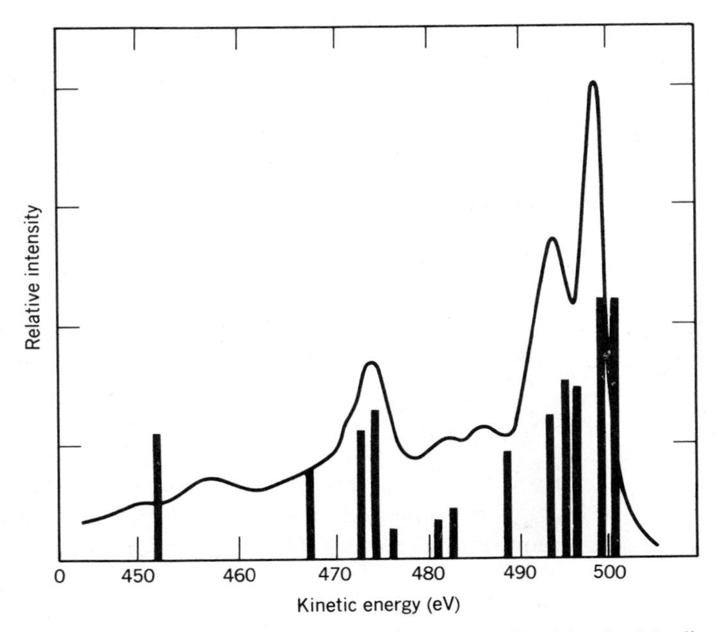

Fig. 2. Spectrum of Auger electrons for water vapor irradiated by the Mg_0 line (Reproduced by permission from Ref. 46).

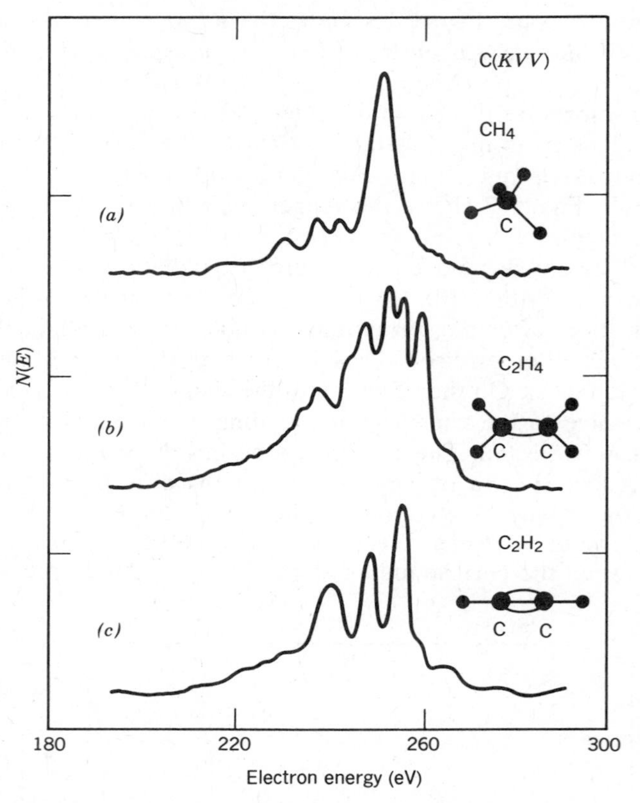

Fig. 3. Shape of the Auger spectrum depending on the type of the chemical bond (Reproduced by permission from Ref. 47).

interaction energy of the formed vacancies. For example, for the Auger transition featured in Fig. 1*b*,

$$E_e = I_K - I_L - I_M - U_{LM} \, ,$$

where I_K, I_L, and I_M are the ionization potentials of the corresponding electron shells, and U_{LM} is the hole–hole interaction term that reflects the fact that the energy required to eject an *M*-shell electron increases with formation of a hole in the *L* shell. Similarly, the total energy of the two Auger electrons in the double Auger transition shown in Fig. 1*c* equals

$$E_{e_1} + E_{e_2} = I_K - I_L - I_M - I_V - U_{LMV} \, .$$

An Auger transition may be regarded as an internal conversion of the energy of the electromagnetic quantum emitted as an electron makes the transition from an upper level to a lower one, into the energy of an external electron. In order to be able to calculate the intensity of the Auger transition, we must know the wavefunction of the initial and the final states of the system, including the wavefunction of an Auger electron in the field of the newly formed ion. The intensity is calculated according to the so-called Wentzel ansatz, which gives

$$w_{\mathrm{fi}} = \frac{2\pi}{\hbar} |\langle \psi_{\mathrm{f}}|\hat{H} - E|\psi_{\mathrm{i}}\rangle|^2 , \tag{3.1}$$

where ψ_{f} and ψ_{i} are the multielectron wavefunctions of the final and the initial states (the former also includes the wavefunction of the Auger electron), and E and $\hat{H}$ are the energy and the Hamiltonian of the system. Since an Auger transition takes place at constant energy, the value of E is the same for both the initial and the final states. The foundation and the discussion of formula (3.1) are presented in Refs. 48 and 49. A precision calculation of the Auger-transition intensity is quite an intricate job—one has to use the modern quantum-mechanical computer methods that make allowance for electron correlation and relaxation of the frame.[50–52]

Besides the Auger effect, there is one more physical mechanism leading to additional ionization after the formation of a core vacancy; it is due to the shake-up of the molecule's electron shell following the ejection of a core electron. The resulting sharp increase of the electromagnetic field in which the external electrons move leads to their excitation or ionization. This is readily seen in photoelectron spectra and in spectra produced by fast electrons, where besides the main maximum corresponding to ionization of the molecule with formation of an ion in the ground electron state, there appear additional peaks of smaller intensity (called satellites) corresponding to excitation or ionization of external electrons (Fig. 4). According to the precision calculations,[53,54] in the case of H_2O, CH_4, and NH_3 molecules, the probability of formation of satellites following the ejection of a K electron is about 0.3, meaning that about 30% of the ions either experience additional ionization or are formed in electron excited states. Only ~70% of the ions are formed in the ground electron state and later on undergo additional ionization via the Auger effect.

This results in a high probability of multiple ionization following the ejection of a K-shell electron. For instance, for Ne ($Z = 10$) the probability of subsequent ionization after the ejection of a K electron is as

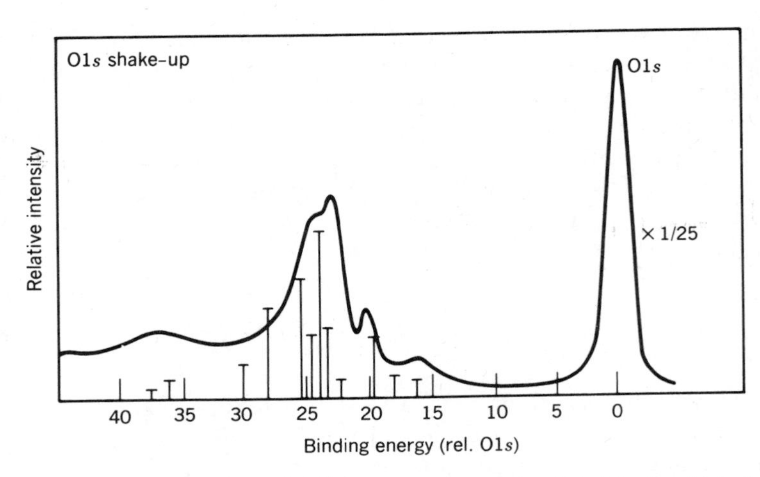

Fig. 4. Photoelectron spectrum of O(1*s*) for water vapor irradiated by the AlK_α line. To left of the main peak are the satellites. Solid line, experimental data[55]; vertical lines, results of calculations.[53] The energy is counted from the main peak corresponding to ejection of 1*s* electrons.

high as 99%, whereas for ejection of *L*-shell electrons it is considerably smaller, amounting only to 13%.[55] In Table III we present the relative yields of Ne ions with different charge multiplicities obtained when the gas was irradiated by the K_αAl X-ray line ($E_{\text{ph}} = 1256$ eV). The total yield of multiple-charge ions amounts to 93.5% (the value smaller than 99% is due to ionization from the *L*-shell and from more highly located shells). A similar distribution in charge multiplicity should be also expected for light molecules. For heavier elements the formation of ions with charge multiplicity ≥ 4 becomes more probable than the formation of double-charge ones.[56]

The total number of ions with a hole in the *K* shell produced by fast electrons per unit absorbed energy is actually relatively small. According to estimates,[57] it is about 10^{-2} per 100 eV, which means that the complete degradation of a 1-MeV electron in a medium results in formation of $\sim 10^2$ ions with a hole in the *K* shell, whereas the average number of all

TABLE III
Distribution of Ions with Different Charge Multiplicities Produced by Irradiation of Ne Gas by an X-Ray Line with $E_{\text{ph}} = 1256$ eV[55]

Ion charge	+1	+2	+3	+4	+5
Yield (%)	6.5 ± 0.7	68.5 ± 2.0	22 ± 2	2.8 ± 0.5	0.3

produced ions is 3×10^4, since the energy needed to produce an ion pair is about 30 eV.[20,21] However, it is the subsequent cascade of ionizations that distinguishes the ionization of the K electrons (and of L and M electrons, for heavier elements) from the whole mass of newlyformed ions.

The formation of a multiply charged ion following the ejection of a core electron often leads to decay of the molecule into several positive fragments, and, in the end, gives different final products than in photochemistry. An ion with a core hole can also make transitions into the highly excited states that cannot be occupied spectroscopically (for instance, into the $C^2\Sigma^+$ state of the N_2^+ ion[58]).

B. The Superexcitation States

The excitation of atoms into discrete states with $E > I_1$ usually leads to autoionization. This is so because the autoionization in atoms takes 10^{-14}–10^{-11} s, and the only competing channel is the radiative one with the time range 10^{-9}–10^{-6} s. Only in rare cases does the emission occur from autoionization states. For example, for an oxygen atom, the autoionization peaks at $\lambda = 878$–879 and 791–793 Å have the ionization efficiency $\eta(E) < 1$. Owing to the constraints imposed by selection rules, the lifetime of such states is 10^{-8} s, which makes the radiative decay possible. For atoms this is an exception. However, the situtation is different in the case of molecules.

Analyzing the data on molecular gases irradiated by vacuum UV emission,[60] Platzman[2] has noted that for certain gases the probability of ionization η (E_{ph}) is smaller than unity when E_{ph} exceeds I_1 by 10 eV or more. This was confirmed in his subsequent study of molecule–noble-gas mixture,[61] done in collaboration with Jesse. They have also observed an isotopic effect: the substitution of deuterium for hydrogen increases the ionization probability. Platzman thus concluded that in such discrete states with $E > I_1$ the predissociation efficiently competes with autoionization. Platzman has named them the *superexcitation* states (SES). The SES were discussed in a special issue of *Radiation Research*[62] (see also Refs. 25 and 63).

Owing to the electron–vibrational interaction in molecules, there is one more possible decay channel for SES. This is the nonradiative relaxation (internal conversion), in which the electron energy is transferred into vibrational energy of molecules (in the condensed phase, into thermal energy of the medium). If the molecule fluoresces, there may also occur fluorescence from the lowest excited state. (According to the empirical rule of Kasha,[64] the molecular fluorescence occurs from the lowest excitation level irrespective of the wavelength of the exciting radiation.)

The reactions of formation and decay of SES are summarized in the following scheme:

$$ABC \xrightarrow{h\nu > I_1} \begin{cases} ABC^+ + e^- & \text{direct ionization,} \\ ABC^{**} & \text{formation of SES,} \end{cases}$$
$$ABC^{**} \longrightarrow \begin{cases} ABC^+ + e^- & \text{autoionization,} \\ AB^* + C & \text{predissociation,} \\ ABC + \Delta E'_{\text{vib}} + h\nu_1 & \text{radiative relaxation,} \\ ABC + \Delta E_{\text{vib}} & \text{nonradiative relaxation.} \end{cases} \tag{3.2}$$

In the optical approximation, the radiation yield of SES for small molecules is estimated[65] to be $g_{\text{SES}} \simeq 0.6\text{–}1$. Similar estimates were obtained by Bednář.[57]

The SES play an important role in radiation chemistry, since they decay into highly excited products (radicals). Such "hot" radicals may take part in endothermal reactions. For instance, in his first basic study,[2] Platzman has pointed out the possibility of the one-stage reaction

$$H^* + H_2O \rightarrow H_2 + OH \tag{3.3}$$

with formation of molecular hydrogen. A number of data indicate that in a similar manner molecular hydrogen is also formed in radiolysis of hydrocarbons.[63]

One can imagine three possible ways for the SES to be formed: (1) the excitation of a core electron into a vacant orbital; (2) the formation of a Rydberg state converging to $I_n \geq I_1$; and (3) the excitation of two valence electrons.

Owing to their short lifetime, the states of the first type for molecules consisting of light elements (and it is these molecules Platzman and Jesse considered) decay according to the Auger mechanism and cannot be considered as SES since the latter are neutral formations. The hydrogen-like excited states, called Rydberg states, live considerably longer, at least if I_n is not too high, that is, if the hole is formed in a valence shell (though the latter may be a deeper valence level). Being sufficiently far from the positive ion, an electron in a Rydberg state moves slower. (The distance between a Rydberg electron and the nucleus may exceed 10 nm.) This favors the exchange between electron and vibrational energies, which leads to predissociation.

The energy levels of atomic Rydberg states are described by a hydrogen-type formula:

$$E_n = \frac{e^2}{2a_0(n - \delta_R)^2}, \tag{3.4}$$

where a_0 is the Bohr radius, e is the electron charge, n is the main quantum number, and δ_R is the Rydberg-shift constant. For molecules, formula (3.4) is valid at small n, when the spacings between the electron levels are much greater than the vibrational and rotational quanta of motion of molecules. In this case, in the adiabatic approximation, the energy of molecular levels can be presented as the sum

$$E_{nvr} = E_n + E_v + E_r , \qquad (3.5)$$

where E_n is the electron energy (3.4), and E_v and E_r are, respectively, the vibrational and the rotational energies of the molecule. For higher n the adiabatic approximation is no longer valid. The interaction of Rydberg electron states with vibrational and rotational motions of the molecule strongly widens the spectral lines, transforming the Rydberg spectrum into an almost continuous one. This results because the lifetime of molecular Rydberg states is many orders of magnitude smaller than in the case of atoms. (For the latter $\tau_{Ry} \sim n^3$ and may be as high as 10^{-3}–10^{-1} s.) For example, at $n = 8$–11, the width of the Rydberg level of an H_2 molecule for the autoionization channel is 1–5 cm^{-1}, which corresponds to $\tau = \hbar/\Delta E \simeq 10^{-12}$ s, that is, on the order of vibrational transitions. However, this lifetime is not too small to rule out predissociation.

The lifetime of Rydberg states with a large angular momentum l is considerably greater. As was shown by Fano[67,68], such states can be formed via the formation and subsequent decay of an intermediate negative complex if the energy of the incident electron exceeds the ionization potential by several electron volts. Owing to the development of a centrifugal barrier, the lifetime of such states becomes considerably longer with respect to both autoionization and predissociation, as well as to the radiative transitions. Apparently, the Rydberg states with $\tau = 10^{-5}$–10^{-4} s discovered in studies[69,70] for molecules N_2, O_2, CO_2, NH_4, and a number of others are of such kind. Let us stress that such long lifetimes have been observed in the rarefied gas phase. In the condensed phase, the situation may be different.

A special kind of SES is the highly excited Rydberg states that converge to I_1. Adding up to the energy of an electron, the vibrational and rotational quanta transfer the latter from a state with $E < I_1$ into a state with $E > I_1$ [see formula (3.5)]. The majority of the autoionization thresholds at E exceeding I_1 by several electron volts are of this electron–vibrational Rydberg kind.[59] In such states the predissociation competes with autoionization, leading to $\eta(E) < 1$. This possible realization of SES was first discussed in a review by one of the authors[25] (see also Ref. 71). The role of Rydberg states in radiation chemistry is discussed in Refs. 72 and 73.

A detailed theoretical study of Rydberg states formed with excitation of one of the valence electrons in a diatomic molecule has been done by Golubkov and Ivanov.[74–76] However, no one yet has made a theoretical study of Rydberg states with a core hole or of Rydberg states in condensed media. In the latter case, the denominator of the right-hand side of formula (3.4) must include the permittivity of the medium ϵ, since, owing to the large radius of Rydberg states (up to hundreds of angstroms), there are many molecules between the electron and the positive ionic residuum. It is still not clear to what extent this will shorten the lifetime of Rydberg states.

A direct confirmation of existence of SES has been obtained only in the gaseous phase. In this case, an efficient method of identifying SES is measuring the luminescence of decay products versus the energy of the ionizing radiation. A series of such measurements has been performed.[77,78] When irradiating hydrocarbons with electrons, the authors have observed the Balmer-series lines of an H atom appearing when the energy exceeded I_1, which indicates the formation of H*. This means that the excitation energy of a hydrogen atom is above 12 eV. In Fig. 5 we show the luminescence of the CH_4 fragments versus the energy of the electron beam. The threshold above which the Balmer series appears is 21.8 eV, which is much higher than the ionization potential $I_1 = 12.51$ eV. On the other hand, H* cannot be formed in a decay of CH_4^+ since in that case the luminescence threshold would have been $I_1(CH_4) + E(H^*) \gtrsim 25.5$ eV rather than 21.8 eV. As is shown in Refs. 77 and 78, the intensity decreases in a strictly regular manner with increase of the line number n,

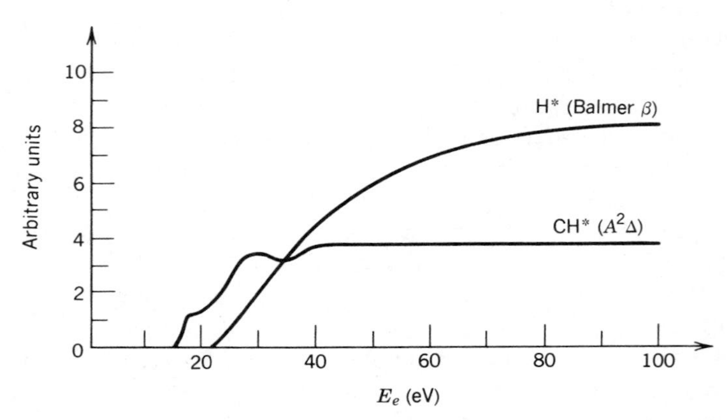

Fig. 5. Energy dependence of the luminescence of CH_4 fragments with irradiation by electrons (Reproduced by permission from Ref. 79). Plotted along the axis of ordinates is the intensity of luminescence in arbitrary units.

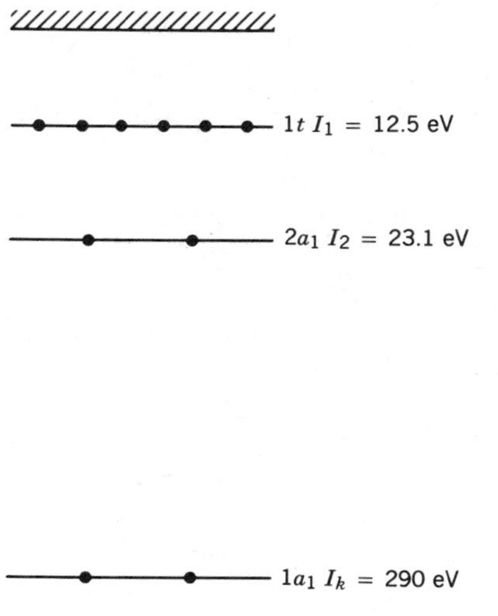

Fig. 6. Scheme of occupation of one-electron level of CH_4 (the level is triply degenerate) and the corresponding experimental values of ionization potentials I_n.

indicating that the SES formed via the excitation of inner valence electrons from nonbound orbitals are of Rydberg nature.

The experimental data presented in Ref. 79 unambiguously prove that the SES are formed in two-electron excitations. According to the measurements of the energy-absorption and ionization probabilities, $\eta(E)$ is below unity for CH_4 at energies from 27 to 80 eV. The scheme of occupation of energy levels in a CH_4 molecule is featured in Fig. 6. Since the highest ionization potential of valence electrons is 23.1 eV, and the next potential, corresponding to ionization of the K shell, is around 290 eV, the SES in this case correspond to excitation of two or more electrons.

Thus, among the three possible ways of SES formation mentioned previously, only the latter two are actually realized.

C. The Collective Excitation States of Plasmon Type

As is well known, in crystalline solids there may be formed collective electron-excitation states called *excitons*.[81,82] Such states are excited only in media with periodic structure and are delocalized over a large volume of atoms (or molecules), their excitation energy being 0.1–0.5 eV lower than the energy of the electron states of isolated molecules that produced them. The nature and spectroscopy of exciton states have been thoroughly studied both experimentally and theoretically. In this section we will dwell on the high-excitation states with energy much greater than the energy of discrete electron states in isolated molecules, and formed not only in periodic crystals but also in amorphous condensed media, including liquids.

The study of polymer and liquid films irradiated by fast electrons has revealed that the energy-loss spectrum has wide intensive maxima around 15–25 eV (see Figs. 7–9). The oscillator strength characterizes the probability of energy absorption; in condensed media it can be expressed through the imaginary part of the inverse permittivity of the medium (for details see Section V.B.2). The curves in Fig. 7 clearly show the pronounced difference in the behaviour of oscillator strengths in gaseous and in liquid phases. The distribution of oscillator strengths in the liquid phase has been found from the experimental optical spectral obtained in Ref. 87. On the other hand, the transition into the solid phase does not lead to any essential changes in the absorption spectrum (cf. the spectrum for polycrystalline ice in Fig. 8). Similar wide absorption peaks have been also discovered in certain molecular liquids (see Refs. 88 and 89). At present these wide absorption bands are believed to correspond to collective excitations of the plasmon type. The *plasmon states* are not

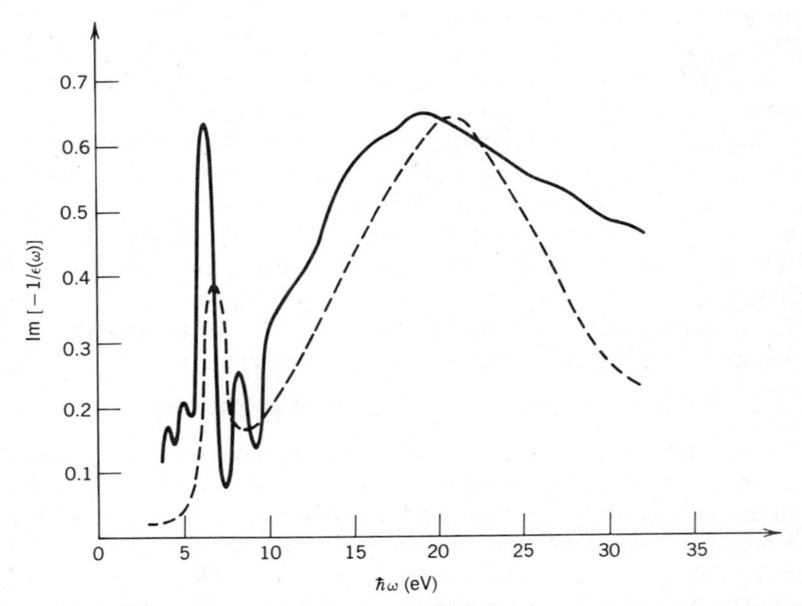

Fig. 7. Spectrum of energy losses in polystyrene films: ---, experiment (Ref. 83); ——, the calculation (Reproduced by permission from Ref. 84).

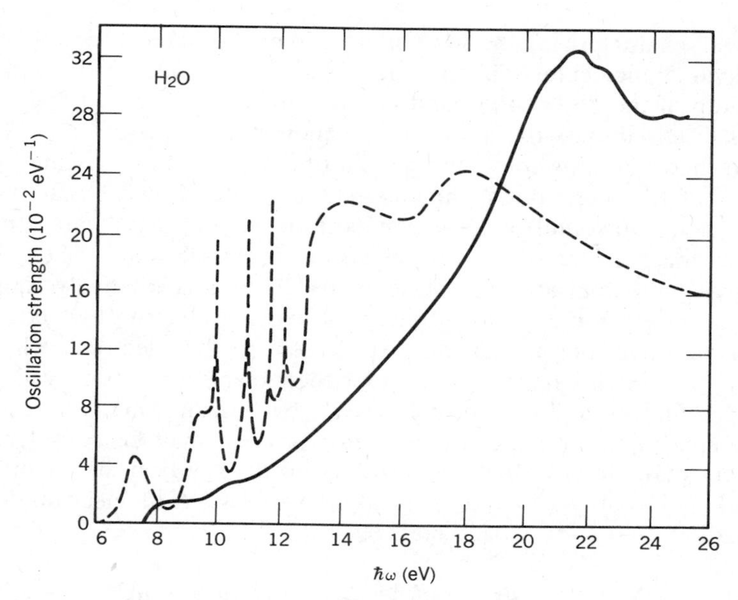

Fig. 8. Spectrum of energy losses in water (taken from Ref. 4): ——, liquid water; – – –, gaseous phase.

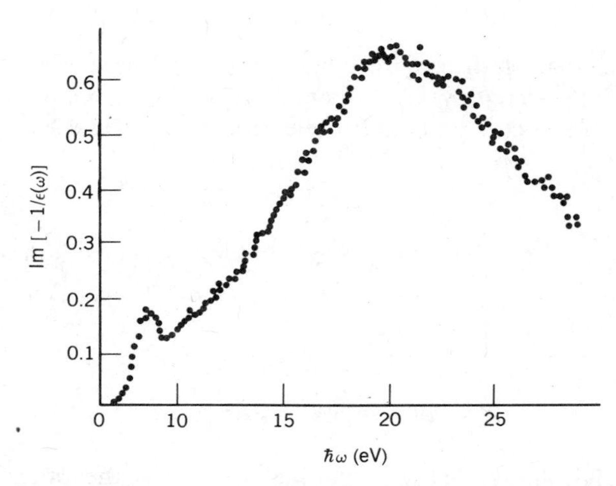

Fig. 9. Spectrum of energy losses in polycrystalline ice (taken from Ref. 85).

localized, characterizing the state of an ensemble of molecules rather than of individual molecules. The lifetime of such states can be estimated from the width of the absorption band and proves to be within 10^{-16}–10^{-15} s. Let us briefly discuss the nature of plasmon states.

The theory of *plasmons*—the quanta of electron-density oscillations in the gas of free electrons in metals—has been developed by Bohm and Pines[90] and is in complete agreement with the experimental data concerning the inelastic losses by fast electrons in metal films. The physics underlying the formation of plasmon oscillations is relatively simple.[90] Imagine that in some region of plasma a passing electron has induced a surplus positive charge. Tending to screen it, the electrons will start moving toward this region, but, under their own momentum, will slip a little bit further and will begin moving backward. This will result in oscillations of the volume-charge density, which may be considered as coherent periodic deviations of the electron density from its mean value at each point of plasma, that is, as a plane wave of electron-density fluctuations

$$\Delta\rho(\mathbf{r}, t) \equiv \rho(\mathbf{r}t) - \bar{\rho}(\mathbf{r}) = \Delta\rho_{\max} \exp\left[i(\omega t - \mathbf{qr})\right] . \tag{3.6}$$

As was shown by Bohm and Pines,[90] under certain conditions the equations of motion for a Fourier component of the electron density can be reduced to harmonic-oscillation equations with the frequency depending only on the density of electrons n_e:

$$\omega_{\mathrm{p}} = (4\pi n_e e^2/m)^{1/2} . \tag{3.7}$$

It is essential that plasmons are longitudinal electric waves, that is, that the electric field strength vector is parallel to the wave vector: $\mathscr{E} \parallel \mathbf{q}$. For a plane wave, the vector potential of electromagnetic field is

$$\mathbf{A} = \mathbf{A}_0 \exp\left[i(\omega t - \mathbf{qr})\right] . \tag{3.8}$$

Expressing the electric and magnetic field strength vectors in terms of **A**, we find that

$$\mathscr{E} = \frac{1}{c}\frac{\partial A}{\partial t} = -i\,\frac{\omega}{c}\,\mathbf{A} = -iq\mathbf{A} , \tag{3.9}$$

$$\mathscr{H} = \mathrm{rot}\,\mathbf{A} \equiv [\nabla\mathbf{A}] = i[\mathbf{qA}] . \tag{3.10}$$

From (3.9) it follows that $\mathscr{E}$ is parallel to **A**. On the other hand, the plasmon oscillations are longitudinal ($\mathscr{E} \parallel \mathbf{q}$), so **A** is also parallel to **q**, and the vector product on the right-hand side of (3.10) is zero, meaning that the magnetic field of plasmon oscillations equals zero. Substituting $\mathcal{H} = 0$ into Maxwell equations for a dielectric medium,[91] we find

$$\text{rot}\, \mathcal{H} = \frac{\epsilon(\omega)}{c} \frac{\partial \mathcal{E}}{\partial t} = 0 . \tag{3.11}$$

and get the following fundamental result: the natural frequencies of plasmon oscillations must obey the equation

$$\epsilon(\omega) = 0 . \tag{3.12}$$

Although as far back as 1960 Fano[92] has pointed out the possibility of existence of collective plasmon-type oscillations in molecular media, this question has been discussed for quite a long time (see Ref. 25), especially as Platzman[17] has shown that Fano's criterion of existence of plasmons, namely,

$$f_1 \omega_p^2 \gg \Delta \omega_{01}^2 \tag{3.13}$$

is not obeyed in real media. In (3.13) f_1 denotes the oscillator strength of the ith mode. In 1974 Brandt and Ritchie[93] have formulated criteria that the dielectric function of a medium must obey in the vicinity of the plasmon-absorption point, while Heller and others[3] have shown that these criteria are obeyed in the case of the wide absorption peak observed for liquid water with a maximum at 21.4 eV (Fig. 8). And though Fano's criterion is not obeyed in real media, the oscillations still can be collective, being realized as *longitudinal polarization waves*.[91,94]

The passage of a particle with the charge e and velocity v through a medium is equivalent to the current $\mathbf{j} = e\mathbf{v}\delta(x - vt)$, which polarizes the medium, and there appear polarization waves. One such wave is schematically featured in Fig. 10. The electric field $\mathcal{E}_{\text{pol}}$ resulting from this polarization interacts with the moving particle, owing to which the latter loses its energy. Solving the Maxwell equations for the medium, one is able to find $\mathcal{E}_{\text{pol}}$ and obtain an expression for the probability of the particle losing the energy $\hbar\omega$ over the path length l following a single scattering into the solid angle $d^2\Omega$ with the deflection angle θ,[94] namely,

$$\frac{\partial^3 W}{\partial(\hbar\omega)\partial^2\Omega} = \frac{l}{(e\pi a_0)^2 q^2} \operatorname{Im}\left[\frac{1}{\epsilon(\mathbf{q}, \omega)}\right], \tag{3.14}$$

where the momentum q transferred to the medium is found from the

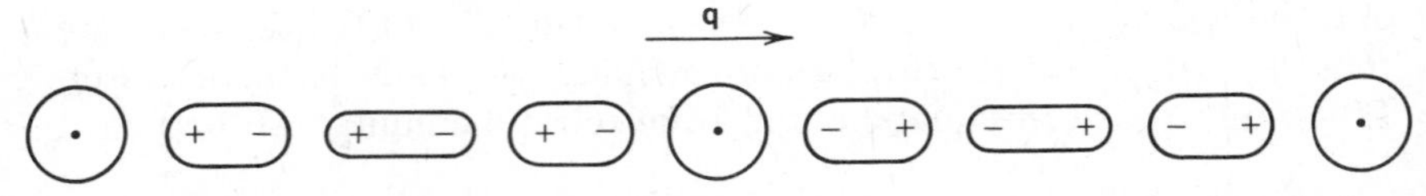

Fig. 10. Schematic diagram of a longitudinal polarization wave in a condensed medium consisting of nonpolar molecules.

energy and momentum conservation laws and depends on the scattering angle θ.

Formula (3.14) for the energy-loss probability contains a typical factor

$$\mathrm{Im}\left[-\frac{1}{\epsilon(\mathbf{q}, \omega)}\right] = \frac{\epsilon_2(\mathbf{q}, \omega)}{\epsilon_1^2(\mathbf{q}, \omega) + \epsilon_2^2(\mathbf{q}, \omega)} .$$

The energy losses are maximum at maxima of $\epsilon_2(\mathbf{q}, \omega)$, which corresponds to discrete transitions in isolated molecules, as well as at zeros of the denominator, that is, for solutions of Eq. (3.12), which corresponds to excitation of collective states. In order to find the frequencies of the latter we must know the explicit form of $\epsilon(\mathbf{q}, \omega)$.

A medium with bound electrons can be regarded as an ensemble of oscillators each of which is characterized by three parameters: its natural frequency ω_i, the oscillator strength f, and the damping constant γ_i, which equals the inversed lifetime of the corresponding state ($\gamma_i = 1/\tau_i$). The well-known expression for the permittivity of the medium[91,95] can be presented in the form[89]

$$\epsilon(\omega) = 1 + \sum_i \frac{f_i \omega_p^2}{\omega_i^2 - \omega^2 - i\gamma_i \omega} . \tag{3.15}$$

Here ω_p is the frequency of plasmon oscillations in a system of free electrons (3.7). The oscillator strengths f_i introduced previously differ from the usual f_{on} (see Section IV) in their normalization ($\Sigma_{i=1}^N f_i = 1$). A method for calculating the thus defined oscillator strengths from experimental values of ϵ_2 is presented in Ref. 89. Since the energy range essential for collective oscillations is $\hbar\omega < 30$ eV, the electrons of inner atomic shells can be disregarded. Thus, the value of n_e is determined by the density of valence electrons only, and only the transitions of these electrons should be taken into account in the sum over i in formula (3.15). A convenient formula for calculating the frequencies ω_p in molecular liquids is presented in ref. 89:

$$\hbar\omega_p = 28.8(N_v \rho / M)^{1/2} \text{ eV} , \tag{3.16}$$

where N_v is the number of valence electrons in one molecule, M is the molecular weight, and ρ is the density in g/cm^3. In Table IV we present the parameters of a number of substances together with the values of $\hbar\omega_p$ calculated according to formula (3.16) and the experimental values of the resonance frequencies ω_r corresponding to the maximum of the energy-loss function $\mathrm{Im}[-1/\epsilon(\omega)]$.

The natural frequencies of a medium must satisfy Eq. (3.12). However, it is very difficult to calculate them in the general form using formula (3.15). So, following Ref. 89, let us consider a number of special cases.

For free electrons in a metal $f_0 = 1$ and $\omega_0 = 0$, while all the other f_i $(i \neq 0)$ are zero. Thus,

$$\epsilon(\omega) = 1 - \frac{\omega_p^2}{\omega^2 - i\gamma_0\omega} . \tag{3.17}$$

If the damping is small, the resonance occurs at $\omega_r \approx \omega_p$, that is, the free plasmons are excited. Indeed, from formula (3.17) we get

$$\mathrm{Im}\left[-\frac{1}{\epsilon(\omega)}\right] = \frac{\omega_p^2\gamma\omega}{(\omega^2 - \omega_p^2)^2 + \gamma^2\omega^2} . \tag{3.18}$$

Function (3.18) has a maximum at $\omega = \omega_p$, where it equals

$$\mathrm{Im}\left[-\frac{1}{\epsilon(\omega)}\right]_{\max} = \frac{\omega_p}{\gamma} . \tag{3.19}$$

TABLE IV
Parameters of Different Substances and the Resonance Frequencies ω_r Corresponding to the Maximum of the Energy-Loss Function

Substance	ρ (g/cm^3)	M	N_v	n_v, (10^{23} cm^{-3})	$\hbar\omega_p$ (eV)	$\hbar\omega_r$, Experimental (eV)
Water	1	18	8	2.675	19.2	21.4[3,89]
						20.4[85]
Glycerine	1.26	92	38	3.135	20.78	20.4[89]
						22[96]
Benzene	0.879	78	30	2.034	16.74	21.3[97]
Polyethylene	0.92	28	12	2.371	18.08	20.6[86]
Polystyrene	1.05	104	40	2.432	18.3	21.7[83]
						22[98]
Poly(methyl methacrylate)	1.187	100	40	2.858	19.84	22[87]

In the vicinity of the maximum the energy-loss function (3.18) is of Lorentz form. With $\gamma \to 0$ it transforms into a delta function. In order to see this, let us use the representation of a delta function for a nonnegative variable (see the Mathematical Appendix A in Ref. 99):

$$\lim \frac{2}{\pi} \frac{\alpha}{\alpha^2 + x^2} = \delta(x) , \tag{3.20}$$

which gives us

$$\lim_{\gamma \to 0} \operatorname{Im} \left[-\frac{1}{\epsilon(\omega)} \right] = \frac{\pi}{2} \omega_p^2 \delta(\omega^2 - \omega_p^2) . \tag{3.21}$$

In many molecules one of the transitions has the largest oscillator strength; let it be f_1. In this case we have the following approximate representation

$$\epsilon(\omega) = 1 + \frac{f_1 \omega_p^2}{\omega_1^2 - \omega^2 - i\gamma_1 \omega} . \tag{3.22}$$

As one can easily check, if the damping is small, the energy-loss function has only one maximum corresponding to $\epsilon(\omega) = 0$ at

$$\omega_r = (f_1 \omega_p^2 + \omega_1^2)^{1/2} . \tag{3.23}$$

At low densities (in the gaseous phase) $\omega_p \approx 0$, and the maximum of the absorption function corresponds to $\omega_r \approx \omega_1$, that is, as for an isolated molecule. In the opposite case of very high densities with

$$f_1 \omega_p^2 \gg \omega_1^2 \tag{3.24}$$

[cf. Fano's condition (3.13)], $\operatorname{Im}[-1/\epsilon(\omega)]$ has a maximum at $\omega_r = \sqrt{f_1}\,\omega_p$. If condition (3.24) were obeyed exactly, the resonance frequency ω_r would be shifted with respect to ω_p toward smaller frequencies ($f_1 < 1$) and would correspond to longitudinal collective oscillations of the gas of valence electrons interacting with the ionic frame. In molecular media one usually observes the opposite: $\hbar\omega_r > \hbar\omega_p$ (see Table IV). So the peak of the energy-loss function can be totally attributed neither to purely collective oscillations of valence electrons nor to the one-electron transitions. The latter situation is typical for molecular liquids.

As an example, let us consider liquid water (Fig. 8). The highest oscillator strength, $f_1 = 0.43$, corresponds to the transition $\hbar\omega_1 = 13.5$ eV. The peak the energy-loss function $\operatorname{Im}[-1/\epsilon(\omega)]$ has around 21 eV is of plasmon nature, that is, corresponds to longitudinal collective oscillations of the whole ensemble of molecules. However, these are not purely plasmon oscillations of free electrons, since condition (3.24) in this case is evidently not obeyed. Indeed, according to (3.16), for water $\hbar\omega_p \simeq$ 19.2 eV. So for the peak with $\hbar\omega_1 = 13.5$ eV and $f_1 = 0.43$ we have $f_1(\hbar\omega_p)^2 = 159$ eV^2 while $(\hbar\omega_1)^2 = 182$ eV^2. Nevertheless, the large deviation from the ionization potential (from 13.5 to 21 eV) indicates that collective effects give a large contribution.

It is very different to calculate the energy-loss function $\mathrm{Im}[-1/\epsilon(\mathbf{q}, \omega)]$ theoretically. At $q \to 0$ one can use the optical data. This way Lindhard[90] has obtained a formula for the permittivity of the electron gas, while the authors of Ref. 100 have constructed a semiempirical formula for $\epsilon(\mathbf{q}, \omega)$ of water.

The energy loss function is related to the dynamical structural factor $S(\mathbf{q}, \omega)$,[90] which describes the scattering of particles in a liquid, namely,

$$S(\mathbf{q}, \omega) = \frac{q^2}{4\pi^2 e^2} \mathrm{Im}\left[-\frac{1}{\epsilon(\mathbf{q}, \omega)}\right]. \tag{3.25}$$

If we find the structural factor from some independent experimental data, say, from those concerning the scattering of neutrons, relation (3.25) will enable us to find the loss function also. Since the structural factor is determined by the density–density electron correlation function, relation (3.25) implies that the excitation of plasmon oscillations is determined by the correlation in electron motion.

The collective excitations may decay in a number of ways. In an electron gas they decay owing to the interaction between electrons. In this case the Coulomb interaction of electrons is strongly screened by the electron cloud and is noticeable only at short distance. Owing to this short-range interaction, the oscillational energy of a plasmon transforms into kinetic energy of individual electrons, that is it is distributed all over the electron gas.[101]

In atomic–molecular media the damping of plasmon states is due to the interaction of plasmon waves with electrons, lattice vibrations, and impurities. The electron–plasmon interaction is a long-range one. With absorption of a plasmon, the momentum q is transferred to the electron, resulting in a decay of the collective state into a single-particle one. The latter process is identical with absorption of a photon with the same energy. Wölff[102] (see also Ref. 103) has shown that in this case the lifetime can be expressed in terms of two optical constants: the absorption coefficient κ and the refractive index n_r, namely,

$$\frac{1}{\tau} = n_r \kappa \omega . \tag{3.26}$$

For water, at frequencies around that of the plasmon state $\hbar\omega_r = 21.4$ eV, we have $n_r = 0.76$ and $\kappa = 0.36$.[3] Using formula (3.26) we find $\tau \simeq 1.1 \times 10^{-16}$ s. The lifetime of a plasmon state can be also estimated using the uncertainty relation. From the energy-loss spectrum in water we find that the half-width of the excitation level $\hbar\omega_r = 21.4$ eV amounts to about 3.4 eV. This gives us $\tau = (1/\Delta\omega_r) \simeq 2 \times 10^{-16}$ s, which is in good agreement with the value given by formula (3.26).

IV. CROSS SECTIONS OF INTERACTION OF FAST CHARGED PARTICLES WITH A MOLECULAR MEDIUM

For studying the effects a charged particle produced in a medium and for simulating the structure of its track, it is necessary to know the probabilities with which a particle loses its energy when exciting and ionizing the molecules of the medium. Such probabilities are conventionally expressed in terms of cross sections. Let J denote the number of particles in an incident beam passing through a unit area of its section per unit time. The number of incident particles scattered through an angle θ into a solid angle $d\Omega$ after exciting a molecule into the nth quantum state is proportional to J:

$$\frac{dJ_n(\theta)}{d\Omega} = \sigma_n(\theta)J\,, \tag{4.1}$$

where the proportionality factor $\sigma_n(\theta)$ depends on the angle θ and on the state into which the molecule was excited. $\sigma_n(\theta)$ is called the *differential scattering cross section*, as it has the dimension of area.

We will present the approximate analytical formulas used for calculating the cross sections of different processes of the interaction of charged particles with molecules, and discuss the limits within which they are valid. In this connection let us first dwell on the general formulation of the problem in the framework of the scattering theory.

A. General Formulas for the Cross Sections of Scattering of a Charged Particle by a Molecule

The scattering of a "structureless" particle with mass M and charge z by a molecule (an atom) is described by the Schrödinger equation

$$[H_0 + T_{\mathbf{r}} + V(\mathbf{r}, \mathbf{r}_1, \ldots, \mathbf{r}_N)]\psi(\xi, \xi_1, \ldots, \xi_N) = E\psi(\xi, \xi_1, \ldots, \xi_N)\,, \tag{4.2}$$

where H_0 is the Hamiltonian operator of an isolated molecule; $T_{\mathbf{r}} =$

$(\hbar^2/2M)\nabla_{\mathbf{r}}$ is the kinetic energy operator of the incident particle; V is the energy of the Coulomb interaction between the particle and the molecule; $\psi(\xi, \xi_1, \ldots, \xi_N)$ is the total wavefunction of the interacting system, ξ_i and ξ being sets of spatial and spin coordinates of the ith electron of the molecule and of the incident particle, respectively; and E is the total energy of the system (the charged particle and the molecule).

The wavefunction of the system must obey the asymptotic boundary conditions[104,105]

$$\psi(\xi, \xi_1, \ldots, \xi_N)_{|\mathbf{r}-\mathbf{r}_i|\to\infty} \to \Phi_0(\xi, \xi_1, \ldots, \xi_N) + \psi^+(\xi, \xi_1, \ldots, \xi_N) \tag{4.3}$$

The first term in this formula describes the initial state of the system before the scattering, the wavefunction of the incident particle being a plane wave:

$$\Phi_0(\xi, \xi_1, \ldots, \xi_N) = v_{\mathbf{k}_0}(\xi)\psi_0(\xi_1, \ldots, \xi_N)\,, \tag{4.4}$$

$$v_{\mathbf{k}_0} = \exp(i\mathbf{k}_0\mathbf{r})\chi(\kappa)\,, \tag{4.5}$$

Here $\mathbf{k}_0$ is the wave vector of the incident particle, $\chi(\kappa)$ is its spin wave function, and κ is its spin coordinate. The second term in (4.3) is a superposition of products of scattered waves and the wavefunctions of all the possible molecular states that obey the energy conservation law:

$$\psi^+(\xi, \xi_1, \ldots, \xi_N) = \sum_n A_{n0}(\mathbf{k}_n, \mathbf{k}_0)\,\frac{1}{r}\,\exp(i\mathbf{k}_n\mathbf{r})\chi(\kappa)\psi_n(\xi_1, \ldots, \xi_N)\,. \tag{4.6}$$

The expansion coefficients in (4.6) are the *scattering amplitudes*, and equal[99]

$$A_{n0}(\mathbf{k}_n, \mathbf{k}_0) = \frac{M}{2\pi\hbar^2}\,\langle\Phi_n|V|\psi^+\rangle\,, \tag{4.7}$$

where, as in formula (4.4),

$$\Phi_n(\xi, \xi_1, \ldots, \xi_N) = \mathbf{v}_{\mathbf{k}_n}(\xi)\psi_n(\xi_1, \ldots, \xi_N)\,. \tag{4.8}$$

The scattering amplitude $A_{n0}(\mathbf{k}_n, \mathbf{k}_0)$ describes a scattering process in which an incident particle with the initial momentum $\hbar\mathbf{k}_0$ is scattered by a molecule into a state with momentum $\hbar\mathbf{k}_n$, while the molecule itself makes a transition from the initial state with the wavefunction ψ_0 into an excited state described by the wavefunction ψ_n. In the particular case where ψ_n coincides with ψ_0, the amplitude (4.7) describes elastic scattering.

The scattering amplitude $A_{n0}(\mathbf{k}_n, \mathbf{k}_0)$ is related to the scattering cross section in a simple way[99,104]:

$$d\sigma_{n0}(\theta) = \frac{k_n}{k_0} |A_{n0}(\mathbf{k}_n, \mathbf{k}_0)|^2 \, d\Omega \,, \tag{4.9}$$

where $d\Omega = \sin\theta \, d\theta \, d\varphi$, and θ and φ are the scattering angles in the center-of-mass system, which determine the direction of $\mathbf{k}_n$ with respect to $\mathbf{k}_0$.

If the final state of the molecule belongs to the continuous part of the spectrum, the corresponding scattering channel describes ionization. In this case the energy level corresponds to the energy of free electrons $\varepsilon_{k'} = \hbar^2 k'/2m$, and the wavefunction is normalized to the delta function $\delta(\mathbf{k} - \mathbf{k}')$ and has an asymptotic form of a plane wave plus a converging wave.[106] Such a state of the molecule may be also regarded as a scattering of one of its electrons in the field of the molecular ion (neglecting the interaction with the incident particle). The corresponding scattering amplitude $A_{k'0}(\mathbf{k}_n, \mathbf{k}_0, \mathbf{k}')$ depends on the momentum $\hbar\mathbf{k}'$ of the electron ejected in the direction (θ', φ') into the solid angle $d\Omega'$, while the differential scattering cross section is

$$d\sigma_{k'0}(\theta, \theta') = \frac{k}{k_0} |A_{k'0}(\mathbf{k}, \mathbf{k}_0, \mathbf{k}')|^2 k'^2 \, d\Omega \, d\Omega' \, dk' \,. \tag{4.10}$$

From the energy conservation law it follows that $\hbar^2(k_n^2 - k_0^2) = 2M\hbar\omega_{0n}$ in the case of excitation, and that $\hbar(k^2 - k_0^2) = 2M(\hbar^2 k'^2/2m - I)$ in the case of ionization, where $\hbar\omega_{0n}$ is the excitation energy of the molecule, I is the ionization potential of the electron, and m is the electron mass.

The scattering amplitude defined in (4.7) characterizes the so-called direct scattering. However, when the scattered electron is slow, there can also occur processes in which the molecule captures the incident electron and emits one of its own. This sort of scattering is described by the exchange amplitude $B_{n0}(\mathbf{k}_n, \mathbf{k}_0)$, the formula for which differs from that for the direct amplitude (4.7) in that in the final state of the system Φ_n the coordinates of the incident electron are transposed with coordinates of molecular electrons, namely,

$$B_{n0}(\mathbf{k}_n, \mathbf{k}_0) = -\frac{m}{2\pi\hbar^2} \sum_i \langle P_{\xi\xi_i} \Phi_n | V | \psi^+ \rangle \,. \tag{4.11}$$

It is due to the exchange scattering amplitude B_{n0} that the cross section of excitation of a molecule into a triplet excited state is nonzero (see Section IV.B.3).

Thus, in order to determine the scattering cross sections we must find the wavefunctions of the system after the scattering for a known interaction potential. This is a very complicated problem in the case of many-electron systems and can be solved only with various approximate methods. We will only briefly discuss the results obtained in the Born approximation and in the quasi-classical impact parameter method. A detailed discussion of various approximate methods can be found in special monographs (e.g. in Refs. 104 and 107) or in reviews (see Refs. 105, 108–112).

B. The Born Approximation

1. *Differential Cross Sections*

If the interaction operator V can be regarded as a perturbation to the Hamiltonian H_0 (this is the case for fast particles the velocity of which is much greater than those of atomic electrons), the function ψ^+ can be found using the perturbation theory. Such an approach was named the *Born approximation*. In the first Born approximation we replace the function ψ^+ by that of the initial state of the scattering system,[48] that is, put $\psi^+ \equiv \Phi_0$, and thereby do not have to solve Eq. (4.2). In this way, for the differential cross section of direct scattering, we get

$$d\sigma_{n0}^{\mathrm{B}}(\theta) = \frac{M}{2\pi\hbar^2} \frac{k_n}{k_0} |\langle \Phi_n | V | \Phi_0 \rangle|^2 \, d\Omega \,. \tag{4.12}$$

A consecutive application of the Born approximation to the problem of calculating the scattering cross sections was first done by Bethe (a detailed discussion of his theory is presented in Refs. 104, 106, 113). Integrating over the coordinates of the incident particle we can obtain simple analytical formulas for the cross sections.

If the energy is transferred is small portions, leading to excitation of a molecule from the ground state into the nth quantum state with transition energy $\hbar\omega_{0n}$ (this sort of collisions are called *glancing*), the cross section is given by Bethe's formula

$$d\sigma_{n0}(q) = \frac{4\pi z^2 e^4}{mv^2} \frac{f_{0n}(q)}{\hbar\omega_{0n}} \frac{dq}{q} \,. \tag{4.13}$$

In this formula v denotes the velocity of the incident particle, which has the charge z; q is the transferred momentum; m and e are the electron mass and charge; and $f_{0n}(q)$ is the *generalized oscillator strength* introduced by Bethe:

$$f_{0n}(q) = \frac{\hbar\omega_{0n}}{\text{Ry}\, a_0^2 q^2} \, |\mu_{n0}(q)|^2 \,, \tag{4.14}$$

where Ry stands for a unit of energy of one Rydberg, equal to $me^4/2\hbar^2 =$ 13.65 eV; $a_0 = \hbar^2/me^4$ is the Bohr radius; and $\mu_{n0}(q)$ denotes the matrix element of the intramolecular transition in question:

$$\mu_{n0}(q) = \langle \psi_n | \sum_{i=1}^{N_e} \exp(iqx_i) | \psi_0 \rangle \,. \tag{4.15}$$

Here we assume that the x axis is directed along the momentum vector $\boldsymbol{q}$, and N_e is the total number of electrons in the molecule (the atom). At small q we can confine ourselves to only the first two terms in the expansion $\exp(iqx_i) \approx 1 + iqx_i$. Owing to the orthogonality of the wavefunctions, the first term gives a zero contribution to the matrix element, and the generalized oscillator strength coincides with the usual optical one:

$$f_{0n} = \frac{\hbar\omega_{0n}}{\text{Ry}\, a_0^2} \left| \langle \psi_n | \sum_{i=1}^{N_e} x_i | \psi_0 \rangle \right|^2 \equiv \frac{\hbar\omega_{0n}}{\text{Ry}} M_{n0}^2 \,, \tag{4.16}$$

where

$$M_{n0}^2 = \frac{1}{a_0^2} \left| \langle \psi_n | \sum_{i=1}^{N_e} x_i | \psi_0 \rangle \right|^2 \tag{4.17}$$

is the conventional notation for the dimensionless squared dipole matrix element.

The differential cross section of a transition into a continuous spectrum can be also expressed in terms of generalized oscillator strengths. However, in this case we must introduce the *spectral density of generalized oscillator strengths* $df_\omega(q)/d\omega = f(\omega, q)$[113,114]:

$$d\sigma_{\omega 0}(q) = \frac{4\pi z^2 e^4}{mv^2} \frac{f(\omega, q)}{\hbar\omega} \, d\omega \, \frac{dq}{q} \,. \tag{4.18}$$

$f(\omega, q)$ is calculated using formula (4.16) in which the final state is now the wavefunction of continuous spectrum corresponding to the system molecular ion plus ejected electron. Introducing the spectral density of oscillator strengths, we are able to study the overlapping bands as well as the discrete transition against the background of continuous spectrum. In the latter case we choose a narrow interval around the maximum of the transition and determine its oscillator strength as

$$f_{0n}(q) = \int_{\omega_{0n}-\Delta\omega}^{\omega_{0n}+\Delta\omega} f(\omega, q)\, d\omega\ . \tag{4.19}$$

If the transferred energy ε is much greater than the ionization potential I_1, the ejected electron can be regarded as free to a very good approximation. Since both the momentum $\hbar\mathbf{q}$ and the energy ε are entirely transferred to the ejected electron, we have $\varepsilon = \hbar^2 q^2/2m$. In this case the differential cross section of ionization is given by the well-known *formula of Rutherford*:

$$d\sigma = \frac{2\pi z^2 e^4}{mv^2}\,\frac{d\varepsilon}{\varepsilon^2}\ . \tag{4.20}$$

Formula (4.20) gives the cross section per one ejected electron. If the transferred energy is much greater than the binding energy of any one of the electrons, the right-hand side of (4.20) must be multiplied by the number of electrons in the molecule (N_e).

The excitation of a molecule may result in a change of its electron and rotational–vibrational quantum numbers. In the adiabatic approximation,[99] the total wavefunction of a molecule can be presented as a product of the electron wave and the rovibrational wavefunction. In those cases where the former is weakly affected by the changes in the relative position of the nuclei (this is usually the case with lower vibrational levels), we can use the *Condon approximation* considering the electron wavefunction only at equilibrium configuration of the nuclei. In this case the oscillator strength factorizes into an electron oscillator strength and the so-called *Frank–Condon factor*, which is the overlap integral of the vibrational wavefunctions of the initial and the final states of the molecule.[115,116]

Thus, in order to calculate the differential cross sections it is enough to know the generalized oscillator strengths. On the other hand, if the cross sections are found experimentally, formulas (4.13) and (4.18) enable us to find the experimental values of the oscillator strengths.[117] We will briefly dwell on the properties of the generalized oscillator strengths.[113,118]

2. *Properties of Generalized Oscillator Strengths*

Knowing the generalized oscillator strengths we are able to predict quite a number of properties of the system concerning the way it is affected by charged particles. For instance, very much information can be obtained from the energy moments of the oscillator strengths, which are defined as

$$S_t(q) = \sum_n (\hbar\omega_{0n}/\mathrm{Ry})^t f_{0n}(q)\,, \tag{4.21}$$

$$L_t(q) = \sum_n (\hbar\omega_{0n}/\mathrm{Ry})^t f_{0n}(q) \ln (\hbar\omega_{0n}/\mathrm{Ry})\,, \tag{4.22}$$

where the sum over n is taken over all the excited states of the system.

At $t = -1$ we have

$$S_{-1}(q) = (qa_0)^{-2} \sum_n |\mu_{n0}(q)|^2\,, \tag{4.23}$$

$$L_{-1}(q) = (qa_0)^{-2} \sum_n |\mu_{n0}(q)|^2 \ln (\hbar\omega_{0n}/\mathrm{Ry})\,. \tag{4.24}$$

$S_{-1}(q)$ characterizes the so-called incoherent scattering, while $L_{-1}(q)$ is related to the total cross section of all the inelastic scattering processes. With $q \to 0$, S_{-1} becomes the sum of squares of dipole matrix elements: $S_{-1}(0) = M^2_{\mathrm{tot}}$.

The zero moments $S_0(q)$ and $L_0(q)$ have the following physical meaning. $S_0(q) = \Sigma_n f_{0n}(q) = N_e$ corresponds to the sum rule for the generalized oscillator strengths,[113,118], while $L_0(q)$ is related to the stopping power of a charged particle. The moment $L_1(q)$ characterizes the statistical fluctuations of the charged particle's energy losses (the straggling) and $S_{-2}(0)$ is proportional to the dipolar polarizability α_d of the molecule: $S_{-2}(0) = \alpha_d / 4a_0^3$.

Many studies have been devoted to calculation of $S_t(0)$ and $L_t(0)$. For example, in Refs. 118 and 119 the authors present the values of $S_t(0)$ and $L_t(0)$ for a large number of molecules and discuss their relation to the properties of the molecules. The authors of Ref. 118 have examined the possibility of calculating $S_t(0)$ and $L_t(0)$ using the additivity rule for molecules starting from the values of similar quantities for atoms. They have shown that with $t \leq -2$ the values of $S_t(0)$ and $L_t(0)$ for molecules calculated this way differ from their exact values by 15–25%, whereas for $t = -1, \ldots, 2$ the deviation is less.

At small q (i.e., at $qa_0 < 1$) we can use the expansion of $f_{0n}(q)$ in powers of qa_0, which gives us[116]

$$f_{0n}(q) = f_{0n} + \sum_{n=1} a_n (qa_0)^{2n}\,, \tag{4.25}$$

where the coefficients a_n are expressed in terms of dipole matrix elements.[117] From (4.25) it follows that for optically allowed transitions

$f_{n0}(q)$ tends to f_{n0} with q tending to zero, while for optically forbidden transitions $f_{n0}(q) \to 0$ at $q \to 0$.

At large q ($qa_0 \gg 1$) the integrand in the matrix element (4.15) contains a rapidly oscillating factor $\exp(iqx_i)$, so the value of the integral is close to zero, owing to which for discrete transition with $qa_0 \gg 1$ $f_{0n}(q) \to 0$. This means that discrete transitions occur mainly at small q.

In order for the integral (4.15) not to be close to zero at large q, the wavefunction ψ_n of the final state must contain a factor $\sim\exp(iqx)$. Such ψ_n correspond to an ionized state of a molecule with the ejected electron having momentum $\hbar q$, that is, of the same type we would have in the case of collisions with free electrons. The cross section in this case is given by the Rutherford formula (4.20).

The function of two variables $f(\omega, q)$ can be pictured as a surface over the plane (ω, q), which is called the *Bethe surface*. The analysis of Bethe surfaces presented in review[113] by Inokuti shows that at $(\hbar q)^2/2m < \hbar\omega$ the function $f(\omega, q)$ has noticeable nonzero values, but falls very rapidly to zero when $(\hbar q)^2/2m$ becomes greater than $\hbar\omega$. In the region $(\hbar q)^2/2m < \hbar\omega$ one can distinguish two subregions. The first one corresponds to small ω and q and describes distant collisions. The behaviour of $f(\omega, q)$ in this subregion depends on the electronic properties of the scattering system.

The second subregion corresponds to large ω and q and is related to close collisions (the knock-on). At very large transferred momenta ($qa_0 \gg 1$) the inelastic scattering by a molecule (an atom) is actually the elastic scattering by a free electron with the cross section given by Rutherford formula. In this case the function $f(\omega, q)$ can be presented analytically as a delta function:

$$f(\omega, q) \approx \delta\left(\hbar\omega - \frac{(\hbar q)^2}{2m}\right). \tag{4.26}$$

Keeping in mind the characteristic properties of generalized oscillator strengths, the authors of Ref. 120 have proposed the following semiempirical formula for $f(\omega, q)$:

$$f(\omega, q) = f(\omega)\theta[\hbar\omega - (\hbar q)^2/2m] + F(\omega)\delta[\hbar\omega - (\hbar q)^2/2m], \tag{4.27}$$

where

$$\theta(x) = \begin{cases} 1, & \text{at } x > 0, \\ 0, & \text{at } x \leq 0, \end{cases}$$

and $F(\omega) = \int_0^\omega f(\omega')\, d\omega$, where $f(\omega)$ is the optical oscillator strength.

Although the actual form of $f(\omega, q)$ is different from formula (4.27), the latter leads to reasonable results when we use it to calculate the cross sections of inelastic collisions and the ionization losses.[120] As one of the reasons for using approximation (4.27), one can consider the fact that the data concerning the Bethe surfaces for molecules are very scant, while there is extensive information about the optical oscillator strengths of molecules both in the discrete and in the continuous regions of the spectrum (see Refs. 119 and 121).

Both theoretical[122–124] and experimental[125] studies of the behavior of f_{0n} versus q show that at $qa_0 < 1$ the oscillator strengths for optically allowed transitions rapidly fall with increase of q, which is in agreement with formula (4.27). However, at $qa_0 > 1$ the behavior of $f_{0n}(q)$ depends on the type of a transition. In the case of Rydberg transitions, $f_{0n}(q)$ has characteristic maxima and minima that are absent in the case of excitation of valence electrons. According to Ref. 123, their appearance is due to the existence of nodes in molecular orbitals.

Since, by definition, both $f_{0n}(q)$ and $f(\omega, q)$ do not depend on the energy of the primary particle, by studying the behavior of f_{n0} versus q at different energies of the electron we can determine the lower limit below which the Born approximation is no longer valid. This sort of experiment shows that for optically allowed transitions, as well as for a number of the optically forbidden ones, $f_{0n}(q)$ becomes independent of the electron energy at least above 300 eV. However, for certain optically forbidden transitions (e.g., those in N_2) the dependence on electron energy has been observed within the whole range of studied energies, from 300 to 500 eV.[116]

3. *Exchange Effects in Scattering*

If the velocity of the incident electron is comparable with those of molecular electrons, the former can be exchanged for one of the latter. Such processes are described by the exchange scattering amplitude, the form of which in the first Born approximation has been found by Oppenheimer.[126] In the *Born–Oppenheimer approximation* the exchange amplitude (4.11) acquires the form

$$B_{n0}(\mathbf{k}_n, \mathbf{k}_0) = -\frac{m}{2\pi\hbar^2} \sum_i \langle P_{\xi\xi_i} \Phi_n | V | \Phi_0 \rangle , \qquad (4.28)$$

where the wavefunctions of the initial (Φ_0) and the final (Φ_n) states of the system are given by formulas (4.4) and (4.8), respectively.

In the simplest case, when the scattering system is a hydrogen atom, after integration over spin variables the exchange amplitude in the

Born–Oppenheimer approximation acquires the form

$$B_{n0}(\mathbf{k}_n, \mathbf{k}_0) = \frac{me^2}{2\pi\hbar^2} \langle \exp(i\mathbf{k}_n\mathbf{r}_1)\varphi_n(\mathbf{r}_2)| - \frac{1}{r_2} + \frac{1}{r_{12}} | \exp(i\mathbf{k}_0\mathbf{r}_2)\varphi_0(\mathbf{r}_1)\rangle . \tag{4.29}$$

Here the incident and the atomic electrons are labeled, respectively, number 2 and number 1, and $\varphi_0(\mathbf{r})$ and $\varphi_n(\mathbf{r})$ are the coordinate wavefunctions of the ground and of the nth excited states of the hydrogen atom. When applied in calculations of exchange cross sections near the threshold, formula (4.29) leads to gross discrepancy,[127] which is due, for one thing, to the fact that the Born approximation is valid only far away from the threshold when the energy of the incident electron is high. In addition, as was shown by Ochkur,[128] the discrepancy becomes greater still owing to the fact that, whereas the Born formula (4.12) for the direct scattering amplitude is the leading term of the asymptotic series in $1/k_0$, in the expansion of (4.29) in powers of $1/k_0$ only the first term has real meaning. The rest of the terms should be omitted being of higher orders of smallness, so they must not be taken into account in the first order of the perturbation theory, which the Born–Oppenheimer approximation actually is. When we extrapolate this formula onto the region of small energies, these terms become large, so that the results have no meaning.

Ochkur[128] has expanded the exchange amplitude in an asymptotic series in powers of $1/k_0$ and has found its leading terms. He has found that the term with the operator $1/r_2$ can be safely neglected, while the leading part of the term describing the interelectronic interaction can be found via the following substitution

$$\int \exp(i\mathbf{k}_0\mathbf{r}_2)\varphi_n^*(\mathbf{r}_2)\frac{1}{r_{12}}\, d^3\mathbf{r}_2 \rightarrow \frac{4\pi}{k_0^2}\exp(i\mathbf{k}_0\mathbf{r}_1)\varphi_n^*(\mathbf{r}_1) . \tag{4.30}$$

As a result, formula (4.29) acquires a form practically coinciding with the formula for the direct amplitude save for the factor, namely,

$$B_{n0}(\mathbf{k}_n, \mathbf{k}_0) = \frac{2me^2}{\hbar^2 k_0^2} \langle \varphi_n(\mathbf{r}_1)| \exp(i\mathbf{q}\mathbf{r}_1)|\varphi_0(\mathbf{r}_1)\rangle = \frac{q^2}{k_0^2} A_{n0}(\mathbf{k}_n, \mathbf{k}_0). \tag{4.31}$$

However, this difference in factors is quite important since it leads to a different dependence on the energy E of the incident electron, namely, the exchange cross section rapidly decreases with decrease of E ($\sim 1/E^3$). Therefore, for fast electrons the exchange scattering can be disregarded.

Since the ground state of a molecule is singlet [the only exceptions are O_2 ($S=1$) and NO ($S=\frac{1}{2}$)], a fast electron interacting with a molecule causes the transitions of the latter into singlet excited states only. However, slow electrons can excite molecules into triplet states as well, owing to the exchange processes. In this case there are no constraints on the spin since it has the same value ($\frac{1}{2}$) for the initial and the final states of the electron plus molecule system. If in the initial state the spin of the incident particle ($s=\frac{1}{2}$) is added to the spin of the molecule ($S=0$), in the final state the particle's spin $s=\frac{1}{2}$ is subtracted from $S=1$. Integrating over the spin variables we get the following formula for the amplitude of the exchange excitation of a triplet electron state of a molecule (or of an atom with a singlet ground state)[128,129]

$$B_{n0}(\mathbf{k}_n\mathbf{k}_0)=\frac{me^2}{\hbar^2}\frac{\sqrt{3}}{k_0^2}\langle\psi_n(\mathbf{r}_1,\ldots,\mathbf{r}_N)|\exp(i\mathbf{q}\mathbf{r}_1)|\psi_0(\mathbf{r}_1,\ldots,\mathbf{r}_N)\rangle\,, \tag{4.32}$$

which is a generalization of formula (4.32). Since, when deriving formula (4.32), we have integrated over the spin variables, the wavefunctions appearing in (4.32) are only coordinate functions and not the total ones, meaning that their permutation symmetry is determined by the corresponding Young tableaux (see Ref. 130). It should be noted further that the matrix element (4.32) contains only one exponential factor rather than a sum of them, as was the case for direct scattering [see formula (4.15)], and it is owing to this fact the right-hand side of (4.32) is not zero. For, if there were a sum of exponential factors, which would have been a symmetric function with respect to permutations of electron cordinates, the matrix element in (4.32) would have been zero since ψ_0 and ψ_n have different permutation symmetries. The largest contribution to the exchange scattering amplitude is given by the region $qa_0 \sim 1$, while for direct scattering the cross section is mainly determined by small qa_0.

If we substitute amplitude (4.32) into formula for the differential excitation cross section (4.9) and integrate it, making the substitution $(1/2\pi)\,d\Omega = q\,dq/k_0k_n$, we will get a factor $(\hbar k_0)^{-6}$ in the total cross section. Thus, while the direct scattering cross section decreases as $1/E$ with decrease of the energy, the exchange cross section behaves as $1/E^3$, meaning that the triplet states can be excited only by the slow electrons.

Rudge[131,132] has modified the approach of Ochkur by correcting the asymptotes of the wavefunctions. The amplitude of exchange scattering in this so-called Ochkur–Rudge approximation is obtained from the Ochkur amplitude by replacing k_0^2 by $[k_n - i(2I_0)^{1/2}]^2$, where I_0 is the ionization energy for the ground state.

The excitations of lower triplet states of benzene in these two approximations have been calculated in Ref. 133. A similar study of excitation cross sections for triplet states of a nitrogen molecule has been done in Ref. 134.

4. *Total Cross Sections of Excitation and Ionization*

In order to obtain the total excitation cross section we must integrate formula (4.13) over the transferred momentum between $q_{\min}$ and $q_{\max}$. The integration limits are determined by the energy and momentum conservation laws, thus being $q_{\min} = k_0 - k_n = \omega_{0n}/v$ and $q_{\max} = k_0 + k_n \approx 2mK/M$. However, at a given transferred energy $\hbar\omega_{0n}$, $f_{0n}(q)$ tends to zero at large q (see Section IV.B.2), so, in the case of excitation, the maximum transferred momentum must be of order of the atomic ones, and we can carry out the integration only between $q_{\min}$ and $q' \sim a^{-1}$ (a is a quantity on the order of the size of a molecule). Such integrations are easily done if we further neglect the dependence of $f_{0n}(q)$ on momentum, taking the optical oscillator strength (this is the optical approximation). As a result, for the optically allowed transitions we have

$$\sigma_{0n}^{\text{opt}} = \int_{q_{\min}}^{q'} d\sigma_{n0}^{\text{opt}} = \frac{4\pi z^2 e^4}{mv^2} \frac{f_{0n}}{\hbar\omega_{0n}} \ln \frac{v}{a\omega_{0n}} . \tag{4.33}$$

The account of the dependence of the generalized oscillator strength on the wave vector q leads to a more complicated behavior of the cross section σ_{0n} (see Refs. 113 and 135). For the dipole-allowed transitions the leading term in σ_{0n}, which is usually used in data processing, is traditionally presented in the form (taking into account relativity)

$$\sigma_{0n} = \frac{8\pi a_0^2 z^2 \text{ Ry}}{mv^2} \frac{f_{0n} \text{ Ry}}{\hbar\omega_{0n}} \left[\ln \frac{2c_{0n} mv^2}{(1-\beta^2)\text{ Ry}} - \beta^2 \right], \tag{4.34}$$

where $\beta = v/c$, and c_{0n} is a dimensionless constant the value of which is found experimentally. In the same approximation, the cross section for the dipole-forbidden transitions has the form[113,136]

$$\sigma_{0m} = \frac{8\pi a_0^2 z^2 \text{ Ry}}{mv^2} b_{0m} , \tag{4.35}$$

where b_{0m} is also a constant, which in the limit $q \to 0$ is determined by the square of the quadrupole-moment matrix element.

For transitions into continuous spectrum (ionization), the differential cross section $d\sigma/d\omega$ is found by integrating formula (4.18) over q, and can be presented in the form

$$\frac{d\sigma}{d\omega} = \frac{8\pi a_0^2 z^2 \text{ Ry}}{mv^2} \frac{f(\omega)}{\hbar\omega} \ln \frac{2c_\omega mv^2}{\text{Ry}} , \tag{4.36}$$

where $f(\omega) = f(\omega, q)|_{q=0}$ is the spectral density of oscillator strengths, and c_ω, like c_{0n}, is found experimentally. Integrating formula (4.34) over the energy, we can find the total ionization cross section. Taking into account relativity, we get

$$\sigma_i = \frac{8\pi a_0^2 z^2 \text{ Ry}}{mv^2} \left\{ M_i^2 \left[\ln \frac{1}{1-\beta^2} - \beta^2 \right] + C_i \right\} , \tag{4.37}$$

where C_i is a constant, and

$$M_i^2 = \int_I^\infty \eta(\omega) \frac{f(\omega) \text{ Ry}}{\hbar\omega} \, d(\hbar\omega) . \tag{4.38}$$

In formula (4.38) I stands for the threshold energy of ionization and $\eta(\omega)$ denotes the ionization efficiency factor. The latter does not have to be unity at $\hbar\omega > 1$. For instance, in the case of SES (see Section II.B) $\eta(\omega) < 1$, while for multiple ionization it can be greater than unity.

In certain cases the ionization cross sections of a many-atom molecule can be presented as a sum of those for simpler molecules. For example, the authors of Ref. 137 have compared the cross sections of ionization of a H_2O molecule by electrons with those obtained according to the formula $\sigma_i(H_2O) = \sigma_i(H_2) + \frac{1}{2}\sigma_i(O_2)$ and have found that the difference between the latter and the experiment is insignificant in the whole range of studied electron energies (from 0.1 to 20 keV).

The authors of Refs. 138 and 139 have shown that the ionization cross section of a complicated molecule can be presented as a sum of ionization cross sections of separate chemical bonds with very good accuracy. Analyzing the experimental data on ionization cross sections of 40 hydrocarbon molecules of different nature, the authors of Ref. 139 have found the effective ionization cross sections of C–H, π(C–C), σ(C–C), C–O, and a number of other bonds, which allows one to approximate the ionization cross section of an arbitrary molecule with given chemical bonds with an accuracy of 2%.

Carrying out the sum in formula (4.34) over all the possible states n, we get the total cross section of excitation from the ground state σ_{ex}. Adding to it the ionization cross section σ_i we obtain the total cross section of inelastic scattering σ_{tot}, which can be presented as

$$\sigma_{\text{tot}} = \frac{8\pi a_0^2 z^2 \,\text{Ry}}{mv^2} M_{\text{tot}} \left[\ln \frac{2C_{\text{tot}} mv^2}{(1-\beta^2)\,\text{Ry}} - \beta^2 \right], \tag{4.39}$$

where C_{tot} is a constant, and M_{tot}^2, like M_i^2 in formula (4.38), is usually determined from experimental data. In Table V we present the values of the constants in formulas (4.38) and (4.39) for a number of atoms and molecules.

According to (4.34), the probability of an optically allowed transition in a molecule from the ground state into the nth state induced by a fast particle is proportional to $f_{0n}/\hbar\omega_{0n}$, meaning that the mainly occupied states must be those that have the largest value of this ratio. These states

TABLE V
Value of Constants M_i^2, C_i, M_{tot}^2, and C_{tot} for Some Atoms and Molecules[a]

Atom/Molecule	M_i^2	C_i	M_{tot}^2	C_{tot}	References
H	0.283	4.037	1	1.567	140
He	0.745	8.0			142
			0.7525	0.6035	136
Ne			1.94	0.41	136
	2.02	18.17			142
Ar			5.086	0.81	136
	3.96	38.14			142
H_2	0.695	8.115			142
N_2	3.74	34.84			142
O_2	4.20	38.84			142
H_2O	3.24	32.26			142
	3.14	30.96			137
	2.657	25.72	3.688	0.7174	141
CH_4	4.23	41.85			142
	4.28	43.17			138
C_2H_4	6.75	68.82			142
	7.32	73.6			138
C_2H_6	6.80	68.93			142
	8.63	88.16			138
C_3H_8	11.93	114.1			142
	13.8	135.3			138
C_6H_6	17.54	162.4			142
	24.2	246.06			138
CH_3OH	6.22	66.4			142
C_2H_5OH	9.94	97.66			142
CH_3COCH_3	11.89	118.0			142
C_6H_{12}	21.97	213.2			142

[a] The sources of ionizing radiation were: electrons and positrons with energy from 0.1 to 2.7 MeV (Ref. 142); electrons with energy from 0.1 to 20 keV (Ref. 137); and electrons with energy from 0.6 to 12 keV (Ref. 138).

do not have to be the first excited state: for example, for aromatic molecules (such as naphthacene or antracene), the oscillator strength of the second or the third excited transition is an order of magnitude higher than that of the first one. In the case of a H_2O molecule, the transition with the highest oscillator strength is the one with $\hbar\omega_{0n} = 13.32$ eV, which exceeds the first ionization potential; therefore, we deal here with a transition into a SES.

However, among secondary electrons generated by a fast charged particle during ionization, the most efficiently produced are the slow electrons, the majority of which are capable of exciting only low-lying singlet states or only the triplet ones. Since the distribution of excited molecules in quantum states in a bulk system is determined by the effect of both the primary particle and the secondary electrons of subsequent generations, it can be noticeably different from the one we would get considering the ratio $f_{0n}/\hbar\omega_{0n}$ (see Ref. 143). Such distributions, as well as the radiolytic yield of ionized states, can be conveniently found using the Monte Carlo method. To use it, we must know the values of σ_{0n}, σ_i, and σ_{ex}. The analytical formulas for the cross sections we have obtained in this section using the Born approximation are very useful in such calculations and have found wide application in many problems (see Section VIII.A.2).

The use of the Born approximation is valid only for fast collisions. According to Ref. 144, the authors of which compared the experimental values of the cross sections for the H_2O molecule with those predicted by formulas (4.38) and (4.39), the agreement is good only at electron energies exceeding 1 keV, whereas at lower energies the predicted values are greater than the actual ones.

A number of methods have been developed for calculating the cross sections at low energies of charged particles (see Ref. 105). However, they do not give good analytical dependences, and it is better to use either the experimental values or the semiempirical formulas for the ionization and excitation cross sections (see, e.g. Ref. 145). Very useful for obtaining estimates are the formulas for the cross sections obtained in the binary-collision approximation (see the reviews by Vriens[146] and Ochkur[147]).

C. The Quasi-Classical Approximation: The Impact Parameter Method

When we study the effect of charged particles on a substance, we often need to estimate the probabilities of excitation or ionization as functions of the distance from the axis of the track (i.e., of the impact parameter b). This is done using the quasi-classical approach, within which we assume that the charged particle moves along a definite trajectory. In the framework of this method, which is justified the most for heavy particles, the effective cross section of a transition is given by the formula

$$\sigma_{0n} = 2\pi \int P_{0n}(b) b \, db \, , \tag{4.40}$$

where $P_{0n}(b)$ is the probability of the transition with the impact parameter b. In order to calculate the probability $P_{0n}(b)$, we have to solve a system of equations of the nonstationary perturbation theory. In the first order the formula for $P_{0n}(b)$ is[99]

$$P_{0n}(b) = \left| \frac{1}{i\hbar} \int_{-\infty}^{\infty} dt \exp(i\omega_{0n} t) \langle \psi_n | \hat{V}(R, t) | \psi_0 \rangle \right|^2 , \tag{4.41}$$

where $\hat{V}(R, t)$ is the perturbation operator acting on electrons of the molecule. In the nonrelativistic limit $\hat{V}(R, t)$ is the Coulomb interaction between the incident particle and the charges inside the molecule.

The trajectory of a fast heavy charged particle is mostly a straight line, since, with the exception of head-on collisions, its interaction with electrons of the system practically does not change its direction. If $\mathbf{R}_0(t)$ is the radius vector connecting the center of mass of the molecule with the charged particle, the trajectory of the latter can be presented as $\mathbf{R}_0(t) = \mathbf{b} + \mathbf{v}t$, where $\mathbf{v}$ is the velocity of the particle and $\mathbf{b}$ is a vector the length of which equals the impact parameter and which is directed perpendicular to the particle's trajectory.

Let $\mathbf{r}_i$ stand for the radius vector of the ith electron with respect to the center of mass of the molecule. Assuming that all $\mathbf{r}_i$ are much less than $\mathbf{b}$, we can expand the Coulomb interaction operator into a multipole series. Keeping only the dipole terms, we get the following expression for $P_{0n}(b)$[99]

$$P_{0n}(b) = \frac{z^2 e^4}{\hbar^2} \left| \int_{-\infty}^{\infty} \frac{M_{0n}^v vt + M_{0n}^b b}{[(vt)^2 + b^2]^{3/2}} \exp(i\omega_{0n} t) \, dt \right|^2 , \tag{4.42}$$

where

$$M_{0n}^v = \sum_{i=1}^{N_e} \langle \psi_n | \mathbf{r}_i \frac{\mathbf{v}}{v} | \psi_0 \rangle \, , \qquad M_{0n}^b = \sum_{i=1}^{N_e} \langle \psi_n | \mathbf{r}_i \frac{\mathbf{b}}{b} | \psi_0 \rangle \, .$$

The integrand decreases as an inverse cube of the distance separating the molecule from the charged particle (i.e., as R_0^{-3}), meaning that the interaction is important only at distances about the distance of closest approach. Using the classical dynamics of collisions,[148] we find the effective collision time to the $\tau_{col} \approx 2b/v$. If the collision time is much greater than the effective time of the quantum transition (the latter may be taken to be the period of oscillations: $T = 2\pi/\omega_{0n}$), the collision is called *adiabatic*. This sort of collisions correspond to the inequality

$$\omega_{0n} b/\pi v \gg 1 . \tag{4.43}$$

In this case, during the time of collision, the integrand in (4.42) oscillates many times and the value of the integral is close to zero. Thus, the adiabatic collisions do not lead to excitation of the molecule.

Moreover, it can be shown that condition (4.43) is stronger than is necessary, that is, that the much greater sign can be replaced by the greater sign and the situation will not change, meaning that the excitation can occur only at effective collision times smaller than the time of the quantum transition. In order to show this strictly we must calculate the integral in (4.41) without any approximations. Presenting the exponent of the integrand as $\exp(i\omega_{0n}t) = \cos(\omega_{0n}t) + i \sin(\omega_{0n}t)$ and integrating each term, we get

$$P_{0n}(b) = \left(\frac{2ze^2\omega_{0n}}{\hbar v^2}\right)^2 \{|M_{0n}^v|^2 K_0^2(\eta_{0n}) + |M_{0n}^b|^2 K_1^2(\eta_{0n})\} . \tag{4.44}$$

Here $\eta_{0n} = \omega_{0n} b/v$, and $K_1(\eta_{0n})$ and $K_0(\eta_{0n})$ are the modified Bessel functions.[146] At large values of their arguments, $K_0(\eta_{0n})$ and $K_1(\eta_{0n})$ behave as $(\pi/2\eta_{0n})^{1/2} \exp(-\eta_{0n})$, so $P_{0n}(b)$ decreases exponentially with increase of b. At $\eta_{0n} = \pi$ both $K_0^2(\pi)$ and $K_1^2(\pi)$ are practically zero [$(0.0296)^2$ and $(0.034)^2$, respectively, see Table 9.8 in Ref. 146]. Putting η_{0n} equal to π we find the effective size of the region of direct excitations to be

$$b_{eff} = \pi v/\omega_{0n} . \tag{4.45}$$

(By direct excitations we mean those that are induced by the primary particle itself and not by the electrons it produces.)

Thus, the excitation has a noticeable probability only at $b \leq b_{eff}$. At small η ($\eta < 0.4$), the functins $K_0(\eta)$ and $K(\eta)$ behave, respectively, as $-\ln \eta$ and η^{-1}. At $b < 0.1 b_{eff}$, the second term in (4.44) is always greater than the first one, and $P_{0n}(b)$ tends to the value

$$P_{0n}(b) = \frac{4z^2e^4}{\hbar^2 v^2} |M_{0n}^b|^2 \frac{1}{b^2} . \tag{4.46}$$

We will get the same result by putting the exponent in the integral (4.42) equal to unity (cf. Ref. 99). In an isotropic medium $|M_{0n}^v|^2 = |M_{0n}^b|^2 = |M_{0n}|^2$. Introducing the optical oscillator strength according to (4.16) instead of $|M_{0n}|^2$, we get

$$P_{0n}(b) = \frac{2z^2e^4}{mv^2} \frac{f_{0n}}{\hbar\omega_{0n}} \frac{1}{b^2} .$$

The total excitation cross section σ_{0n} is found by integrating formula (4.40) over b between $b_{\min} = a$ and $b_{\max} \simeq b_{\text{eff}}$:

$$\sigma_{0n} = \frac{4\pi z^2 e^4}{mv^2} \frac{f_{0n}}{\hbar\omega_{0n}} \ln\left(\frac{\pi v}{a\omega_{0n}}\right) . \tag{4.47}$$

This formula exactly coincides with the one obtained by integrating the Bethe's cross section (4.13) over q, provided we take $q_{\min} = \omega_{0n}/\pi v$ [when deriving (4.33) we took $q_{\min} = \omega_{0n}/v$]. This fact is no coincidence, since there is an indirect relation between q and b, namely, small q corresponds to large b, and vice versa: $q_{\max} \sim b_{\min}^{-1} \approx a^{-1}$. Fano[150] has made the transformation from momentum representation to a representation in terms of the impact parameter b in the Bethe's formula (4.13) and has obtained an expression for the differential cross section coinciding with (4.44).

In 1924, before the creation of modern quantum mechanics, Fermi[151] had obtained a general formula (not requiring the interaction to be weak) for the probability of excitation or ionization of an atom by a nonrelativistic charged particle passing by at a distance b from the atom. In doing so, Fermi assumed that the effect the charged particle has on the molecule is equivalent to the effect of light waves of the continuous electromagnetic emission the particle produces with the frequency varying from $\omega = 0$ to some maximum value $\omega_{\max}$ determined by the energy of the particle. In this case the probability of excitation of the atom $P_{0n}(b)$ is equal to the probabiity of absorption of a light quantum corresponding to the absorption line. For the probabiity of absorbing an energy $\hbar\omega$, Fermi obtained the expression

$$P(\omega, b) = 1 - \exp\left[-\int \frac{\alpha(\omega)}{2\pi\hbar\omega} J(\omega)\, d\omega\right] , \tag{4.48}$$

where $\alpha(\omega)$ is the photoabsorption coefficient, and $(1/2\pi)\, J(\omega)\, d\omega$ is the total intensity of light in the frequency range $d\omega$.

In order to determine $J(\omega)\, d\omega$, Fermi performed the harmonic analysis of the electric field of a particle. Fermi's method can be generalized to the case of relativistic particles. Under the condition $ze^2/\hbar v \ll 1$, coinciding with the requirement necessary for the Born approximation to be valid, the obtained expression for $P_{0n}(b)$ coincides with the one we have obtained previously [formula (4.44)]. For a relativistic particle, the component of its electric vector parallel to its direction (let us denote it as E_v) undergoes Lorentz contraction, while the perpendicular component correspondingly becomes greater, meaning that the field of a relativistic particle flattens in the direction of its motion[152]:

$$E_v = ze\gamma(x - vt)/R^{*3}, \qquad E_b = ze\gamma b/R^{*3}, \tag{4.49}$$

where $\gamma = (1 - \beta^2)^{-1/2}$, $R^{*2} = \gamma^2(x - vt)^2 + b^2$, and the x axis is directed along $\mathbf{v}$. Performing the derivation according to Fermi[148] we get

$$J(\omega) = \frac{2cz^2e^2\omega^2}{\pi v^4\gamma^4}\,[K_0^2(\eta') + \gamma^2 K_1^2(\eta')], \tag{4.50}$$

where $\eta' = \gamma\omega b/v$.

When carrying out the integration in (4.48), let us keep in mind that in the resonance case $\alpha(\omega)$ is nonzero only in a very narrow region around the resonance frequency ω_{0n}, and so

$$\int \frac{J(\omega)\alpha(\omega)\,d\omega}{2\pi\hbar\omega} = \frac{J(\omega_{0n})\alpha_{0n}}{2\pi\hbar\omega_{0n}} \equiv B(\omega_{0n}, b), \tag{4.51}$$

where $\alpha_{0n} = \int \alpha(\omega)\,d\omega = \sigma_{0n}^{\text{ph}}$ can be interpreted as the cross section of photoabsorption. The latter can be expressed in terms of the optical oscillator strength using the formula[120]:

$$\sigma_{0n}^{\text{ph}} = (2\pi^2e^2\hbar/mc)f_{0n}. \tag{4.52}$$

As a result, we get

$$B(\omega_{0n}, b) = \frac{2z^2e^4f_{0n}\omega_{0n}}{mv^4\gamma^4}\,[K_0^2(\eta'_{0n}) + \gamma^2K_1^2(\eta'_{0n})], \tag{4.53}$$

where $\eta'_{0n} = \gamma\omega_{0n}b/v$.

It can be shown that at $ze^2/\hbar v \ll 1$ we have $B(\omega_{0n}, b) \ll 1$. Under this condition the probabiity $P_{0n}(b) \simeq B(\omega_{0n}, b)$ coincides with the value of $P_{0n}(b)$ in the relativisitc case obtained in the first Born approximation [see formula (14) in Ref. 153], while at $\gamma = 1$ it coincides with (4.44).

Putting, as we have done earlier, the argument of the Bessel functions equal to π, we find the effective size of the region of direct excitations in the relativistic case

$$b'_{\mathrm{eff}} = \pi\gamma v/\omega_{0n} . \tag{4.54}$$

Owing to the Lorentz factor in formula (4.54), when v approaches c, we have $b'_{\mathrm{eff}} \to \infty$, meaning that a molecule can be excited by a very distant passing particle, which is in contradiction with reality. This is a consequence of the fact that in our derivation (as in Ref. 150) we made no allowance for the weakening of the interaction between a charged particle and molecule when $b \gg a$, which is due to the polarization of the medium. The account of dielectric properties of the medium should lead to finite values of b'_{eff} even at $v \approx c$.

V. ENERGY LOSSES OF CHARGED PARTICLES IN A MOLECULAR MEDIUM

A. The Molecular Stopping Power

1. Methods of Calculating the Molecular Stopping Power and Its Properties

A fast charged particle moving in a substance loses its energy in elastic and inelastic collisions with atoms (or molecules) of the medium and also by emitting electromagnetic (bremsstrahlung and Vavilov–Čerenkov) radiation and in nuclear processes. For the greater part of its path length, the retardation of a particle is mainly due to inelastic (ionization) energy losses. The retardation owing to elastic collisions is efficient only for the slowly moving heavy charged particles (i.e., at the end of their path).[154]

It is very unlikely for a charged particle to excite the nuclei or to induce nuclear reactions, and, as a rule, the contribution of the latter processes to the total energy losses of the particle can be neglected. For heavy particles the losses of energy on the bremsstrahlung radiation (S_{rad}) are also small in comparison with ionization losses. Only for electrons may they become comparable with the latter, but even then the energy of electrons must be sufficiently high (for water it must be about 90 MeV).[154]

A charged particle emits Vavilov–Čerenkov radiation (VCR) when its velocity becomes greater than the speed of light in the medium, that is, when $v > c/n_{\mathrm{r}}$, where n_{r} is the refractive index of the medium. This is so when the energy of the particle is sufficiently high—for electrons in water it must be $\geq$270 keV. And though the contribution of VCR to the energy losses is small, it can nevertheless be essential in photosensitive media (see Section VIII.C).

Thus, the ionization losses supply the main part of the energy absorbed by the medium and are responsible for most of radiation effects. There are a large number of theoretical and experimental studies of ionization losses [see, for example, the monographs, Refs. 148, 154–156, and the reviews, Refs. 113 and 157], so here we will discuss only some of the aspects of the problem.

If a medium is considered as an ensemble of individual molecules, the average energy losses of a charged particle on electron retardation per unit path length can be calculated according to the formula

$$-\frac{dE}{dx} \equiv S_e = \sum_i n\hbar\omega_{0i}\sigma_{0i} , \tag{5.1}$$

where n is the density of particles in the medium, and the summation sign also implies integration over the continuous states. Using formulas (4.34) and (4.36) for σ_{0i}, which we have obtained within the Bethe theory, we get the following expression for S_e:

$$S_e = \frac{4\pi z^2 e^4}{mv^2} nN_e\left[\ln \frac{2mv^2}{(1-\beta^2)I_M} - \beta^2\right] , \tag{5.2}$$

Here N_e is the number of electrons in a molecule and I_M denotes the average excitation potential of a molecule defined according to the relation

$$N_e \ln I_M = \sum_i f_{0i} \ln (\hbar\omega_{0i}) . \tag{5.3}$$

The ratio $S_e/n = s_e$ characterizes the stopping power corresponding to one molecule and is called the *molecular stopping power* (it is measured in eV cm^2 per molecule). For electrons, the expression for S_e is usually presented in the form[5]

$$S_e = \frac{2\pi e^2}{mv^2} nN_e\left\{\ln \frac{Emv^2}{2(1-\beta^2)I_M^2} - [2(1-\beta^2)^{1/2} - 1 + \beta^2]\ln 2 + (1-\beta^2) + \tfrac{1}{8}[1-(1-\beta^2)^{1/2}]^2\right\} , \tag{5.4}$$

where $E = mc^2[(1-\beta^2)^{-1/2} - 1]$ is the relativistic kinetic energy of an electron.

The principal parameter characterizing the molecular stopping power in Bethe's theory is the average ionization potential I_M, which depends only on the properties of the molecule. There are different ways of determining I_M experimentally and theoretically.[157] A direct method is to measure the energy losses ΔE in a thin layer of an absorbent. Putting $S_e = \Delta E/\Delta x$ (Δx is the thickness of the layer), we can determine I_M using formula (5.2). Another way of measuring I_M is to use the pathlength energy dependence. To this end one measures the path lengths of particles with different initial energies (for details see Ref. 157).

In the case of molecules, the method for determining I_M is based on Bragg's additivity rule, according to which[22]

$$nN_e \ln I_M^{\mathrm{ad}} = \sum_a n_a z_a \ln I_a , \tag{5.5}$$

where I_a is the average ionization potential of an atom of the ath kind, Z_a is its charge, and n_a is the density of such atoms in the medium. When using formula (5.5) to calculate I_M, we suppose that the energy losses on electron retardation in a molecular medium are a sum of ionization losses of individual atoms forming a molecule. Bragg's rule is a very convenient way of calculating I_M, since the values of I_a are known for a large number of atoms (see Ref. 157).

Apparently, Bragg's rule is only approximate. When atoms combine into a molecule, their oscillator strengths and the energy levels of valence electrons change essentially. Consequently, the true value of I_M must also be different from the one given by formula (5.5). An obvious example is the difference between the value $I_{\mathrm{H}_2} = 19$ eV calculated by Platzman[158] and the value $I_{\mathrm{H}_2} = 15$ eV predicted by formula (5.5).

Each electron of an atom contributes to I_a. However, the chemical bonds are formed only by valence electrons. Thus, for heavy atoms, containing a large number of electrons, the changes in the energy levels of valence electrons when such atoms combine into a molecule must give only a small contribution to the deviation of I_M from its additivity-rule value. This is indeed the case: the effect of chemical bonds on the value of I_M becomes smaller with increase of the nuclear charge of the atoms forming a molecule.[157] In Table VI we compare the values I_M^{ad} obtained according to the additivity rule [formula (5.5)] with the values I_M^{dip} obtained by direct calculations using the dipole oscillator strengths according to the formula[118]

$$I_M^{\mathrm{dip}} = 2\,\mathrm{Ry} \exp\left[L_0(0)/S_0(0)\right] , \tag{5.6}$$

where the quantities $L_0(0)$ and $S_0(0)$ are given by formulas (4.21) and (4.22) at $t = 0$ and $q = 0$. As one can see in Table VI, the values I_M^{ad} are smaller than I_M^{dip} obtained by direct calculation, the difference between

TABLE VI
Values of Average Ionization Potentials for a Number of Molecules[118]

Molecule	I_M^{ad} (eV)	I_M^{dip} (eV)	Deviation (%)
H_2	14.98	19.26	−22
NH_3	47.88	53.69	−11
H_2O	64.91	71.62	−9
N_2	78.77	81.84	−4
NO	86.38	87.82	−2
N_2O	83.88	84.9	−1
O_2	93.65	95.02	−1

the two rapidly becoming smaller with decrease of the nuclear charge of the molecule's atoms.

In formulas (5.2) and (5.4) the quantity I_M enters the logarithmic term, which changes only slightly when we vary the value of I_M if $2mv^2/(1-\beta^2) \gg I_M$. Thus, in the case of the *fast* particles, the chemical bond has a weak effect on molecular stopping power, and Bragg's rule works well.

If the charged particle is a nucleus with no electron shells around it, the molecular stopping power depends on the velocity of the particle in the following way. At relativistic velocities $v \approx c$, the dependence of S_e on v is determined by the logarithmic term in the brackets, because at $v \to c$ the factor preceding the brackets tends to some finite limit. So the molecular stopping power experiences the so-called relativistic rise. As for S_e, its relativistic rise in real dense media is slowed down by the density effect (see Section V.B.2).

Thus, if starting at ultrarelativistic values we go down the energy scale, the molecular stopping power will first decrease, then will reach a minimum, and then will begin growing again. This latter increase is due to the fact that the factor preceding the brackets now grows more rapidly than the expression inside the brackets decreases. As an example in Fig. 11 we present the curves $s_e = S_e/n$ for water as functions of the energy of the proton and the electron. For values of S_e for the proton we have used the data of Janni,[159] while for the electron we have taken the values of Pages et al.[160] within the energy range 10^4–10^8 eV, and the Ashley[161] at $E < 10^4$ eV.

When the velocity of a particle becomes smaller than those of molecular electrons of a given shell, the latter no longer takes part in the retardation. This slows down the rise of the retardation power occurring at small velocities. At a certain velocity this effect becomes dominant, and as v decreases further, the stopping power also decreases. In the

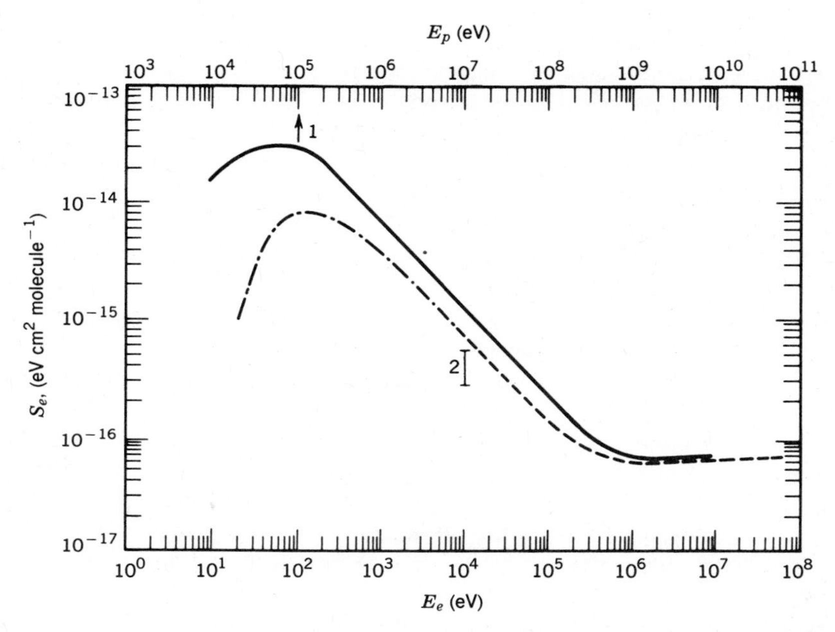

Fig. 11. Molecular stopping power of water for protons (curve 1, data of Ref. 159) and electrons (curve 2: –·–·–, data of Ref. 161; – – –, data of Ref. 160).

vicinity of this value of v, the curve $s_e(v)$ has a maximum called the Bragg's peak (Fig. 11).

At velocities near the Bragg's peak it is no longer correct to calculate s_e using formulas (5.2) and (5.4) with I_M given by formula (5.3) in which the sum is carried out over all electron shells. A number of methods have been developed to deal with this situation. The most widely used among them is the following. Leaving the definition (5.3) of I_M unaltered, one introduces the inner shell corrections into formulas (5.2) and (5.4). Such corrections have been calculated, for instance, in Refs. 22 and 155–159.

Another method has been proposed by Neufeld.[162] Within this method N_e and I_M are no longer considered to be parameters of a molecule (an atom) of the medium independent of the charged particle's velocity. The presentation of N_e and I_M as functions of the particle's energy (or more precisely, of the maximum transferred energy) is widely used for calculating the stopping power (see Refs. 163 and 164).

If the energy of the incident particle is near Bragg's maximum, the core electrons do not contribute to the stopping power, and, consequently, the relative contribution of valence electrons becomes greater. In this case the logarithmic term in (5.2) and (5.4) becomes very sensitive to variations of I_M, the role of chemical bonds becomes greater, and Bragg's rule no longer works. As was shown in Ref. 165, for protons, the deviation from Bragg's rule becomes noticeable when their energy falls below 150 keV. The influence of chemical bonds on the stopping power has been studied experimentally in a number of works (e.g., in Refs. 166–169). In particular, the authors of Ref. 164 have shown that at low energies of the incident particle the molecular stopping power is dependent on the nature of the chemical bond.

At low energies of the ion ($v < v_0 Z^{2/3}$), the Born approximation becomes invalid, and the consistent theoretical analysis of the retardation process becomes a very complicated problem. For this reason many authors have tried to develop a theory using various models (see Ref. 170). For instance, according to the model of Firsov,[171] an ion loses its energy by capturing an electron of a molecule and thus acquiring a momentum proportional to the velocity of the ion v. As a result, there appears a retarding force acting on the ion. As the ion moves away from the molecule, the electron it has captured returns to the molecule, but now no momentum is transferred since the electron occupies a high-lying level. The formula of Firsov and its modification (the LSS theory[172]) are the most widely used formulas for calculating S_e in this region of energies (see Ref. 164).

According to Refs. 171 and 172, at low energies of the incident particle, the energy losses s_e are proportional to v. The recent experimental[173] and theoretical[174] studies indicate that at $v \sim v_0 Z^{2/3}$ the dependence of s_e on v becomes nonlinear. However, at present we have no strict theory describing the retardation of slow ions. And since Firsov's and the LSS formulas provide sufficient accuracy (10–15%) necessary to study the retardation effects, they are widely used in many problems.

2. *Particular Features of Retardation of Multicharged Ions*

A heavy multicharged ion (with its electron shell "stripped" to a considerable extent) is capable of capturing the electrons of a medium even at high velocities. In doing so it forms its own electron shells, thus acquiring a certain electronic structure. As a result, besides excitation and ionization, there appear additional channels of interaction, such as the transfer of electron excitation, the emission of Auger electrons, and, at slow collisions, different types of chemical reactions. The role of elastic collisions also becomes greater. For fission fragments, such collisions may become dominant at the end of their path.

The knowledge of the ion's charge is very important for calculating the stopping power. Owing to loss and capture charge processes, after a beam of ions passes through a sufficiently thick layer of matter, its ions acquire different charges. The distribution of the ions in charge is characterized by a certain mean (effective) charge z_{eff} and by a certain half-width. z_{eff} can be calculated from the cross sections of an electron being lost and being captured by an ion. However, this method is not an easy one and does not provide high accuracy, since the charge-exchange cross sections are known only approximately. For this reason the methods that are widely used today for calculating z_{eff} are those that do not need preliminary calculation of the cross sections. These are the methods based on criteria of Bohr, Lamb, Brannings, and others. The problems concerning the charge state of ions are considered in detail in reviews[157,175,176] and, partially, in monographs.[156,177]

In the case of fission fragments Bohr[177] has obtained the following expresion for z_{eff}:

$$z_{eff} = Z_1^{1/3} v/v_0 , \qquad (5.7)$$

where Z_1 is the charge of the ion's nucleus. He used a criterion which states that at each point of its path a fission fragment retains only those electrons the velocity of which is greater or equal to its own. Formula (5.7) is valid when $1 < v/v_0 < Z_1^{2/3}$. And though its predictions are 20–30% higher than the actual values of z_{eff}, it gives a correct dependence of z_{eff} on velocity.

Analyzing the experimental data concerning the dependence of z_{eff}/Z_1 on $v/v_0 Z_1^{\gamma}$, a number of authors have obtained the universal curves

$$z_{eff}/Z_1 = 1 - c \exp(-v/v_0 Z_1^{\gamma}) . \qquad (5.8)$$

The authors of Ref. 176 propose setting the parameters c and γ equal to 1.032 and 0.69, respectively. Some authors[156] propose the values $c = 1$ and $\gamma = \frac{2}{3}$.

The values of the parameter γ corresponding to experimental dependences of z_{eff} on the charge Z_1 have been found in Ref. 178. Using the statistical Fermi–Thompson model of an atom, the authors have calculated the velocity distribution of external electrons for atoms of heavy elements. Using the Bohr criterion, they have obtained a formula for z_{eff} of the form (5.7) with the exponent of Z_1 equal to $\frac{1}{4}$. For this value of γ the calculated values of z_{eff} for heavy fission fragments in a hydrogen medium practically coincide with the experimental ones.

For media more complicated than hydrogen, where the electrons may have different orbital velocities, the Bohr criterion alone is not enough. At present we have no good theory that would explain the influence of a medium on z_{eff}, although there have been many attempts in this direction.[176] For instance, in Ref. 178 the authors proposed to introduce into formula (5.7) a correcting factor η that depends on the effective potential and on the atomic number of the target's nuclei.

At $v < v_0 Z^{2/3}$ the effective charge z_{eff} is proportional to v/v_0. So the dependence of s_e on velocity in this case is solely due to the logarithmic term in formula (5.2), which decreases as the ion's velocity falls. Consequently, as the ion slows down, s_e must also decrease, that is, its behavior is exactly opposite to the case of protons and alpha particles, where s_e increases as the particle's velocity falls, until it approaches Bragg's peak. In Section VIII.D we will show how this particuliarity affects the structure of the track of a multicharged ion.

B. Specific Features of the Interaction of Charged Particles with a Condensed Medium

1. *The Ionization Potential in the Condensed Phase*

A noticeable difference between the condensed and the gaseous phases is the much greater density of matter in the former case, and, consequently, the need to make allowance for the interaction between the molecules (or atoms) of the medium. In fact, it is owing to this interaction that liquids and solids exist at all. The intermolecular interaction results in energy levels in the condensed phase being shifted with respect to those in the gaseous phase. There also appears a new, qualitatively different type of excited states—the collective ones (excitons and plasmons). Another difference between the two phases, which is very important for radiation chemistry, is the change (usually, a decrease) in the ionization potential. Let us now consider the physics underlying this change.

In the gaseous phase, an electron ejected from a molecule becomes free, and so for each filled electron level we have only one ionization potential. However, in the condensed phase an ejected electron can be in three different states: free, quasi-free, and solvated. So the definition of the ionization potential becomes ambiguous.

We can find the potential at which a free electron appears by measuring the threshold of external photoemission E_e^{ph}. However, the ionization is not always accompanied by electron emission. We can consider the ionization event to have occurred if the electron is transferred to the conductivity band. The corresponding ionization potential I_c equals the energy needed to transfer an electron to the bottom of the conductivity band. It is found experimentally by measuring the threshold of photoconduction current. In crystalline insulators I_c can be found from the limit to which the energy series for the Wannier–Mott[179] exciton converges.

An electron in the conductivity band is quasi-free, since for it to escape from the solid or the liquid we must supply the electron with the so-called work function, which equals the energy an electron has at the bottom of the conductivity band taken with an opposite sign. (This energy is measured from the energy of an electron in vacuum.) Denoting it as V_0, we can write the relation between the ionization potential I_c and the external emission threshold as

$$I_c = E_e^{ph} + V_0 . \tag{5.9}$$

The diagram in Fig. 12 shows the energy levels in a dielectric for negative V_0.

A large number of studies have been devoted to measuring the ionization potential in the liquid and the solid phases (see Refs. 179–181, 189, 190). Some of these results are presented in Table VII, from which one can see that for most substances V_0 is negative, and so the ionization potential in the condensed phase is smaller than the photoemission threshold E_e^{ph}. However, for some substances (for instance, for *n*-pentane, *n*-decane, and neon), V_0 is positive, meaning that in this case it is more advantageous, from the energy point of view, for an electron to make a transition into vacuum than to remain in a quasi-free state.

The difference between E_e^{ph} and the ionization potential I_g in the gaseous phase equals the energy required for reorganizing the medium after one of the molecules has been replaced by a positive ion.[193,194] (Approximately, we can take this energy to be the polarization energy P_+

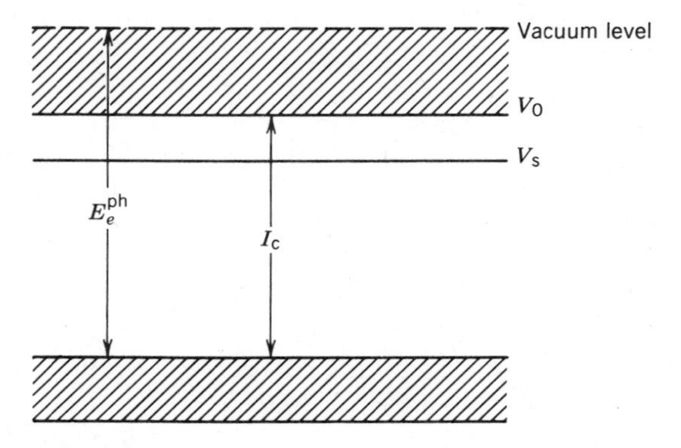

Fig. 12. Scheme of energy levels in a dielectric. I_c is the ionization potential in the condensed state, E_e^{ph} is the photoemission threshold, V_0 is the energy of the bottom of the conductivity band, and V_s is the solvation energy. (V_0 and V_s are negative since they are measured from the energy of the vacuum level.)

TABLE VII
Ionization Potentials for the Condensed and Gaseous Phases

Substance	I_g (eV)	I_c (eV)	State	$I_g - I_c$ (eV)	V_0 (eV)
Water	12.56	8.76[a]	Liquid	3.8	−1.3[188]
		8.7[180]	Solid	3.86	
Toluene	8.82	7.75[181]	Liquid	1.07	−0.22[182]
Cyclopentane	10.53	8.82[181]	Liquid	1.71	0.17,[183] 0.28[184]
n-Pentane	10.35	8.86[181]	Liquid	1.49	0.01[185]
Cyclohexane	9.88	8.43[181]	Liquid	1.45	0.01[186]
n-Hexane	10.18	8.7[181]	Liquid	1.48	0.02[185]
2,2-Dimethylpropane	10.35	8.55[181]	Liquid	1.80	−0.43,[187] − 0.35[183]
2,2-Dimethylbutane	10.06	8.49[181]	Liquid	1.57	−0.24,[184] − 0.15[182]
2,2,4-Trimethylpentane	9.86	8.38[181]	Liquid	1.48	−0.15,[184] − 0.18[187]
Tetramethylsylane	9.79	8.05[181]	Liquid	1.74	−0.55,[183] − 0.62[187]
Helium	24.58		Liquid		1.05[179]
Neon	21.56	21.69[179]	Solid	−0.13	1.1[179]
			Liquid		0.67[179]
Argon	15.68	14.20[179]	Solid	1.48	0.3[179]
		13.40[179]	Liquid	2.28	−0.2[179]
Krypton	13.92	11.60[179]	Solid	2.32	−0.3[179]
		11.00	Liquid	2.92	−0.45[179]
Xenon	12.08	9.28[179]	Solid	2.80	−0.46[179]
		9.20	Liquid	2.88	−0.61[179]

[a]The value calculated according to formula (5.9) with $E_e^{\text{ph}} = 10.06$ eV.[189]

of the medium. Then, the relation between the ionization potentials in the condensed and in the gaseous phases can be written down as[191,192]

$$I_c = I_g + P_+ + V_0 \,. \tag{5.10}$$

For estimating P_+ in liquids one often uses the following simple formula for the energy of polarization of a medium by a single charge[184]

$$P_+ = -\frac{e}{2r_+}\left(1 - \frac{1}{\epsilon}\right). \tag{5.11}$$

Here ϵ is the dielectric permittivity of the medium (since electron polarization is a rapid process taking about 10^{-15} s, one should take the high-frequency value of ϵ) and r_+ is the effectve radius of the ion, the experimental values of which can be found only indirectly. Messing and Jortner[192] have done more detailed calculations of the polarization energy, making allowance for the electrostatic interaction between the ion and its surrounding, which they considered as a continuum. Their final formula coincides with (5.11), where r_+ equals σ,

$$\sigma(\rho) = 2(4\pi\rho/3\eta)^{-1/3} \equiv 2r_s/\eta^{1/3}\,, \tag{5.12}$$

where ρ is the density, η is a parameter that characterizes the closeness of packing in a liquid, and r_s is the Wigner–Seitz radius. The authors of Ref. 180 have calculated P_+ taking into account not only the polarization of the medium as of a continuum, but also the direct electrostatic interaction of the ion with the molecules in the closest coordination sphere.

The polarization energy is always negative. Since in most cases V_0 is also negative, the ionization potential of a medium lowers with transition from the gaseous state to the condensed one. The only exceptions are those rare cases where V_0 is positive and, at the same time, is greater than $-P_+$, as it is for neon.

For most liquids the ionization potential is 1–2 eV lower than it is in the gaseous phase. However, for such a polar liquid as water, the difference is considerably greater, amounting to 3.86 eV. This value is given by formulas (5.9) and (5.10) if we take the experimental value of the photoemission threshold $E_e^{\text{ph}} = 10.06$ eV presented in Ref. 186 and $V_0 = -1.3$ eV (see Ref. 188).

As was pointed out by Byakov,[195] it is very probable that molecules in a liquid form dimers and clusters, the ionization energy of which can be expressed through their dissociation energy using the Born cycle. In the case of a dimer we have

$$\begin{array}{ccc} M_2 & \xrightarrow{I_{M_2}} & M_2^+ + e \\ {\scriptstyle D_{M_2}}\uparrow & & \downarrow{\scriptstyle D_{M_2^+}} \\ M + M & \xleftarrow{I_M} & M + M^+ + e \end{array} \tag{5.13}$$

and the energy conservation law gives us

$$I_{M_2} = I_M - (D_{M_2^+} - D_{M_2})\,. \tag{5.14}$$

For liquids with a high dimerization probability, the ionization potential I_g in formula (5.10) should be replaced by the right-hand side of Eq. (5.14). For most dimers the value of $D_{M_2^+}$ is unknown; however, on general grounds, we can expect that owing to the stronger electrostatic interaction, the dissociation energy of an ion exceeds that of a neutral dimer (the only exception is the H_2 molecule, which is more strongly bound than the ion H_2^+). This results in additional lowering of the ionization potential of liquids. For complexes consisting of atoms of alkali metals, the ionization potential of the cluster lowers with increase of the number of atoms in it, approaching gradually the value of the work function for a planar metal surface.[196]

As we have mentioned at the beginning of this section, in condensed media, besides the free and quasi-free states, an ejected electron can also be in the solvated state. The energy released in this case is the solvation energy V_s. From the thermodynamical point of view, the ionization with solvation of the ejected electron is more advantageous since the required energy equals $I_c + V_s$ (V_s is always negative and is of the order of 1–2 eV). Physically, such an ionization may correspond to preionization from an excited state as a result of the electron tunneling from the excited molecule to the nearest trap (see the discussion in Refs. 197 and 198).

Apparently, it is this type of preionization that explains the appearance of solvated electrons in water at photon energy $E_{ph} = 6.5$ eV observed in Refs. 192 and 193. Thus, besides the ionization with the potential I_c given by formula (5.10), in polar media with high probability of rapid solvation of an ionized electron there may also occur preionization with solvation of the ejected electron with the potential $I_c + V_s$.

2. *The Energy Loss Spectrum in Condensed Media*

The spectrum of energy losses of the fast electrons that were deflected at small angles after passing a thin layer of a substance in the gaseous state is similar to the optical absorption spectrum. This is so because the inelastic scattering at small angles corresponds to small momentum transfers and the probability of energy losses in this case is determined by the optical oscillator strength $f(\omega)$ [see formulas (4.34) and (4.36)]. The latter is related to the photoabsorption cross section $\sigma_{ph}(\omega)$ as $f(\omega) = (mc/2\pi^2\hbar e^2)\sigma_{ph}(\omega)$,[120] and it is this correspondence that is the basis of electron spectroscopy methods[117,201] for determining the optical oscillator strengths. For this reason the maxima in the energy loss spectrum in gaseous molecular media correspond to maximum values of the oscillator strengths and are located at energies below 10–15 eV.

However, the energy loss spectrum of fast electrons that have passed through thin films and molecular liquids is different from the energy loss spectrum in gases (see Section III). Besides the maxima corresponding to transitions of individual molecules into excited states, there have been discovered maxima of absorption at energies 20–22 eV (see Table IV). These maxima are supposed to correspond to excitation of collective states of the plasmon type. Their location corresponds to the maxima of the energy loss function $\mathrm{Im}\,[-1/\epsilon(\omega)]$ for electrons scattered at small angles.

In the Born approximation, the differential cross section of a fast electron loosing the energy $\hbar\omega$ and transferring the momentum q in an infinite medium per one molecule is given by the formula[202]

$$\frac{d^2\sigma}{d\omega\, dq} = \frac{2e^2 N_e}{\pi n \hbar v^2} \operatorname{Im}\left(\frac{-1}{\epsilon(\omega, q)}\right)\frac{1}{q}, \tag{5.15}$$

where n is the concentration of electrons and N_e is the number of electrons in a molecule. Following Fano,[203] for characterizing the probability of transitions in a medium we introduce the function $F(\omega, q)$, which plays the role of a "macroscopic" differential oscillator strength and is defined as

$$F(\omega, q) = \frac{2\omega N_e}{\pi\omega_{\text{p}}} \operatorname{Im}\left(\frac{-1}{\epsilon(\omega, q)}\right), \tag{5.16}$$

where ω_{p} is given by formula (3.7). Substituting (5.16) into formula (5.15) we transform the latter to the form completely coinciding with expression (4.18).

For the relatively narrow peak of energy losses with the maximum at $\hbar\omega_{0n}$, the transition cross section σ_{0n} integrated over q and ω is given by formula (4.34) in which f_{0n} is replaced by the quantity F_{0n} equal to

$$F_{0n} = \int_{\omega_{0n}-\Delta\omega_{0n}}^{\omega_{0n}+\Delta\omega_{0n}} d\omega\, F(\omega),$$

where $F(\omega) = F(\omega, q=0)$. In dilute media $\epsilon_1(\omega) = 1$, $\epsilon_2(\omega) \ll 1$, and $\operatorname{Im}[-1/\epsilon(\omega)] \cong \epsilon_2(\omega)$, so F_{0n} coincides with the usual oscillator strength determining the probability of a molecule undergoing a transition from the ground state to the nth excited state with transition energy $\hbar\omega_{0n}$ (see Ref. 91).

In condensed media, the transition with the largest value of F_{0n} usually does not correspond to the transitions of an individual molecule into an excited state. For a number of substances the largest F_{0n} corresponds to excitation of collective states of the plasmon type, such as the $\hbar\omega_{\text{p}} = 21.4$ eV transition in water that has $F_{\text{p}} = 2.03$.[4]

Thus, the excitation of collective plasmon-type states is one of the important mechanisms of energy loss by fast particles in thin layers of condensed matter, the thickness of which is much smaller than the path length of the particles. However, this is not necessarily so with thick absorbents, as it is sometimes erroneously stated (see, for instance, Ref. 3).

The primary energy lost by an initial particle in a thick absorbent is also equally divided between excitation and ionization. The ionization results in formation of secondary electrons, the energy spectrum of which is shifted toward the low-energy end. So most of them are not capable of exciting the high-lying collective states, and their energy is spent on excitation of low-lying one-particle states. This results in that the distribution of the energy lost by a charged particle in an infinite medium over the quantum states of the absorbing system is different from the one given by the energy-loss spectrum. This was quantitatively confirmed by the results of computer experiments on the primary stage of radiolysis performed in Ref. 143. These results are discussed in Section VIII.A.2.

At high frequencies we have $\epsilon_1(\omega) \approx 1$ and $\epsilon_2(\omega) \ll 1$. For light media such frequencies lie above the far end of the ultraviolet region,[91] for example, at $\hbar\omega > 50$ eV in the case of water.[204] In this frequency range $\mathrm{Im}\,[-1/\epsilon(\omega)] \approx \epsilon_2(\omega)$, and the energy-loss spectrum in the condensed phase is practically the same as it is in the gaseous phase. Consequently, the state of aggregation of the medium has little effect on the probability of large energy losses. This is so because a particle can lose large quantities of energy only in close collisions with individual molecules located near the particle's track, and since the collision time is very short, the polarization of the medium is not essential.

3. *The Influence of the State of Aggregation on the Ionization Losses of Fast Charged Particles*

Following from formula (4.54), the transfer of energy on excitation of molecules has a noticeable probability even in the case where the impact parameter is much greater than their size d. Since the intermolecular spacings in a condensed medium are of order of d, a charged particle interacts with many of its molecules. The polarization of these molecules weakens the field of the particle, which, in its turn, weakens the interaction of the particle with the molecules located far from the track. This results in that the actual ionization losses are smaller than the value we would get by simply summing the losses in collisions with individual molecules given by formula (5.1). This polarization (density) effect was first pointed out by Swann,[205] while the principles of calculation of ionization losses in a dense medium were developed by Fermi.[206]

At present the density effect has been quite thoroughly studied both theoretically and experimentally. There are different ways of obtaining the calculation formulas for S_e. In particular, we can make allowance for the effect the surrounding medium has on the electromagnetic field of a particle by making the substitution $c^2 \rightarrow c^2/\epsilon(\omega)$ in the formula for the relativistic differential cross section of energy and momentum transfer and taking into account the complexity of the dielectric permittivity $\epsilon(\omega)$.[120] This way we get the Bethe–Bloch–Sternheimer formula, which is similar to formulas (5.2) and (5.4) except that the correction δ is now negative and the average ionization potential is given by the formula

$$\ln \tilde{I}_M = \frac{2}{\pi \omega_p^2} \int_0^\infty \omega \, \mathrm{Im} \left(\frac{-1}{\epsilon(\omega)} \right) \ln (\hbar \omega) \, d\omega \, . \tag{5.17}$$

The methods for calculating the correction δ for the density effect (the so-called Sternheimer correction) are largely presented in the literature (see, for example, Refs. 22, 155, and 157). The values of the parameters entering the calculation formulas for a large group of substances and materials are presented in Ref. 207. According to this study, the relative decrease of the stopping power owing to the density effect, that is, $\Delta = -100(S_e' - S_e)/S_e$ (S_e' and S_e are the stopping powers, respectively, with and without correction for the density effect), becomes greater with increase of the charged particle's energy. For example, for water at the energy of the electron about 1 MeV and the energy of the proton ~2 GeV, Δ is about 1–1.5%, while for a 1 GeV electron it amounts to 29.3%. At ultrarelativistic energies the density effect also manifests itself in dilute media. In air at electron energy 1 GeV, Δ is about 11.8%.

Thus, the density effect plays an important role at very high energies. However, the particles studied in radiation chemistry usually have smaller energies, so after replacing I_M by $\tilde{I}_M$ determined by (5.17) we can use formulas (5.2) and (5.4) for calculating the ionization losses in condensed media with sufficiently high accuracy without introducing the correction δ. For a large number of substances, the values of $\tilde{I}_M$ are presented in Ref. 207.

At nonrelativistic energies the influence the state of aggregation has on molecular stopping power manifests itself first of all in the change of I_M. In comparison with the gaseous state, the oscillator strengths in the condensed state are shifted toward higher energy, and, consequently, the value of $\tilde{I}_M$ has to be higher than that of I_M. Indeed, in water $\tilde{I}_M =$ 75.0 eV, whereas in steam $I_M = 71.6$ eV[207] (according to some estimations,[23] it may even be as low as 65 eV). Since the value of $\tilde{I}_M$ is higher, the molecular stopping power must be lower in condensed media than it is in the gaseous ones.* With decrease of energy, the ratio s_e^g/s_e^c grows (see, for example, the experimental data presented in Ref. 208). At small energies the magnitude of this effect in media consisting of light elements, for instance in hydrocarbons,[209] may be as high as 25%, whereas in media

* We are speaking of the stopping power per *single molecule*.

of heavy elements it does not exceed 10%. Actually, the state of aggregation affects the value of s_e via the change in I_M, similar to the effect of chemical bonds on the additivity rule (see Ref. 209).

I_M is not the only thing affected by the change of the aggregation state of the medium. Another is the effective charge of slow ions. There are certain data indicating that such ions have greater effective charges in solid targets than they do in gaseous ones.[156] It has also been noted that the charges become somewhat greater with considerable increase of gas pressure. However, according to data of Ref. 208, the effective charges in water are, on the contrary, smaller than they are in stream. Since the data concerning the influence the state of aggregation has on effective charges of ions often contradict one another, it is still difficult to say how all this affects the value of s_e.

For calculating the stopping power of condensed media for nonrelativistic particles, one often uses nowadays the so-called *dielectric* approach[90] based on formula (5.15). Using this method the authors of Ref. 161 have calculated the stopping power of water for electrons with energy below 10 keV. At energies above 10 keV the values of the stopping power obtained in Ref. 161 coincide with those given by the Bethe–Bloch formula 160 (see Fig. 11).

4. *The Influence of Dielectric Properties of a Medium on the Distribution of Energy in the Track*

The retardation of a particle in a condensed medium can be considered from the macroscopic point of view. In particular, the energy the particle loses over unit length of its path can be calculated either as the work done by the particle against the retarding force of the electric field induced in the medium by the particle's charge,[91] or as the flux of the Poynting vector through a cylindrical surface with radius b (b is the impact parameter) surrounding the particle's path. The latter method has been first employed by Tamm and Frank[210] for analyzing the losses of energy on Čerenkov radiation, and later, by Fermi[206] who used it for calculating the ionization losses. The advantage of this method is that the formulas one obtain for the energy losses depend on the impact parameter, which is important for analyzing the spatial distribution of energy in the particle's track. There are also other methods, which give the same results for the energy losses as the method of Fermi.[206]

In particular, the energy lost by a charged particle can be found by measuring the change in the energy of the electromagnetic field of the medium in which it moves. For nonrelativistic particles this method was used by Fröhlich and Platzman.[212] This method can be generalized to the case of relativistic energies in the following way. The field of a moving particle changes the internal energy of a unit volume of a dielectric. According to Ref. 212, this change in the internal energy can be presented as

$$\Delta E(b) = (4\pi)^{-1} \int_{-\infty}^{\infty} \mathscr{E}(b, t) \frac{\partial \mathscr{D}(b, t)}{\partial t} \, dt , \tag{5.18}$$

where $\mathscr{E}(b, t)$ is the strength of the electric field at the distance b from the particle at the moment t, and $\mathscr{D}(b, t)$ is the corresponding electric induction. Performing the Fourier transformation from t to ω in $\mathscr{E}(b, t)$ and $\mathscr{D}(b, t)$ and some other simple transformations, we can rewrite expression (5.18) in the form

$$\overline{\Delta E(b)} = (2\pi)^{-1} \int_{0}^{\infty} \omega \epsilon_2(\omega) |\mathscr{E}(\omega, b)|^2 \, d\omega . \tag{5.19}$$

The Fourier transform $\mathscr{E}(\omega, b)$ is found by solving the Maxwell equations. In the case of a fast particle with charge ze moving with constant speed, the components $\mathscr{E}_v(\omega, b)$ and $\mathscr{E}_b(\omega, b)$ of its electric field ($\mathscr{E}_v \| \mathbf{v}$ and $\mathscr{E}_b \| \mathbf{b}$) are given by the formulas[148]

$$\mathscr{E}_v = -\frac{ize\omega}{v^2} \left(\frac{2}{\pi}\right)^{1/2} \left[\frac{1}{\epsilon(\omega)} - \beta^2\right] K_0(\lambda_\omega b) ,$$

$$\mathscr{E}_b = \frac{ze}{v} \left(\frac{2}{\pi}\right)^{1/2} \frac{\lambda_\omega}{\epsilon(\omega)} K_1(\lambda_\omega b) , \tag{5.20}$$

where $\lambda_\omega = (\omega/v)[1 - \beta^2 \epsilon(\omega)]^{1/2}$, and $K_0(x)$ and $K_1(x)$ are the modified Bessel functions. The sign of λ_ω is chosen in such a way that we would get delayed potentials. In the general case with complex $\epsilon(\omega)$, the sign of λ_ω in such that $\text{Re}\, \lambda_\omega > 0$.

The energy losses per unit path length in collisions with impact parameters exceeding some given value b can be presented as

$$S_{>b} = 2\pi \int_{b}^{\infty} \Delta E(b') b' \, db' \tag{5.21}$$

Substituting (5.20) into (5.19) and carrying out the integration, we reproduce the results obtained by Fermi[206]:

$$S_{>b} = \frac{2(ze)^2}{\pi v^2} \text{Re} \int_{0}^{\infty} i\omega \left[\frac{1}{\epsilon(\omega)} - \beta^2\right] \lambda_\omega^* b K_0(\lambda_\omega b) K_1(\lambda_\omega^* b) \, d\omega . \tag{5.22}$$

At large values of their arguments, both $K_0(x)$ and $K_1(x)$ behave as

$(\pi/2x)^{1/2}\exp(-x)$. So in absorbing media [where $\epsilon_2(\omega)\neq 0$), the energy losses decrease exponentially at large b, as Re $\lambda_\omega > 0$. Thus, even in the case of ultrarelativistic particles, for every medium there is a certain effective value b_{eff} of the impact parameter that determines the size of the region of direct excitations. The value of b_{eff} depends on the dielectric properties of the medium via the parameter λ_ω.

If at certain frequencies ω_s the dielectric permittivity of the medium ϵ is zero, formula (5.22) can be rewritten as[211]

$$S_{>b} = \frac{(ze)^2}{v^2}\sum_s \frac{\omega_s}{|\epsilon'(\omega_s)|}\lambda_s b K_1(\lambda_s b)K_0(\lambda_s b) + \frac{(ze)^2}{c^2}\int_{\beta^2\epsilon(\omega)>1}\left[1-\frac{1}{\beta^2\epsilon(\omega)}\right]\omega\, d\omega\,, \tag{5.23}$$

where $\lambda_s = \omega_s/v$ and $\epsilon'(\omega_s) = d[\epsilon(\omega)]_{\omega=\omega_s}$. The first term in formula (5.23) gives the so-called polarization losses—the losses on excitation of longitudinal waves (collective states). The effective size of the region where such excitations occur can be found from the condition $\lambda_s b \leq \pi$ and equals

$$b_{\text{eff}} = \pi v/\omega_s\,. \tag{5.24}$$

The second term in formula (5.23) is nonzero only if $\beta^2\epsilon(\omega) > 1$, which is the condition of existence of transversal waves in the medium. These waves are the Vavilov–Cherenkov radiation,[211] so that second term gives us the losses of particle's energy on Vavilov–Cherenkov radiation. This radiation is observed in the frequency range where $\epsilon(\omega)$ is real [i.e., $\epsilon_2(\omega) = 0$],[148] and the absorption is negligible. In this case the second term is independent of b, and the particle's field has the form of waves propagating into infinity. If the absorption is large, the Vavilov–Cherenkov radiation is absorbed in the nearest vicinity of the point where it was produced.

VI. RETARDATION OF SLOW ELECTRONS

A. The Subexcitation Electrons

By tradition, the electrons with energy below 100 eV are called slow electrons. At such energies the analytical formulas we have obtained in Section IV.B in the first Born approximation are no longer valid, and we have to use some semiempirical expressions. Nevertheless, the principal channel of energy loss is still the ionization and excitation of molecules of the medium.

The situation becomes different, however, when the energy of an electron falls below the lowest electron-excitation level $\hbar\omega_{01}$ (we take the energy of the ground state $\hbar\omega_0$ to be the zero point of the energy scale). Such electrons, called by Platzman[213] the *subexcitation* electrons, are no longer capable of transferring their energy to electronic degrees of freedom of the medium.

The subexcitation electrons lose their energy in small portions, which are spent on excitation of rovibrational states and in elastic collisions. In polar media there is an additional channel of energy losses, namely, the dipole relaxation of the medium. The rate with which the energy is lost in all these processes is several orders of magnitude smaller than the rate of ionizaton losses (see the estimates presented in Section II), so the thermalization of subexcitation electrons is a relatively slow process and lasts up to 10^{-13} s or more. By that time the fast chemical reactions, which may involve the slow electrons themselves (for example, the reactions with acceptors), are already in progress in the medium. For this reason, together with ions and excited molecules, the subexcitation electrons are active particles of the primary stage of radiolysis.

As a rule, the lowest excitation level in molecules is a triplet level. Since such states are efficiently excited by slow electrons (see Section IV.B.3), it is the energy of the lowest triplet level that one should take as a boundary energy for subexcitation electrons rather than the energy of the first singlet excitation level, as was done in Ref. 23.

The role of subexcitation electrons is most important when the irradiated medium contains small amounts of impurity molecules the excitation energy $\hbar\omega_{0j}^{i}$ (or the ionization potential I^i) of which is below $\hbar\omega_{01}$. Such additive molecules can be excited or ionized by the subexcitation electrons the energy of which is between $\hbar\omega_{0j}^{i}$ and $\hbar\omega_{01}$, and, consequently, the relative fraction of energy absorbed by an additive will be different from what it should be if the distribution of absorbed energy were solely determined by the relative fraction of valence electrons of each component of the mixture.[213,214] According to estimates of Ref. 215, this effect is observed when the molar concentration of the additive is of the order of 0.1%. This selective absorption with ionization of additives has been first pointed out by Platzman as an explanation for the increase in the total ionization produced by alpha particles in helium after small amounts of Ar, CO_2, Kr, or Xe were added (the so-called Jesse effect).[216]

The subexcitation electrons are characterized by a certain distribution function $\eta(E)$ defined in the range $0 < E < \hbar\omega_{01}$. The first speculations concerning the form of $\eta(E)$ where made by Magee and Burton,[217] who analyzed the energy distribution of electrons ejected during photoionization of atoms. Later, Platzman has proposed the following general form of the function $\eta(E)$ for a helium medium:

$$\eta(E) = A + B(E + I_0)^{-3} , \qquad (6.1)$$

where I_0 is the ionization potential for helium, and constants A and B are determined by the requirements

$$\int_0^{\hbar\omega_{01}} \eta(E)\, dE = 1 , \qquad \int_0^{\hbar\omega_{01}} E\eta(E)\, dE = \bar{E}_{\text{sub}} ,$$

where $\bar{E}_{\text{sub}}$ is the average energy of subexcitation electrons. Still later, El Komoss and Magee[218] have shown that the function

$$\eta(E) = \frac{(1 + \hbar\omega_{01}/I_0)^2}{\hbar\omega_{01}(1 + \hbar\omega_{01}/2I_0)} \left(1 + \frac{E}{I_0}\right)^{-3} , \qquad 0 < E < \hbar\omega_{01} , \qquad (6.2)$$

is in good agreement with Platzman's theoretical and experimental analysis of the spectrum of subexcitation electrons. The subsequent calculations of the degradation spectrum of electrons in helium[219] gave similar results.

The energy distribution of subexcitation electrons in liquids has not been studied thoroughly, and there is only a small number of papers on the matter.[143,220] In Ref. 220 the authors have found $\eta(E)$ considering the liquid as a dense gas and using the formulas for the cross sections obtained within the theory of binary collisions.[146] In Ref. 143 the spectrum of subexcitation electrons was calculated using the Monte Carlo method. Apparently, it was in this study the influence the state of aggregation of water has on the energy distribution of subexcitation electrons was considered for the first time.

According to Ref. 143, with transition to the liquid state, the spectrum of subexcitation electrons shifts toward lower energies. This may be due to the fact that the lower part of the energy-loss spectrum in water is shifted toward higher energies, and, consequently, each event of excitation in water requires, on the average, greater energy than it does in water vapor. For this reason the energy of subexcitation electrons in water must be smaller than it is in steam.

It is convenient to present the energy spectrum of subexcitation electrons in dimensionless units as a function $\hbar\omega_{01}\eta(E)$ of the variable $E/\hbar\omega_{01}$. In Fig. 13 we show the spectrum for helium obtained in Ref. 219 presented in such a way. The dots in the figure show the corresponding data of Ref. 143 for water and water vapor: each dot represents the average value for a small energy interval. One can note that the spectra

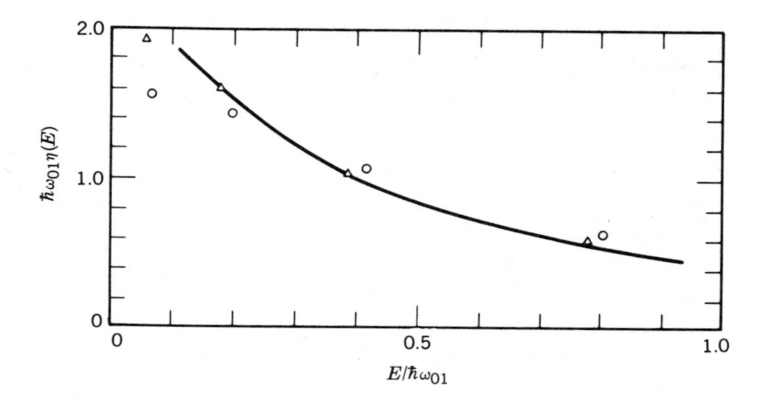

Fig. 13. Spectrum of subexcitation electrons in helium[219] (solid line), in water vapor (⊙), and in liquid water (△).[143] N_i is the number of subexcitation electrons.

for water and helium are very much alike. Although this may certainly be a coincidence, it seems that the energy spectrum of subexcitation electrons presented in such a way is weakly sensitive to the specific kind of the medium.

Regarding the formation of subexcitation electrons, many authors consider only two possibilities: they assume that each subexcitation electron either has been ejected during ionization (and happened to have the energy below $\hbar\omega_{01}$), or it is one of the fast electrons that has slowed down to energies below $\hbar\omega_{01}$. However, there may be other possible ways for subexcitation electrons to be formed. One of them is via a decay of a superexcitation state according to the ionization channel. Such a process may occur both in gaseous and in condensed media, and the majority of electrons produced in this case are the subexcitation ones.

In condensed media the slow electrons may be additionally formed via the decay of collective excited states of the plasmon type. However, initially, the energy of such electrons is still high enough to excite or even ionize the molecules. For instance, the energy of an electron produced with decay of the $\hbar\omega_p = 21.4$ eV plasmon-type state in water is $E = \hbar\omega_p - I_c = 12.64$ eV. An electron with such an energy is still capable of exciting or ionizing one molecule, and only after that it becomes a subexcitation electron.

Thus, it is not easy to calculate the energy spectrum of subexcitation electrons taking into account all these processes, and, apparently, it can only be done using computer simulation.

B. The Main Processes of Slow Electron Interaction with Molecules

Let us briefly discuss the principal channels by which the energy of slow electrons can be transferred to molecules. A more detailed survey of these problems is presented in a number of reviews (see, for example, Refs. 105, 221, 222).

Vibrational excitation. There are two processes leading to vibrational excitation of a molecule when it collides with an electron: the direct excitation and the resonance excitation where the electron is captured by the molecule. The direct vibrational excitation occurs owing to the dependence of the potential of interaction between an electron and a molecule on the internuclear spacings in the molecule. The cross section of direct excitation varies smoothly and is on the order of 10^{-17}–10^{-16} cm^2. The corresponding interaction time is much smaller than the period of nuclear vibrations. For describing this sort of excitation it is sufficient to use the momentum or the adiabatic approximation.[105]

Near the resonances the cross section of vibrational excitation increases by several orders of magnitude. From the physical standpoint we can divide each resonance process into three stages. At the first stage the incident electron is captured by the electron shell of the molecule, forming an intermediate negative molecular ion. The second stage is the vibrational motion of the nuclei of the newly formed ion, which eventually leads to the third stage of the process—the decay of the intermediate ion.

Depending on the kind of the intermediate molecular ion, all resonance processes can be divided into two groups.[116] The first group is the so-called *shape resonances*, where the electron is trapped in a potential well formed in the ground electron state of the molecule by centrifugal or polarization forces. The lifetime of such states is between 10^{-15} and 10^{-10} s.[223]

The other group consists of the so-called *Feshbach resonances* that are formed when the incident electron first excites the molecule directly and then is captured by the molecule which is now in an excited state, provided that by doing so the molecule benefits in energy. Very often this excited state of the molecule is one of the Rydberg states, since in that case a positive electron affinity is preferable.

The maxima near the resonances on the curve representing the dependence of the vibrational excitation cross section on the energy of the electron can be very different in shape: they can be wide (up to 10 eV) or of medium size (several electron volts) or very narrow. Sometimes these maxima are arranged in groups. For the most thorougly studied molecule, N_2, the maxima of different structures are observed throughout the energy range from 2 to 30 eV (Fig. 14). The wide maxima correspond to short lifetimes of the intermediate ion ($<10^{-14}$ s). Such short-lived shape resonances were discovered for H_2 and N_2O molecules around 2 ev, for N_2 near 8 and 14 ev,[221] and for H_2O near 7 eV.[105] After the decay of this short-lived resonance, the molecule still remains in an excited state, which is usually one of the low-lying vibrational ones.

The lifetime of a resonance with a maximum of medium width is on order of 10^{-14} s. During this time the molecule can make only one vibration and decays via autoionization. It provided to be very convenient to describe such resonances using the theory of Herzenberg originally proposed for explaining the existence of a resonance at 2.3 eV in a N_2 molecule[224] (see also Ref. 225). This theory is also known as the *boomerang model*. This model is very suitable for describing the shape resonances and allows one to determine the cross sections of vibrational excitation, including the differential ones (with respect to the scattering angle).[222]

The resonances with long lifetime have narrow maxima. In this case the lifetime of the intermediate state is long enough to allow the molecule to make a large number of vibrations. The decay of the intermediate ion leads to occupation of high-lying vibrational levels of the neutral molecule (the corresponding theoretical study is presented in Ref. 226). Such long-lived resonances have been observed for such molecules as O_2,[227] NO, and N_2 [221] (in the latter case there are two isolated peaks between 11.4 and 12 eV).

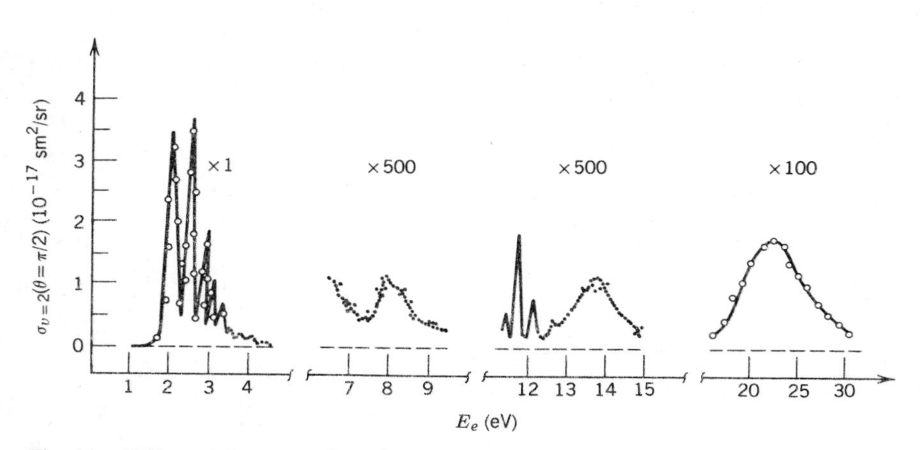

Fig. 14. Differential cross section of excitation of the vibrational level $v = 2$ in a nitrogen molecule within different energy ranges (Reproduced by permission from Ref. 221).

Rotational excitation. The rotational excitation of molecules by electron impact has been studied less experimentally. At present there are no direct methods of measuring the cross sections of rotational transitions, and the latter are found by measuring the mobility of electrons in gases.[228] So the available data are mostly theoretical. A favorable circumstance here is the fact that we are able to obtain sufficiently accurate analytical formulas for the cross sections of rotational excitation that do not require detailed knowledge of the molecule's structure.[223]

As was shown,[229] at energies below the threshold of vibrational excitation the main role in excitation of rotations is played by the interaction of electrons with multipole electric moments of molecules. This interaction is long ranged, and large distances are essential for scattering. At such distances the multipole potentials are small, so we can consider the interaction as a perturbation, and use the Born approximation for calculating the amplitudes of scattering by molecules with fixed axes. (The exceptions are the polar molecules, which have a large dipole moment.)

The scheme of calculation we have outlined has been widely applied to diatomic molecules. In this way the authors of Ref. 107 have obtained formulas for the cross sections of excitation of rotational levels owing to the charge–dipole (for heteronuclear molecules) and the charge–quadrupole (for homonuclear molecules) interactions. These results are in satisfactory agreement with experimental data at small velocities of the incident electron.

The Born approximation was used[230] for calculating the cross sections of rotational excitation for multiatomic molecules of symmetrical and antisymmetrical top types. In the second study the authors have obtained very large cross sections of excitation of the first rotational levels: at the energy of the electron of 0.01 eV, the cross section for H_2O molecules was about 3×10^{-13} cm^2, and about 10^{-12} cm^2 for the H_2CO molecule. The cross sections for diatomic molecules are much smaller: for electron energies from the threshold to 0.01–0.1 eV, the cross sections for molecules H_2, N_2, O_2, and CO are about 10^{-17}–10^{-16} cm^2.[230,231]

When the energy of the electron is below 10 eV, the most probable process is elastic scattering. The cross sections of elastic scattering are usually much greater than those of inelastic processes (except for the resonance ones) and range from 10^{-15} to 10^{-14} cm^2.

The experimental results on electron scattering for a number of molecules are presented in Refs. 232 and 233. The data on cross sections of momentum transfer are summarized in Ref. 234. Among the recent experimental studies let us note Refs. 235–237 (molecules CO, CO_2, CH_4, and H_2O).

The elastic scattering of electrons by CO_2, OCS, and CS_2 molecules has been studied theoretically.[238] The authors of Ref. 239 have calculated the differential and the total cross sections of scattering of 10-eV electrons by an acetylene molecule. The total cross section proved to be $54.5a_0^2$, which is greater than that for a N_2 molecule $(34.4a_0^2)$. The authors believe this to be an illustration of how larger geometrical size of a molecule results in a greater scattering cross section.

Among the different types of collisions of slow electrons with molecules we have considered previously, the most efficient with regard to energy losses is the vibrational excitation. Although the elastic scattering has a larger cross section, it is less efficient in this respect in the region of energy where vibrational excitation can exist. This is due to the fact that in elastic collisions an electron transfers only a small part of its energy, which, on the average, is proportional to the ratio of the electron mass to the mass of the molecule, which is always below 10^{-4}–10^{-3}.

Among the processes leading to annihilation of free electrons, the most efficient is the dissociative recombination of an electron with a molecular ion. At small electron energies the cross section of such processes exceeds 10^{-13} cm^2.[240] The cross sections for other types of recombination are much smaller. The cross section of dissociative attachment of an electron to a neutral molecule can vary within broad limits from 10^{-23} to 10^{-14} cm^2,[223,243,244] and is the largest for halogen-containing molecules.

The dissociative attachment of electrons is a resonance process that also occurs via formation of an intermediate negative molecular ion. It takes place when the electron affinity of a molecule's fragment is greater than the dissociation energy.[245] Halogen atoms have a high electron affinity, and this is the reason why the dissociative decay of a negative halogen-containing molecular ion with formation of an atomic negative halogen ion has a high probability (see Refs. 241, 246, 247).

C. The Theory of Retardation of Subexcitation Electrons in Condensed Media

In the problem of retardation of subexcitation electrons, the two important characteristics are the thermalization time and the thermalization path length. In condensed media the key role is played by thermalization path length, which determines how far can an electron travel away from its parent ion when it is thermalized. The thermalization path length determines the probability of formation of a free ion.

A broad survey of studies of thermalization of slow electrons in gases is presented in Ref. 248. Recently, there has been published a series of papers by Koura[249] devoted to simulation of electron thermalization in gases using the Monte Carlo method. So we will not discuss the case of gaseous media and will concentrate on condensed media alone.

The thermalization path length of subexcitation electrons has been the object of many discussions from the time the first track models appeared up to this day. The reason is that for quite a long time there were no direct methods of measuring the path lengths of slow electrons, while the corresponding theoretical analysis is very difficult owing to the need to take into account all the processes relevant to retardation of subexcitation electrons.

Samuel and Magee[250] were apparently the first to estimate the path length l_{th} and time τ_{th} of thermalization of slow electrons. For this purpose they used the classical model of random walks of an electron in a Coulomb field of the parent ion. They assumed that the electron travels the same distance l between each two subsequent collisions and that in each of them it loses the same portion of energy ΔE. Under such assumptions, for electrons with energy 15 eV and for ΔE between 0.025 and 0.05 eV, they have obtained $\tau_{th} \sim 2.83 \times 10^{-14}$ s and $l_{th} = 1.2$–1.8 nm. At such short l_{th} a subexcitation electron cannot escape the attraction of the parent ion and in about 10^{-13} s must be captured by the ion, which results in formation of a neutral molecule in a highly excited state, which later may experience dissociation. However, the experimental data on the yield of free ions indicated that a certain part of electrons nevertheless gets away from the ion far enough to escape recombination.

The subsequent theoretical calculations of the rate of energy loss and of the path lengths of subexcitation electrons were based on formula (5.22). Since the velocity of subexcitation electrons is much smaller than the speed of light, from (5.22) we get the following expression

$$-\frac{dE}{dt} = \frac{2e^2}{\pi v} \int_0^\infty \omega \operatorname{Im}\left(\frac{-1}{\epsilon(\omega)}\right) x K_0(x) K_1(x)\, d\omega , \tag{6.3}$$

where $x = \omega b_{min}/v$ and b_{min} is the minimum value of the impact parameter, which is usually taken to be equal either to the intermolecular spacing d or to the de Broglie wavelength of electrons $\lambda\!\!\!^{-}$ if the latter is smaller than d. This formula was used by Fröhlich and Platzman[212] to calculate the energy-loss rate in the spectral range corresponding to dipole relaxation. (This is the SHF region with the wavelength of order of 1 cm.) The reason they considered only this range of frequencies was that at that time the form of the function $\epsilon(\omega)$ was not known in a wide range of frequencies. The functions $\epsilon_1(\omega)$ and $\epsilon_2(\omega)$ were presented in the form

$$\epsilon_1(\omega) = \epsilon'(\omega) + \frac{\epsilon_s - \epsilon_{ir}}{1 + \omega^2\tau^2} ,$$

$$\epsilon_2(\omega) = \epsilon''(\omega) + \frac{(\epsilon_s - \epsilon_{ir})\omega\tau}{1 + \omega^2\tau^2} ,$$

where τ is the relaxation time and ϵ_s is the static dielectric constant. The functions $\epsilon''(\omega)$ and $\epsilon'(\omega)$ describe, respectively, the absorption and dispersion. ϵ_{ir} is the value of the dielectric permittivity at a frequency ω_{ir} which is smaller than all the leading frequencies of IR absorption. In this way they have obtained the following formula for the rate of energy loss on dipole relaxation:

$$-\frac{dE}{dt} = \frac{\pi e^2}{4b_{min}} \frac{\epsilon_s - \epsilon_{ir}}{n_r^4 \tau_d} , \tag{6.4}$$

where n_r is the refractive index and τ_d is the time of dipole relaxation.

This formula is valid for kinetic energies of the electron of 1–10 eV, and according to it, the loss rate is independent of electron energy. For $\epsilon_s = 80.4$, $\epsilon_{ir} = 4.9$, $n_r = 1.3$, $\tau_d \sim 1.01 \times 10^{-11}$ s, and b_{min} equal to the intermolecular spacing, Fröhlich and Platzman have obtained $-dE/dt \approx 10^{13}$ eV/s, which they believed to be comparable with the rate of energy loss on vibrational excitation. Following from (6.4), in the case of water the change in $-dE/dt$ under the change of the aggregation state or of the temperature is determined by the corresponding change in τ_d. Since for ice near the melting point τ_d is about 2×10^{-5} s, the loss rate $-dE/dt$ is about 10^7 eV/s, which is six orders of magnitude smaller than in liquid water. However, according to the measurements of solvation time in liquid water and in ice,[251] the difference in the energy-loss rate must be considerably smaller, which means that in the case of ice the losses on dipole relaxation are not crucial.

Formula (6.3) gives the rate of energy loss in the case of straight-line motion of electrons. The influence the curvature of the electron trajectory has on the rate of energy loss of dipole relaxation has been studied by the Monte Carlo method in Refs. 252 and 253 and analytically in Ref. 254. According to Ref. 252, the energy-loss rate on dipole relaxation for a straight-line trajectory calculated using the Monte Carlo method is 15% higher than is predicted by formal (6.4), and becomes larger still (about 1.3 times as large) when we make allowance for the curvature of the trajectory owing to the random walks of the electron in the medium. If, with the random walk method, we also simulate the resonance capture of the electron by the molecule, the resulting $-dE/dt$ will be 1.6 times greater than that for a straight-line trajectory.

As in the Monte Carlo calculations,[252,253] the analytical calculations of Ref. 254 show that the loss rate on dipole relaxation for a curved trajectory is $4/\pi = 1.26$ times higher than for a straight-line motion.

Nevertheless, according to Ref. 254, the value of $-dE/dt$ is independent of the specific way the electron moves in the medium, which contradicts the results of Ref. 252.

Besides the losses on dipole relaxation of the medium (which are sometimes called continuous losses), a subexcitation electron can lose its energy on excitation of intra- or intermolecular vibrations. This usually happens in the IR region of frequencies. The authors of Refs. 252 and 253 have calculated the rate of energy loss on excitation of intramolecular vibrations for benzol and polyethylene, and have found it to be $(2–3) \times 10^{-12}$ eV/s. As for polar media, the authors supposed that the rate of energy loss on dipole relaxation can be as high as 10^{13} eV/s. However, since neither benzol nor polyethylene are polar molecules, they had no means of comparing the contributions of the two mechanisms.

Such a comparative study has been made by Byakov and his collaborators.[29,255] They have shown that in the case of water the main contribution to the loss rate given by formula (6.3) comes from excitation of intramolecular vibrations rather than from dipole relaxation. This is all the more so in nonpolar media where the main channel of continuous losses is not the relaxation of constant dipole moments (which are zero) but the polarization losses due to the electron-inducing dipole moments in molecules. The possible exceptions are the media consisting of molecules with a high degree of symmetry, such as methane and neopentane, which have no active vibrations in the IR region.

According to Ref. 29, the rate with which a 7.5-eV electron loses its energy on dipole relaxation in liquid water is about 10^{12} eV/s, while the total loss rate on excitation of vibrations is $\sim 4 \times 10^{13}$ eV/s. The main contribution to losses in the IR range comes from maxima in the region of libration of molecules ($\nu \sim 700\ \text{cm}^{-1}$) and from valence intramolecular vibrations ($\nu \sim 3400\ \text{cm}^{-1}$). Since in D_2O the reduced masses are approximately twice as large, the loss rate on valence vibrations there must be smaller than in H_2O and, accordingly, the relaxation time and the path length must be greater.

As was noted in Ref. 29, the absorption in the IR region in alcohols and other polar liquids has the same characteristic features as in water. So in such liquids the IR losses due to excitation of intramolecular vibrations usually must give a crucial contribution to retardation of electrons. If in hydrogen-containing substances the retardation time depends on those intramolecular vibrations for which the reduced mass is determined by hydrogen atoms, there must exist an isotopic effect.

Formula (6.3) makes no allowance for the energy losses in close collisions with molecules, including those in which a negative intermediate ion is formed. So the conclusions of Ref. 29 are correct only if the losses in distant collisions play a crucial role in retardation. The contribution of close collisions to energy losses was estimated by Magee.[253] For a 1-eV electron in water the contribution of these collisions to the rate of energy losses in about 5×10^{13} eV/s, while for hexane-type hydrocarbons, it is about 7×10^{13} eV/s. The authors of Refs. 252 and 253 believe that for electrons with energy above 2 eV the losses in close collisions are greater than they are in distant ones. The experimental data[256–260] concerning the passage of slow electrons through thin films indicated that close collisions play an important role in retardation of subexcitation electrons. For instance, in Ref. 258 the peaks in spectra of characteristic losses in alkanes at electron energy around 0.6 eV were attributed to resonance losses due to formation of a negative ion. In Ref. 260 a similar explanation was given to the peak at ~1.3 eV in benzene.

It seems impossible at present to calculate the thermalization path length theoretically. Formula (6.3) is of no use for that matter, since it makes no account for energy losses in close collisions. Besides, the subexcitation electrons experience quite efficient elastic scattering. The latter can be disregarded when we calculate the energy losses, since the corresponding transferred energy is very small. However, the contribution of elastic scattering to thermalization path length is comparable and may even be greater than that of nonelastic losses. The authors of Ref. 261 have made an attempt to estimate the thermalization path length of electrons in ice, water, and benzene. To this end they first integrated formula (6.3) over ω using the experimental curves for $\mathrm{Im}[-1/\epsilon(\omega)]$, thus obtaining the path length of an electron with account of losses in distant collisions only (l_{dist}), and then found the effective thermalization path length versus the electron energy assuming that the former obeys the relation

$$l_{\mathrm{th}} = (2 l_{\mathrm{dist}} l_{\mathrm{el}})^{1/2} , \tag{6.5}$$

where l_{el} was taken to be the free path length of an electron in the medium. l_{el} in formula (6.5) represents the contribution of elastic processes to the thermalization path length. For water the authors have used the dependence of l_{el} on electron energy for the gaseous phase. In the case of benzene, l_{el} was taken to be equal to the intermolecular spacing (i.e., about 0.5 nm).

According to calculations performed in Ref. 261, an electron with an energy of 7.5 eV in water travels $l_{\mathrm{th}} = 30$ nm away from the ion before thermalization. In benzene approximately the same distance (≃33 nm) is traveled by a 3.4-eV electron. Using the energy distribution of electrons presented in Ref. 217, the authors of Ref. 261 have obtained the distribution of electron–ion pairs in distance and then have calculated the yield of free ions. And though their results concerning the yield of ions are in good agreement with experiment, this does not mean that the values they have obtained for l_{th} are reliable, since, in the first place, the yield of free ions is only weakly sensitive to variation of thermalization path length. The other reason is that when calculating l_{th} the authors did not take into account the energy losses in close collisions, which might have considerably lowered the values of the thermalization path length.

In order to find the thermalization path length of subexcitation electrons in *n*-hexane, Mozumder and Magee[262] have considered the retardation of electrons with $E > E_v = 0.4$ eV as a two-stage process. They supposed that in the first stage such electrons rapidly lose the surplus energy $\Delta E = E - E_v$ exciting intramolecular vibrations, and in that way manage to get about 1.7 nm away from the ion, which is only a small part of their total thermalization path length. At the second stage these electrons (called *subvibrational* electrons since now $E < E_v$) lose their energy in small portions on excitation of intermolecular vibrations and in elastic collisions. Since these processes have low efficiency, an electron with $E \approx E_v$ manages to get far away from the ion, its total thermalization path length at 300 K amounting to ~8 nm, which is much smaller than the value calculated in Ref. 261. According to Mozumder and Magee, the main part of the total thermalization path length is made up by the distance the electron travels when its energy falls below the level of excitation of intramolecular vibrations, that is when it becomes a subvibrational electron.

D. Experimental Studies of Thermalization Path Lengths

1. *Radiation-Chemical Methods*

The first experimental data concerning the thermalization path lengths of electrons have been obtained by a radiation-chemical method,[363] which consisted of the following. One measures the yield of electron–ion pairs G_{fi} that have escaped recombination in the track in the presence of an external electric field with strength $\mathscr{E}$. If $\eta(E)$ is the energy distribution of electron–ion pairs, and $g(E, r)$ gives the probability of observing an electron–ion pair with energy E separated by a distance r, the measured yield of ions G_{fi} obeys the relation[264]

$$G_{fi} = G_i \int_0^{E_{\max}} \eta(E)\, dE \int_0^{\infty} g(E, r)\varphi(r)\, dr\ , \tag{6.6}$$

where $\varphi(r)$ is the probability of a pair separated by a distance r escaping recombination.

If we know the form of $g(E, r)$, the value of r average over this function would give us the average thermalization path length l_{th}. So in order to find l_{th} knowing the experimental values of G_{fi}, it is necessary to know the explicit form of $\eta(E)$, $g(E, r)$, and $\varphi(r)$ as well as the yield of ion pairs G_i. The function $\varphi(r)$ is found by solving the diffusion equation for electrons in the field of a positive ion and in an external electric field $\mathscr{E}$. The solution was found by Onsager.[265] In the limit of a weak external field the function $\varphi(r, \mathscr{E})$ has the form[264]

$$\varphi(\mathscr{E}, r) = \exp\left(\frac{r_c}{r}\right)\left\{1 + \exp\left[-\frac{\mathscr{E}e^3 r}{2\epsilon(k_B T)^2}\right]\right\}, \tag{6.7}$$

where $r_c = e^2/\epsilon k_B T$ is the Onsager radius, k_B is the Boltzmann constant, T is the temperature of the medium, and ϵ is the dielectric constant.

However, we have no way of knowing the form of $\eta(E)$, to say nothing of $g(E, r)$. So in order to determine the thermalization path length from radiation-chemical experiments, one usually performs the following procedure. One introduces the function $F(r)$ according to the equation

$$F(r) = \int_0^{E_{\max}} \eta(E)g(E, r)\,dE, \tag{6.8}$$

which is then substituted into formula (6.6). $F(r)$ characterizes the distribution of electron–ion pairs in separating distance, and its analytical form is found by fitting different trial functions in order to obtain the best correspondence with the actual dependence of G_{fi} on $\mathscr{E}$. The most probable value r_0 of the separating distance given by the distribution function $F(r)$ is taken to be the average thermalization path length of an electron in liquid. It equals the average distance the electron travels away from its parent ion during thermalization. When justifying this approach, Schmidt and Allen,[263] appeal to the results of Ref. 262 according to which the thermalization path length is mainly made up by the distance the electron travels when its energy falls below 0.4 eV, and so one can expect that its total thermalization path length weakly depends on its initial energy.

Different types of functions have been used as trial functions for determining $F(r)$.[263,266–269] The best agreement with the dependence of G_{fi} on $\mathscr{E}$ has been obtained with the exponential function $F(r) = (1/r_0)\exp(-r/r_0)$ and with the Gauss function $F(r) = (4r^2/\pi^{1/2}r_0^3)\exp(-r^2/r_0^2)$, and in certain cases with a combination of the latter with a power function.[268–273]

The total yield of ions G_i formed in a liquid at the primary stage of radiolysis cannot be measured experimentally. So the value of G_i is either taken to be equal to the yield of ions produced by radiation in a gaseous medium, or, like r_0, is considered as a parameter whose value is chosen so as to get the best correspondence with the experimental dependence of G_{fi} on $\mathscr{E}$.

So, as one can see, this method of determining the thermalization path length is not straightforward and involves many assumptions and suppositions. First, the value of r_0 depends on the specific choice of the function $F(r)$ (see, e.g., Ref. 269) and is too uncertain to enable us to determine the processes that are responsible for retardation of subexcitation electrons in a medium (see the discussion in Ref. 271). However, by comparing the values of r_0 for different substances we are able to determine some of the factors that affect the path length of slow electrons. Since the value of r_0 depends on the density ρ of the liquid, it is reasonable to compare the products ρr_0 rather than the values of r_0 themselves.

The influence of different properties of a liquid on the value of ρr_0 was considered in Refs. 268, 272, and 273. There it was shown that for polar liquids the product ρr_0 is greater than it is for nonpolar ones. Among liquids of the same type, ρr_0 is the greatest for those with symmetric molecules, such as CH_4, CF_4, and neopentane. The parameter ρr_0 changes only slightly with variation of temperature, and r_0 is smaller when the dipole moments are larger. All these general features are well explained by modern theoretical conceptions concerning the dependence of the rate of energy loss on the properties of a liquid (see Section VI.C). The influence of structural factors, such as isomerization, the number of carbon atoms and the number of chains in hydrocarbon molecules, on the path length of electrons has been studied in Ref. 268.

The procedure we have described previously is not the only radiation-chemical method of determining the average path length of subexcitation electrons in liquids. In Refs. 274 and 275 r_0 was found using a method based on measuring the yield of ions versus the concentration of acceptors. For polar media Mozumder[276] has developed a theory of neutralization of isolated electron–ion pairs, which enables one to calculate r_0 knowing the ratio $G(e_s)/G_i$. However, the introduction of additional parameters [such as diffusion coefficients, relaxation time, and the ratios $G(e_s)/G_i$ themselves] for which we have no definite information does not guarantee the reliability of the values one obtains for r_0. In particular, the values of r_0 Mozumder has obtained for alcohols were too high, the value for ethanol being almost twice as large as for methanol. In order to somehow lessen this sharp difference in r_0, Mozumder had to assume that the relaxation time is the same for all alcohols.

2. *Photoemission Methods*

The electrons that take in the formation of a certain distribution of electron–ion pairs in separating distance under ionizing radiation may have different energies. In particular, their energy can be above the threshold of excitation of electron states of molecules. So using the radiation-chemical methods we practically cannot obtain the dependence of the thermalization path length of a slow electron on its energy. In this respect the most efficient are the methods using the photoemission of electrons out of a metal into electrolytic solutions.[188,277–279] In that case, from the very beginning, the electrons introduced into the solution have energies below the excitation threshold, which enables us to study the solvation and thermalization of slow electrons with no regard for excitation and ionization of the medium by fast electrons.

An important advantage of the photoemission method is the fact that the electrons emitted into the solution have a known or the studied energy distribution with a strictly definite boundary energy E_{max}[188]:

$$E_{max} = \hbar\omega - \hbar\omega_0 + e\varphi \, , \tag{6.9}$$

where ω is the frequency of the incident light, $\hbar\omega_0$ is the work function of electrons for the given metal, and φ is the potential of electrodes. The energy E_{max} can be changed either by varying the energy $\hbar\omega$ or by varying the potential φ (within 1.5 eV). The transformations that the emitted electrons undergo in the solution can be presented as a series of stages (see the scheme in Fig. 15).

In photoemission experiments one measures either the photocurrent or the photopotential when an electrode is placed in a solution and is illuminated by light that the solution does not absorb. The photocurrent of emitted electrons depends on the every $\hbar\omega$ and the potential φ according to the so-called five halves law[277–280]:

$$j_m = cE_{max}^{5/2} = c(\hbar\omega - \hbar\omega_0 - e\varphi)^{5/2} \, , \tag{6.10}$$

where the constant c is usually weakly dependent of ω. The subsequent thermalization and solvation of emitted electrons transform their energy distribution into a spatial distribution $F(x)$ in the coordinate x normal to the surface of the electrode.

It is practically impossible to reconstruct the form of $F(x)$ in detail by measuring the photocurrent. However, its first two moments can be measured in photoemission experiments. The first moment of the function $F(x)$ is the projection of the average thermalization path length of electrons on the normal to the surface of the cathode:

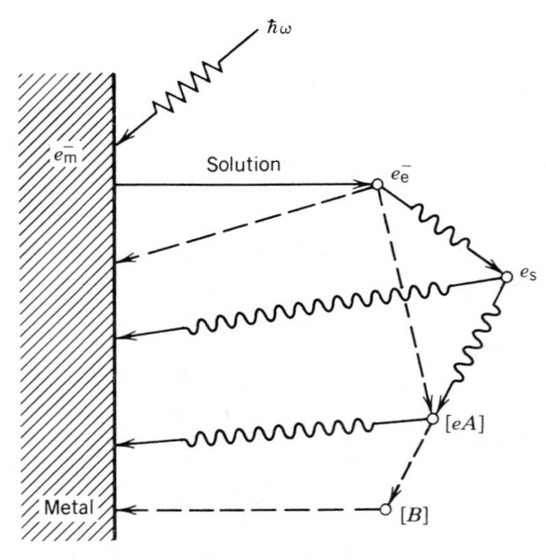

Fig. 15. Scheme of transformations of an emitted electron in a solution (according to Ref. 279), e_m^- is the electron in a metal, e_e^- is the emitted electron, e_s is the solvated electron, $[eA]$ and $[B]$ are reaction products, and A is the acceptor. The possible homogeneous reactions and the return of "dry" electrons and products to the electrode are plotted by dashed lines.

$$x_0 = \int_0^\infty xF(x)\,dx\,. \tag{6.11}$$

Several approaches have been employed for measuring x_0 by the photoemission method.

In their experiments Holroyd et al.[183] have studied the dependence of the photocurrent on the strength of the electric field between the cathode and the anode. The corresponding theoretical dependence was calculated using the function $F(x) = (2/x_0\sqrt{\pi})\exp(-x^2/x_0^2)$ obtained by variation of x_0 and was in agreement with the measured one. This approach is very close to the radiation-chemical method; however, it allows one to vary the energy of electrons. Most of the values of x_0 obtained for organic liquids are smaller than the corresponding values of r_0 obtained in radiation-chemical experiments.[263] However, on the average, the difference between them does not exceed 1.7 nm, which Holroyd et al.[183] believe to be in agreement with the theoretical predictions of Ref. 262. Using the same method, Tauchert et al.[281] have obtained, for a number of cryogenic liquids the values of x_0 depending on the energy of the emitted electrons and on the temperature of the liquid.

More precise methods of determining x_0 are based on the analysis of the dependence of the photocurrent either on the concentration of acceptors or on time in the absence of external fields. In experiments of the first type, one measures the stationary photocurrents at different concentrations of acceptors.[282] At low concentrations one is able to obtain the values of x_0 without using any specific model for the function $F(x)$. This method was used in Refs. 283 and 284 for studying the dependence of x_0 in water on the maximum energy of emitted electrons E_{max}. It was shown that at $0.3 < E_{max} < 2$ eV this dependence is approximately linear, namely, $x_0 = a + bE_{max}$, where $a = 0.5$ nm and $b = 1.5$ nm/eV.

An opposite result was obtained in Refs. 285–287: at $0.7 < E_{max} <$ 1.8 eV the value of x_0 is independent of electron energy. When the range of studied energies E_{max} was increased to 2.9 eV, it was found[288,290] that in water solutions there are two regions of energy with different dependence of x_0 on E_{max}: at $1 < E_{max} < 2$ eV, x_0 is independent of electron energy (being about 2.5 nm), while at $E_{max} > 2$ eV, x_0 grows with increase of E_{max}. References 288 and 290 have attributed this independence of x_0 to resonance energy losses (in close collisions). However, as was pointed out by one of us,[73] this explanation is unconvincing. The subsequent studies of Raitsimring et al.[30,291] gave a more precise dependence of x_0 on E_{max} in the region of small energies from 0.2 to 1.3 eV, and the range of E_{max} studied was extended to 3.8 eV.

According to Ref. 30, in D_2O, x_0 monotonically grows as E_{max} increases from 0.2 to 2.9 eV. However, in water the picture is different. When E_{max} is varied from 0.2 to 1.3 eV, x_0 increases monotonically, as it does in D_2O, but on reaching $E_{max} = 1.3$ eV, x_0 stops growing and remains practically constant until the energy reaches 2.3 eV, after which it begins to grow again. A very unusual result was obtained for $E_{max} = 3.8$ eV: at that energy the thermalization path length rapidly falls both in H_2O and in D_2O. At present we have no explanation of this fact, as well as to the independence of x_0 in H_2O at $E_{max} = 1.3$–2.3 eV. Konovalov et al.[291] suppose that the sharp decrease of x_0 at $E_{max} = 3.8$ eV is due to the change of the energy distribution of emitted electrons at high photon energies. Since the energy distribution of photoelectrons appearing when a photocathode is irradiated by light with $\lambda = 222$ nm and 193 nm has not yet been studied, it is possible that it may be different from the one predicted by the five halves law. Another argument in favor of this explanation is the fact that the values of x_0 depend on the type of photon source used, even at $E_{max} < 2$ eV (see, e.g., Fig. 6 in Ref. 277). Therefore, La Verne and Mozumder[248] may be right in asserting that, owing to the high density of traps in polar media, what one actually measures is not the thermalization path length but the average distance traveled by a hot electron until it is captured in one of the traps.

The second approach to measuring the thermalization path length of emitted electrons is based on analyzing the kinetics of e_s returning to the electrode, that is, the ratio $j(\tau)/j_m$. This approach is more convenient both in its experimental realization and in data processing,[292] since acceptors are not supposed to be present in the solution. In particular, approximate estimates of x_0 can be obtained using the relation between x_0 and the time τ_e of diffusive return of e_s to the surface of the electrode, namely, $x_0 \simeq (D_e \tau_e)^{1/2}$.[279] Since $\tau_e \sim 10^{-9}$–10^{-8} s, such experiments can only be performed with pulsed light sources having pulse duration smaller than 10^{-9} s. This is why the experiments performed in Ref. 293, in which the authors used a nitrogen laser, as well as those of Ref. 294 where the second harmonic of a ruby laser was used, failed to detect the diffusive return of e_s, since the duration of light pulses in both cases was 10^{-8} s. However, the results of these experiments gave an upper limit for the thermalization path length, since they mean that τ_e is below 10^{-8} s, so for $D_e = 5 \times 10^{-5}$ cm^2/s we get $x_0 < 7$ nm.

For the same reason, Konovalov et al.[292] have used excimer lasers with pulse duration ≤4.2 ns. The values obtained for x_0, which correspond to a Gauss function, are presented in Table VIII. In deuterated water x_0 is approximately 1.3 times greater than it is in normal water, which is in qualitative agreement with the rate of energy loss on intramolecular vibrations calculated in Ref. 29, which in the former case is two times lower. Besides, at electron energies ~1.6–1.7 eV the cross section of resonance energy loss in close collisions in D_2O is also about two to three times smaller than in H_2O.[295] If in formula (6.5), we take l_{dist} for D_2O two times greater than for H_2O, and take the same l_{el} for both media, we will get $l_{th}(D_2O) \simeq \sqrt{2} l_{th}(H_2O)$.

In recent years the energy dependence of thermalization path lengths has been studied experimentally using a method based on photoionization by light of different wavelengths of a molecule-additive introduced into the matrix being studied. The molecule-additives often used are, for instance, those of pyren or of N,N,N',N'-tetramethyl-p-phenylene-

TABLE VIII
Thermalization Path Length of Slow Electrons in Water Electrolytic Solutions[292]

Excimer	λ(nm)	E_{max} (eV)	x_0[a] (nm)	
			H_2O	D_2O
KrF	248	2.3	2.5 ± 0.5	3.2 ± 0.5
KrCl	223	2.9	4.0 ± 0.8	5.0 ± 1.0

[a] In the case of isotropic scattering, the thermalization path length is $r_0 = \sqrt{3} x_0$.

diamine.[296] In such experiments one studies the dependence of the yield of ion pairs on the strength of the external electric field. The following data processing is equivalent to the approach used in the radiation-chemical method. Other methods of introducing slow electrons into a solution and of measuring the thermalization path lengths have been used in Refs. 297 and 298.

According to theoretical and experimental studies of path lengths of slow electrons, the distance separating an electron pair is mainly determined by the path length of a subexcitation electron. Indeed, according to Table VIII, the average thermalization path length of an electron with $E_{max} = 2.9$ eV in H_2O is $r_0 \simeq \sqrt{3}x_0 \simeq 6.8$ nm, while the ionization path length of a 100-eV electron in water is on the order of 1.5–2 nm, that is, considerably smaller than the thermalization path length. However, owing to the difference in values of thermalization path lengths obtained using different methods, we are not able at present to obtain a sufficiently reliable energy dependence of path lengths of slow electrons. One of the principal obstacles is the absence of reliable information about the energy distribution of slow electrons that are introduced into the medium. While the absence of data on the absolute values of cross sections of resonance processes and of elastic collisions makes it impossible to carry out a complete theoretical study of retardation of slow electrons.

VII. DELOCALIZATION OF THE ENERGY OF IONIZING RADIATION IN A MOLECULAR MEDIUM AND ITS RADIATION-CHEMICAL MANIFESTATION

A. Formulation of the Problem

The passage of a charged particle through a medium results in the formation of disturbance areas along the particle's trajectory that contain excited molecules, positive ions, and knocked out electrons and atoms. These disturbances areas make up the track of the particle (see Section VIII). An important role in the process of formation of the track and in the following radiation-chemical transformations is played by the degree of delocalization of the initially absorbed energy.

As far back as 1960, Fano[299] had pointed out the two possible reasons for delocalization of the energy absorbed by a medium. The first one has to do with the quantum-mechanical nature of microparticles; the second one is connected with the possibility of excitation of plasmon-type collective oscillations.[92]

The quantum-mechanical nature of microobjects manifests itself in the Heisenberg uncertainty principle. It was common belief that the limitations imposed by this principle are not essential. In Ref. 10 this was substantiated by the fact that during the period of time $\sim 10^{-16}$ s the excitation is localized on one of the molecules. However, this substatiation cannot be considered correct. As was stressed in Ref. 73, since the subsequent localization is of stochastic nature, the initial delocalization makes it impossible to determine the point where the energy absorption took place more precisely than is allowed by the uncertainty principle.

As for delocalization due to collective oscillations, the very existence of the latter in molecular media has been the subject of discussion for quite a long time (see Refs. 4 and 25). Later it was established that, in a molecular medium, too, fast moving charged particles can induce states of collective nature (see Section III.C). Let us now consider both of these reasons in more detail.

B. Delocalization and the Uncertainty Principle

Consider a particle with the energy E transferring part of its energy ΔE to the medium. At the moment when the energy is transferred, there is an uncertainty ΔE in the energy of the particle and a corresponding uncertainty Δx in its location. The relation between these two uncertainties has been found in Ref. 10 for ultrarelativistic particles and in Ref. 73 for nonrelativistic ones. Let us now consider the general case.[300]

As in Refs. 10 and 73, we will start with the uncertainty principle formula connecting the uncertainties in the coordinate and the momentum of a microparticle[106]

$$\Delta x \, \Delta p_x \gtrsim \hbar \, . \tag{7.1}$$

According to relativistic mechanics, for the energy of a particle moving along the x axis and its momentum p_x, we have the following relations[152]:

$$E^2 = c^2(p_x^2 + m_0^2c^2) \, , \tag{7.2}$$

$$p_x = Ev_x/c^2 \, . \tag{7.3}$$

From (7.2) we have $E\,\Delta E = c^2 p_x \, \Delta p_x$ or $\Delta p_x = (E/c^2 p_x)\,\Delta E$. Substituting this into (7.1) and using (7.3), we get the following formula for the uncertainty in the coordinate of the particle when it is transferring the energy ΔE:

$$\Delta x \gtrsim \hbar v_x / \Delta E \, . \tag{7.4}$$

This will also be the uncertainty in the location of the point where the energy was absorbed. Since at small velocities all relativistic relations tend to their nonrelativistic analogs, formula (7.4) holds for the whole range of velocities of the bombarding particle.

The preceding derivation has been performed for a free microparticle. However, the electron moves in a medium, and it is owing to its interaction with molecules that its energy acquires an uncertainty ΔE. The conditions under which we can consider the electron as a free particle coincide with the conditions of applicability of the first Born approximation (within which we have derived all the main formulas for cross sections of interaction in Section IV), namely, it is necessary that the interaction energy be small in comparison with the kinetic energy of the particle, which is just the case for fast particles.

Let us also note that relation (7.4) can be obtained from formula (2.2), since the uncertainty in the coordinate of energy transfer is related to the uncertainty in the transfer time ($\Delta x = v\,\Delta t$), so the former is larger the higher the particle's velocity. The larger the transferred energy, the smaller the delocalization is. When we are knocking out $1s$ electrons, the energy absorption is actually localized. Indeed, according to (7.4), even in the case of ultrarelativistic particles ($v = c$) when the energy transfer is 300 eV [$\Delta E \gtrsim I_s(c)$], we have $\Delta x = 0.66$ nm, and for $\Delta E = 500$ eV [$\Delta E \geq I_s(0)$], $\Delta x = 0.4$ nm.

However, when the energy losses are small (and this is the case for most of the inelastic scattering processes) delocalization can be considerably large. The most probable channel of energy losses in condensed media is the excitation of collective states of the plasmon type. In the fourth column of Table IX we present the values of Δx calculated at different energies of the incident electron for the case where the energy is transferred on excitation of plasmon states in water. As one can see from this table, for relativistic electrons the delocalization along the x axis due to excitation of a plasmon state is no less than 9.2 nm, that is, quite considerable. The degree of uncertainty in the location of the point where the energy transfer took place is greater for excitation of low-lying electron states of molecules and may be more than 20 nm (see the third column of Table IX).

Thus, the area of delocalization may contain up to several dozens of medium-size molecules, and though later on the energy is localized on one of them, this localization is stochastic, and thus, the coordinates of the points of ionization or excitation cannot be determined more precisely than to within Δx.

According to (7.4), the delocalization depends on the velocity of the particle, so for heavy particles the values of Δx presented in Table IX correspond to greater energies, namely, $E = (M/m_e)E_e$. For example, a proton has the velocity $v \approx c$ at $E \gtrsim 2$ GeV.

TABLE IX
Delocalization due to Energy Transfer to the First Singlet State $\hbar\omega_{01} = 8.4$ eV and to the Plasmon State $\hbar\omega_r = 21.4$ eV in Water by an Electron with Energy E_e

		$\hbar\omega_{01}$	$\hbar\omega_r$	
E_e (eV)	v (10^8 cm/s)	Δx (nm)	Δx (nm)	b_{pl} (nm)[a]
10^2	5.93	0.464	0.18	0.565
10^3	18.7	1.465	0.575	1.8
10^4	58.5	4.582	1.8	5.65
10^5	164	12.85	5.0	15.7
10^6	282	22.1	8.61	27.0
10^7	300	23.5	9.2	29.0
10^8	300	23.5	9.2	29.0

[a] b_{pl}-effective size of collective excitations, see below Section VIIC.

C. The Delocalization due to Excitation of Collective States

The second reason for delocalization of energy losses is the collective nature of excited states.* This collectivity may exist even for excited electronic states of a single molecule. The simplest example is the excitation of π-electron states, which are delocalized along the molecule. When a fast electron excites such a molecule, it transfers its energy to the whole ensemble of π electrons. As a result, the energy absorption is delocalized along the molecule, and the latter can be a long (e.g., a polymer molecule).

Because of periodic symmetry, the electronic excitation states in a molecular crystal are also of collective nature. These are the well-studied exciton states.[81,82] Their energy is close to that of discrete electronic states of isolated molecules (4–8 eV), but the excitation envelops a large group of molecules, migrating efficiently up to 100 nm along the crystal.[82] In the same manner, because of efficient migration, the excitation of a fragment of a polymer chain rapidly spreads over the whole molecule.[37]

Thus, the excitation of discrete electronic states ($\Delta E_n < I_1$) in polymer molecules and molecular crystals is certainly delocalized.

Collective excited states exist not only in crystals but also in liquids and polymer films, when the latter are irradiated by fast charged particles. Such excitations are of plasmon nature, their energy, as a rule, being much higher than the first ionization potential, that is, somewhere about

*Here we do not discuss the delocalization of the energy initially localized on one of the molecules owing to its nonradiative transfer to molecule-acceptors[30] (see, e.g., Ref. 36). A detailed analysis of dissipation of excitation energy of a molecule-donor owing to concentration quenching by molecule-acceptors is presented in a recent review by Burstein.[301]

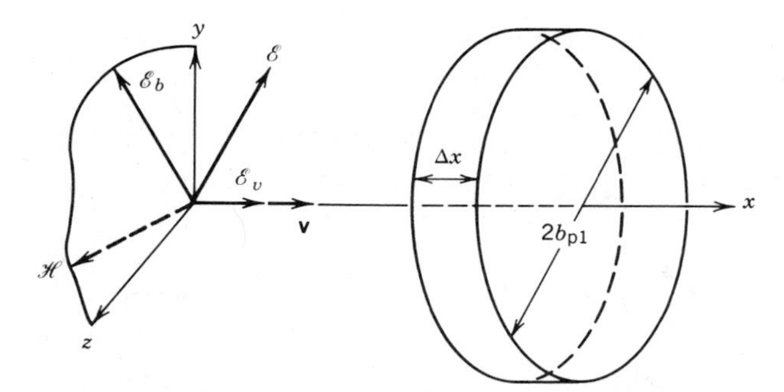

Fig. 16. Schematic picture of the region of delocalization of energy losses of a fast charged particle. Δx is the uncertainty in the coordinate of the point of energy transfer, b_{pl} is the radius of collective excitations, and $\mathscr{E}$ and $\mathscr{H}$ are the strengths of the electric and magnetic fields.

20–25 eV (see Section III.C). It is important to estimate the size of the region enveloped by collective excitation.

The electromagnetic field created by a moving charge is isotropic at small velocities. For relativistic particles we have a sharp anisotropy: the electric vector lies within a narrow angle near the plane perpendicular to the the direction of the particle[152] (see Fig. 16), and so collective oscillations are excited in the same plane. Since the group velocity of polarization waves is small, the period of time $\sim 10^{-16}$ s (the average lifetime of collective oscillations) is too short for the energy to migrate outside the region of initial excitation. The effective size of the region enveloped by this excitation is given by formula (5.24), so $b_{pl} = \pi v/\omega_r$, where ω_r is the resonance frequency (its values are presented in Table IV). In the fifth column of Table IX we present the values of b_{pl} for water versus the energy of the electron. The maximum value $b_{pl} = 29$ nm corresponds to ultrarelativistic electrons with $v \simeq c$.

D. The Radiation-Chemical Manifestation of Delocalization

Thus, part of the energy transferred to a molecular medium by a charged particle is certainly delocalized. And though later this energy is localized on one of the molecules, this localization is stochastic, and thus the coordinates of the points of ionization and excitation cannot be determined more precisely than to within the magnitude of b_{pl} or Δx we have presented previously. This circumstance is important, first of all, when one simulates tracks of charged particles using the Monte Carlo method, where the track is presented as a set of points where the interaction took place.[302,303] Even if the plasmon states are not formed in the system, the quantum-mechanical uncertainty alone imposes limitations on the quantities in the relation (7.4). For instance, if we want to find to a certain precision the coordinates of a point where an event of inelastic loss occurred, then at a fixed energy E of the particle a lower limit will be placed on the transferred energy ΔE; while, if we fix ΔE, we will get an upper limit on the particle's energy.

The delocalization of energy losses is particularly important in radiobiology. Biosystems are very nonhomogeneous in sensitivity to radiation, so the effect the latter produces in them, as a rule, is due to affects on their individual microstructures. In microdosimetry[304,305] one calculates a function $f_{t,a}(\Delta E, s)$ which determines the probability of the energy ΔE being transferred (index t) or absorbed (index a) when a beam of charged particles passes through a layer with thickness s. Since the uncertainty in energy localization equals the thickness of the layer, for the energy transfer from formula (7.4) we get

$$\Delta E \gtrsim \hbar v / s\,. \tag{7.5}$$

As an answer to the question of how essential this limitation is, let us consider a number of typical examples from radiobiological studies. In Ref. 304 the function $f_t(\Delta E, s)$ was calculated for a beam of protons with energy $E = 7$ MeV and $s = 2$ nm. According to (7.2), ΔE cannot be smaller than ~12 eV. The same thing applies to the function $f_a(\Delta E, s)$ presented in Ref. 305 for 1-keV electrons, which determines the probability of energy absorption in a spherical volume with diameter $d =$ 1 nm. Here, also, we have $\Delta E \geq 12$ eV. Since the losses of energy on excitation of electron–vibrational states may as well be considerably smaller, the use of microdosimetry characteristics for regions of order of several nanometers in size, that is, in nanodosimetry, is quite problematic.

As was shown previously, the linear dimensions of the area enveloped by collective oscillations of plasmon type may be as large as 30 nm. Since in the presence of collective peaks, the oscillator strengths are concentrated near the maximum of the peak, the yield of plasmon absorption is rather large. Since it follows from the behavior of the energy loss function in water,[4] the existence of plasmonic absorption diminishes the relative fraction of oscillator strengths corresponding to transitions into discrete states in comparison with their distribution in steam. Since a plasmon-type collective oscillation decays mostly by transferring its energy to a single molecule with subsequent ionization, the redistribution of energy of discrete excited levels to collective ones must lead to enhancement of the relative fraction of ionization in liquid systems in comparison to the gaseous ones. (For how this fact is taken into account, see the calculation of the yield of intermediate products of water radiolysis in Section VIII.A.) At the same time, as was stressed by Voltz,[84] the decay of collective excitations may be an additional source of SES (Section III.B). For a SES to be excited with decay of a collective exctation, it is necessary that the energy released would be in the region of SES formation for the considered molecules.

In conclusion, let us dwell on the consequences of delocalization of electronic excited states in long molecules and of the effective migration of electronic excitation along a polymer chain.[37] These processes lead to rupture of mostly the weaker chemical bonds, as well as of those near embedded atoms and near the defects of the chain, since it is there the excitation energy is localized. In such a polymer matrix there can be no track effects caused by the energy transferred to electrons of the medium, since the absorbed energy rapidly dissipates throughout the electron subsystem. The situation here is similar to that in metals, where the energy absorbed by electronic degrees of freedom dissipates over the volume of the metal and no linear tracks (which are due to electronic degrees of freedom) are formed.

VIII. TRACKS OF CHARGED PARTICLES AND THEIR STRUCTURE

A. Tracks of Fast Electrons

1. *Classification of Track Structures*

When a fast electron passes through a medium and loses its energy exciting and ionizing the molecules, it does so in discrete portions and not in a continuous manner. If the energy it transfers to the medium is on the order of or smaller than the ionization potential of a molecule, there appears an electron–ion pair or an electronically excited molecule. If, on the other hand, the energy transferred in a single event exceeds the ionization potential by several times, there forms a region of energy localization containing several electron–ion pairs and excited molecules.

In early studies[306,307] such regions of a medium containing one or several ion pairs were called clusters. However, later on, following Samuel and Magee,[250] they started calling them *spurs*. It is the spurs formed around the trajectory of a primary electron that make up the latter's track.

According to the data on energy-loss spectrum of fast electrons in thin layers of substance (see Section V.B.2), the most probable amount of lost energy is small, being 10–15 eV in dilute media and 20–25 eV in condensed ones. Therefore, it is most probable that a fast electron generates a spur with one or two ion pairs. However, we have no certain information about the initial size and shape of spurs for the reason that there are no reliable data on path lengths of slow electrons (see Section VI). So one usually assumed that at the beginning of the chemical stage of radiolysis a spur has a spherical form with a Gaussian distribution of chemically active particles in it.[23,308]

During the period of time between two consecutive events of energy transfer, an electron travels a distance l_{in}, which is called the free path length with respect to inelastic scattering. Its average value equals

$$\bar{l}_{in} = (n\sigma_{tot})^{-1}, \tag{8.1}$$

where σ_{tot} is the total cross section of inelastic scattering of an electron by a molecule [formula (4.39)] and n is the density of molecules in the medium. For relativistic electrons in water, $\bar{l}_{in}$ is about 200–300 nm, while for a 10-keV electron it is an order of magnitude smaller. If we take the initial size of a spur to be equal to the average thermalization path length of a subexcitation electron (according to Table VIII, for water $r_0 \lesssim 7$ nm), then comparing the values of $\bar{l}_{in}$ and r_0 we see that for electrons with $E_e > 10$ keV, $\bar{l}_{in} > r_0$, meaning that the main part of the track of a relativistic electron is a set of nonoverlapping (isolated) spurs.

For electrons with $E_e \leq 1$ keV, the value of $\bar{l}_{in}$ is 1–10 nm, that is, smaller than r_0. Therefore, in the track of such an electron, the regions containing excited molecules and electron–ion pairs are no longer isolated from each other, forming track structures consisting of overlapping spurs. Following Mozumder and Magee,[309] the energy required to form a given track structure is taken to be the criterion for classification of different track structures.

According to this classification, the energy of a fast electron in water is spent on formation of the spur itself, with energy between 6 and 100 eV, as well as on formation of blobs, with energy 0.1–0.5 keV, and of short tracks, whose energy is 0.5–5 keV. The upper limit for the energy of a short track (5 keV) is determined by the condition of spurs overlapping each other. According to Ref. 309, short tracks are cylindrical regions of high concentration of ionizations and excitations.

Originally, a blob was defined as a formation generated by a secondary electron, the energy of which exceeds 100 eV but is not high enough to allow it to escape the attraction of the parent ion. For water this energy is about 625 eV at room temperature. With this in view, Mozumder and Magee[309] have taken the upper limit of the blob energy to be 500 eV. However, this was quite an arbitrary choice, and later Magee and

Chatterjee[310] have proposed to take the upper limit to be 1.6 keV, which is the energy of an electron below which elastic scattering becomes efficient. According to Magee and Chatterjee, a blob can be considered as an ellipsoid.

In Ref. 309 the initial "radius" of a spur with energy E was estimated to be proportional to $r_0 \sim E^{1/3}$, and for $E = 100$ eV, r_0 was taken to be 1.7 nm. Magee and Chatterjee[308] determined the size of a spur in a different way. They considered an electron–ion pair as a Rydberg state of a molecule, and have estimated the size of the electron's orbit in this state to be about 3–3.5 nm by the time it is hydrated. The initial size of a spur containing only one electron–ion pair was estimated to be 3 nm, while the size of the electron orbit was believed to be smaller in spurs containing several ion pairs since in that case the electron moves in the field of several ionic centers. This led Magee and Chatterjee[308] to a nontrivial result; the size of a spur is the smaller the more ion pairs it contains. For instance, the initial size of a spur with six ion pairs ($E = 100$ eV) is $r_0 \simeq 2$ nm.

In Ref. 309 the authors have found the distribution of spurs in size in the case of water using the Monte Carlo method. They have used the energy-loss spectrum for water vapor assuming that, since the stopping power of water in the gaseous and in the liquid phases is approximately the same, the energy-loss spectra in both cases must be practically identical. Later this idea was used in Ref. 308. As was shown in Section V.B.2, in reality the spectrum of energy losses in liquid water is different from the one in water vapor. This difference leads to different average ionization potentials $\bar{I}_M$ (the difference is 5–10 eV); however, this has little effect on the stopping power since the latter depends on $\bar{I}_M$ only logarithmically. And so the weak variation of the stopping power under the change of the aggregation state cannot be considered as an argument in favor of the energy-loss spectrum being independent of the state of aggregation. Thus, the distribution of spurs in size calculated in Refs. 308 and 309 needs to be determined more precisely.

The distribution of the energy needed to form different track structures has been calculated in Ref. 309 using the Monte Carlo method in a wide range of primary electron energies, from 2 keV to 10 meV. According to these results, with increase of the energy of the primary electron, the fraction of this energy spent on formation of isolated spurs becomes larger, while the relative fraction of the energy needed to form blobs reaches a wide maximum between 20 and 100 keV.

The track structures considered in these papers were only those formed by primary electrons. Santar and Bednář[311] have carried out the calculations of Ref. 309 in more detail and have taken into account the formation of track structures by secondary electrons. According to their results, the account of the total energy degradation spectrum of secondary electrons essentially changes the distribution of energy among different track structures, increasing the relative fraction of blobs. This is due to the fact that an electron which forms a short track may also generate additional short tracks and blobs. Besides, each short track end with a blob.

Such a presentation of a fast electron track as a set of spurs, blobs, and short tracks is widely used in radiation chemistry for describing the processes that occur in a condensed medium exposed to electron or gamma radiation.[7] However, this presentation is not the only one there is. Other possible approaches are discussed in Ref. 305, where, in particular, the authors note that the most general description of track structures is the one using correlation functions.

2. *Simulation of a Fast Electron Track*

The tracks of fast electrons are widely studied using the Monte Carlo method, which allows one to simulate the passage through a medium of primary electrons and of electrons of all subsequent generations. With such simulation one can obtain a spatial picture of the track not only by the end of the physical stage but also at subsequent moments of time. The possibilities of the Monte Carlo method regarding simulation of radiative transformations have been demonstrated most spectacularly in Ref. 16. Using the previously developed method,[100] the authors have calculated the coordinates of ionization and excitation events in the track of a 5-keV electron in water. Then, they have chosen the mechanism according to which the primary active particles H_2O^+, H_2O^*, and e transform in the following physicochemical stage of radiolysis and have obtained the distribution of chemically active particles, such as H_3O^+, e_{aq}, OH, and H, at the beginning of the chemical stage, that is, at times about 10^{-11} s. After that, they followed the transformation of these particles owing to diffusion-controlled reactions between them up to the time $\sim 10^{-6}$ s, when the track period ends. An example of the picture of evolution of an electron track between 10^{-11} and 2.8×10^{-7} s is shown in Fig. 17.

Studying the electron tracks with the Monte Carlo method, the authors of Refs. 302 and 303 have used the so-called "stochastic" approach, within which one fixes a simultaneous picture of the spatial distribution of excitation and ionization events. The tracks found this way are sets of spatial points where the inelastic scattering events took place. With this at hand it proves to be possible to calculate the energy absorption spectrum in sensitive volumes of the irradiated medium[303] and to calculate the shape of the line and the slope of electronic spin echo signals.[302] Such a

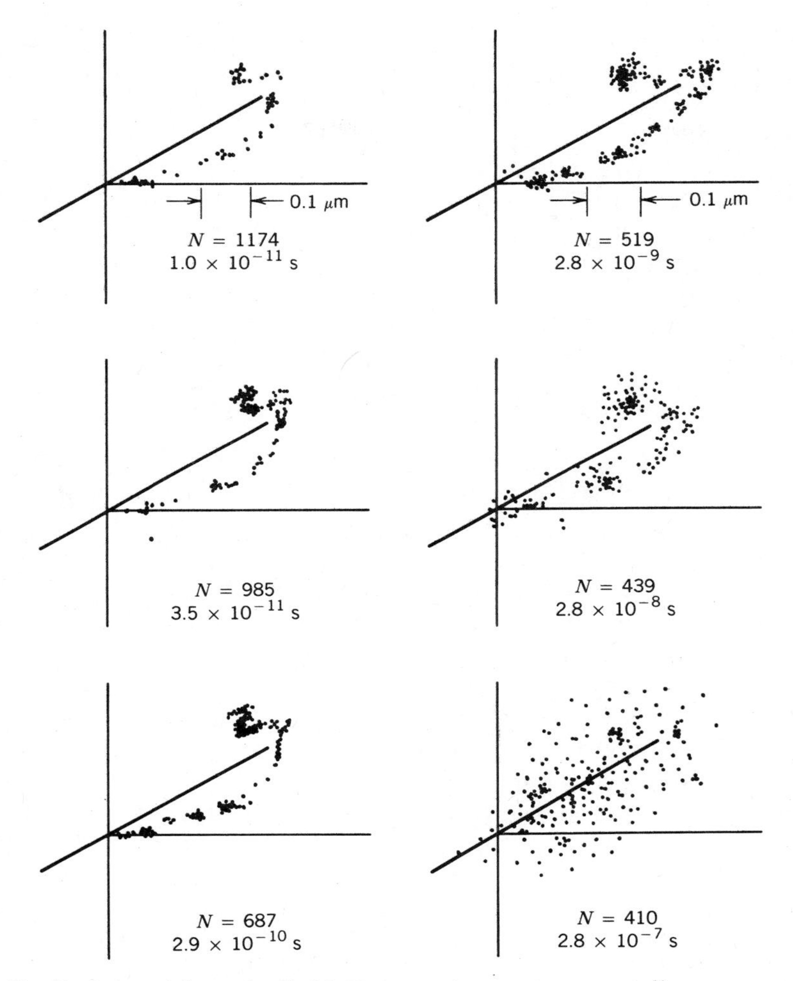

Fig. 17. Evolution of the track of a 5-keV electron in water between 10^{-11} and 2.8×10^{-7} s (Reproduced by permission from Ref. 16).

determinate approach is justified only for electrons with relatively low energies at which the uncertainty in localization of energy is smaller than the size of a molecule or the size of biological-sensitive volumes. (We have discussed these problems in Section VII.) Unfortunately, the latter fact is usually not taken into account, which may lead to incorrect results obtained in this manner.

To use the Monte Carlo method on a computer, we need to know the cross sections of elementary events. Such cross sections are known for a number of substances in the gaseous state (see Section IV). However, with liquids the situation is not so favorable. At the same time, it is the liquid state that is more interesting to us from the point of view of studying the radiation-chemical processes in tracks of particles of different nature. Owing to the lack of data on elementary processes of interaction of ionizing particles with liquid media, for quite a long time the simulation of the primary stage of radiolysis in liquids was carried out the same way as in gases, that is, the liquid was considered as a dense gas. The only exceptions were studies[4,14,16,143] in which the authors took into account the specific features of the condensed state. Presenting a liquid as a dense gas, we omit from consideration the peculiarities in the distribution of energy over the quantum states in the liquid, in particular, the existence of collective excited states (see Section V.B).

B. The Influence the State of Aggregation of a Medium Has on the Primary Stage of Radiolysis

In Section V.B we have pointed out the principal differences between processes of interaction of charged particles with a medium in the condensed and in the gaseous phases. Owing to the lowering of the ionization potential and to the change in the spectrum of energy losses with transition from the gaseous state to the condensed one, we should, first of all, expect a change in yields of primary active particles. The first studies of this kind were reported in Ref. 4, where the main attention was paid to the role of plasmon states in distribution of the energy of a fast electron over the quantum states of the system.

According to these results, despite the fact that plasmon states absorb about 30% of the energy of a primary electron in water, this has little effect of the yield of primary active particles after the plasmon states have decayed. The authors also did not observe any noticeable decrease in the average energy spent on formation of an ion pair W_i with transition from steam to liquid water.

An opposite result has been obtained in Ref. 14, where, using the Monte Carlo method, the authors made a comparative study of the influence the state of aggregation of water has on yields of ionized and excited states and on the way electron energy is absorbed. According to Ref. 14, with transition from vapor to liquid, the number of ionizations induced by a fast electron increases, while the value of W_i (at E_e = 10 keV) lowers from 30.0 to 24.6 eV. Both in liquid water and in water vapor the yield of ions exceeds that of excited molecules. In our study,[143] which we have made in collaboration with Sukhonosov, we simulated the primary stage of water radiolysis and have obtained a somewhat different result.

We have found (see Table X) that the transition from water vapor to liquid water changes the ratio of ionization and excitation yields: whereas in steam the yield of excitations is somewhat higher, in liquid the yield of ionizations g_{ion} is more than two times greater than the yield of excitations g_{exc} (without taking into account collective states). If we take into account that the decay of plasmon states ($g_{pl} \simeq 1.12$) mainly results in ionization, the ratio g_{ion}/g_{exc} will be about 3. In Table X we compare the calculated values of g_{ion}, g_{exc}, and W_i. The value $W_i = 29.5$ eV obtained from Ref. 143 for water vapor is in better agreement with the experimental value $W_i = 29.6 \pm 0.3$ obtained from Ref. 312. Our results on the total yield g_{tot} are also in better accord with the known mechanism of water radiolysis, from which it follows that in steam $g^{g}_{tot} = 7.3$, while in liquid $g^{l}_{tot} \leq g^{g}_{tot}$.[313]

In Ref. 143 we have also studied the occupation of quantum states, in steam and in water, induced by a fast electron. We have found that secondary electrons play an important role in the distribution of the energy of a primary electron over the quantum states of the absorbing system. Since the secondary electrons, and, more so, the electrons of subsequent generations, produced with ionization of molecules are most likely to have small energies, most of them are not capable of exciting high-lying excited states. Therefore, it turns out that the large value of the oscillater strength of a high-lying excitation level (for instance, of diffusion bands in water vapor or of plasmon states in liquid water) does not mean that this state will be the main occupied one. As one can see in Table XI, the yield of the collective plasmon-type state is not at all dominating, amounting to about 20% of the total yield, which is not much higher than the yield of the lowest-excited state, although the oscillator strength of the latter is more than 100 times as small and the ratio $F_{0n}/\hbar\omega_{0n}$, which determines the probability of fast electrons exciting the nth excited state, is 40 times smaller for the lowest-excited state than

TABLE X

Yields of Ionized g_{ion} and Excited g_{exc} States and the Values of W_i in Liquid Water and in Water Vapor Irradiated by 10-keV Electrons

	Water Vapor			Liquid Water		
Quantity	Ref. 4	Ref. 14	Ref. 143	Ref. 4	Ref. 14	Ref. 143
g_{ion}	—	3.33	3.39	—	4.07	3.65
g_{exc}	—	3	3.82	—	3.6	2.76
$g_{tot} = g_{ion} + g_{exc}$	—	6.33	7.21	—	7.67	6.4
W_i (eV)	30.71	30.0	29.5	28.48	24.6	27.4

TABLE XI

Radiation Yield per 100 eV of Absorbed Energy for Electronically Excited and Ionized States Produced by Electrons with Initial Energy E_e in Water and in Water Vapor[143]

Electron–Excitation State	Water Vapor					Water				
	Excitation Energy $\hbar\omega_{0n}$ (eV)[a]	Oscillator Strength f_{0n} [a]	E_e (keV)			Excitation Energy $\hbar\omega_{0n}$ (eV)[a]	Oscillator Strength F_{0n} [a]	E_e (keV)		
			1	5	10			1	5	10
$\tilde{A}\,^1B_1$	7.4	0.04	1.25	1.2	1.19	8.4	0.018	0.79	0.81	0.79
$\tilde{B}\,^1A_1$	9.67	0.0665	0.53	0.48	0.48	10.1	0.039	0.39	0.39	0.38
Rydberg $(A+B)$	10.69	0.0624	0.34	0.31	0.31	11.26	0.0089	0.05	0.05	0.05
Rydberg $(C+D)$	11.46	0.02006	0.86	0.78	0.78	11.93	0.0536	0.22	0.21	0.21
Diffusion bands	13.32	0.411	1.16	1.08	1.06	14.1	0.103	0.22	0.21	0.21
Collective excitations						21.4	2.03	1.19	1.10	1.12
Ionization	12.6		3.28	3.38	3.39	11.3[b]		3.53	3.68	3.65

[a] Values of $\hbar\omega_{0n}$, f_{0n}, and F_{0n} are taken from Ref. 4.

[b] At present we have performed calculations with $I_i = 8.76$ eV. The results will be published elsewhere.

TABLE XII
Yields of Products per 100 eV of Absorbed Energy, $E_e = 5$ keV

Products	H_3O^+	OH	H	H_2	O	H*	e^-_{dry}	e^-_{aq}
Results of Ref. 143	5.16	5.85	0.61	0.39	0.39	0.086		5.16
Data of Ref. 16	6.3	8.4	2.1	0.3	0.3			6.3
Experimental data of Refs. 314–316	4.8	5.9	0.7	0.45			5.4	4.7

it is for the collective excitation. The production of plasmon states takes about 25% of the total energy, which is in agreement with the results of Ref. 4 and contradicts those of Ref. 3.

Using the values of yields of primary active particles and retracing their transformation according to the scheme proposed in Ref. 16, in Ref. 143 we estimated the radiation-chemical yields of primary radiolysis products appearing by the time $\sim 10^{-11}$ s. The results obtained are presented in Table XII together with results of calculations of Ref. 16. Comparing these results with experimental data,[314–316] we see that the results of Ref. 143 are in better agreement with experiment than those of Ref. 16.*

C. Vavilov–Cherenkov Radiation and Photoradiation Reactions in Tracks

As it follows from the classical work of Tamm and Frank,[210] the electromagnetic field of a charged particle moving through a medium with speed greater than the phase velocity of light in this medium (i.e., $v > c/n_r$, n_r is the refractive index) lags behind the particle and excites a certain kind of radiation that was named after its discoverers—the Vavilov–Cherenkov radiation (VCR).[317,318] The VCR has found wide application in high-energy physics in Cherenkov detectors.[319] Since the relative fraction of particle's energy spent on VCR is very small (tenths of a percent), until recently the VCR has usually been neglected in radiation-protection physics and in radiation chemistry. However, a sufficiently intensive beam of VCR quanta must induce light-sensitive photochemical reactions in the medium, the yield of which depends differently on the energy of primary electrons and on the depth of penetration than in the

* At present we have performed calculations with the scheme of primary stage of radiolysis which differs from that in Ref. 16.

case where these reactions are induced by ionizing radiation. The latter fact should permit one to distinguish the effect caused by VCR from the whole mass of radiation-chemical transformations.

Before the role of VCR in photoradiation processes had been realized, the VCR itself was used as a detector of short-lived active particles produced in a medium under irradiation in pulsed radiolysis apparatus. The detection of the signal of optical density variation within picosecond time intervals, that is, within 10^{-12}–10^{-11} s, meets great difficulties connected with synchronization of the signals, with exact measurement of their time delay and with equipment sensitivity. In the University of Toronto in 1967 a pulsed set was designed for studying the radiation-chemical processes of picosecond duration in which the VCR pulses were used as analyzing light.[320,321]

When irradiating solutions, it is important to distinguish the contribution of the VCR-induced luminescence from that of radioluminescence induced by ionizing radiation. In this connection a sufficiently convincing series of experiments was carried out by Steen,[323] who measured the luminescence of indole in a solution of water with glycol using two different ionizing sources: ^{60}Co and an X-ray source with $E_{ph} = 220$ keV. Since the threshold energy of VCR production in water is $E_e = 280$ keV, the second source did not produce VCR. This enabled Steen to show that about 80% of the singlet luminescence of indole irradiated by ^{60}Co is excited by VCR quanta.

The contribution of VCR to radioluminescence has been also studied in Refs. 324 and 325. The results close to those of Ref. 323 have been obtained by Zhuravleva[325] in a comparative study of radioluminescence of 3-aminophthalamide (AP) in ethanol solutions irradiated by electrons with $E_e = 3.5$ MeV and by X rays with $E_{ph} = 50$ keV.

As has already been mentioned, picosecond pulsed radiolysis offers great possibilities for studying the short-lived transient processes. In Ref. 326 the solutions of 2,5-diphenyloxazol (DPO) in different solvents were irradiated by picosecond electron pulses obtained from an accelerator. The authors have found two types of excitations of DPO, which they have named the fast and the slow excitations. With fast excitation the luminescence appears during the electron pulse and stops growing at the end of the pulse, after 10 ps. With slow excitation the luminescence is formed within ~1 ns. At small DPO concentrations the observed intensity of fast luminescence cannot be explained by direct excitation by electrons (cf. data of Ref. 325). Analyzing the results of experiments with different solvents and different types of additives, Katsumura et al.[326] conclude that the main part of the fast luminescence of DPO is due to VCR absorption.

The chemical transformations induced by simultaneous exposure to light and to ionizing radiation are of great interest to radiation chemistry. For a number of materials one observes the so-called synergy effect, that is, the nonadditivity (the mutual increase) of the effects produced by different types of radiation. For instance, the simultaneous irradiation by light from a lamp and by accelerated electrons leads to a considerable increase of the efficiency of marcoradical formation (see Chapter V of Ref. 327).

Papers[328,329] by Polyansky and one of the authors contain calculations of the spectral and energy yields of VCR in tracks of electrons with energies attained in accelerators used in radiation chemistry. It is emphasized that the VCR can induce photoradiation processes in tracks of fast electrons, that is, the inner photoradiation effect.

The total number of photons within the spectral range (λ_1, λ_2) emitted with retardation of an electron with energy E is given by the integral

$$N^{\mathrm{ph}}(\lambda_1, \lambda_2) = \int_{\lambda_1}^{\lambda_2} d\lambda \int_{E_0(\lambda)}^{E} dE' \, \frac{d^2N^{\mathrm{ph}}}{dx\, d\lambda} \left(\frac{dE'}{dx}\right)^{-1}, \tag{8.2}$$

where $E_0(\lambda)$ is the threshold energy of emission. For a transparent isotropic medium, the number of photons emitted per unit electron path length within a unit spectral interval is given by the Tamm–Frank formula[317]

$$\frac{d^2N^{\mathrm{ph}}}{dx\, d\lambda} = \frac{2\pi\alpha}{\lambda^2} [1 - \beta^2(E) n_{\mathrm{r}}^{-2}(\lambda)], \tag{8.3}$$

where $\alpha = \frac{1}{137}$. The function $n_{\mathrm{r}}(\lambda)$ is well described by the Cauchy formula

$$n_{\mathrm{r}}(\lambda) = A + B/\lambda^2, \tag{8.4}$$

where A and B are constants characterizing the material. The function $\beta(E)$ is determined by the relativistic expression for velocity and can be written

$$\beta^2(E) = \frac{E(E + 2mc^2)}{(E + mc^2)^2}. \tag{8.5}$$

The refractive index for water in the spectral range (200–900 nm) is approximated in Ref. 330 as

$$n_{\mathrm{r}}(\lambda) = 1.2998 + 2.291 \times 10^3/\lambda^2 \tag{8.6}$$

in accord with formula (8.4). Figure 18 shows the results of calculation of dN^{ph}/dx and N^{ph} for water in the spectral range (200–600 nm),[328] while in Fig. 19 we compare the losses on VCR per unit path length with total energy losses of an electron $dE/\rho\, dx$, that is, with the mass stopping power. Unlike the stopping power, the losses on VCR show a clear tendency for saturation, although they do not exceed 0.17% of the total energy losses. This means that in the overall energy balance the losses on VCR cam be ignored for electron energies used in radiation studies ($E_e < 30$ MeV). However, the total number of VCR quanta is in no way small and may have considerable effect on photosensitive radiation processes.

This is due to the large electron fluxes in modern accelerators. For a current of 10 mA we have an electron flux $J_e = 10^{17}\ s^{-1}$. The photon flux equals the product

$$P = N^{ph} J_e \,. \tag{8.7}$$

As a result, for a beam with energy $E_e = 10$ MeV, we have $P \approx 10^{20}$ photons per second. Table XIII shows the number of VCR quanta in the spectral range 200–600 nm emitted with complete retardation of an electron beam in water for average parameters of accelerators used in radiation-chemical studies. It follows from Table XIII that the number of

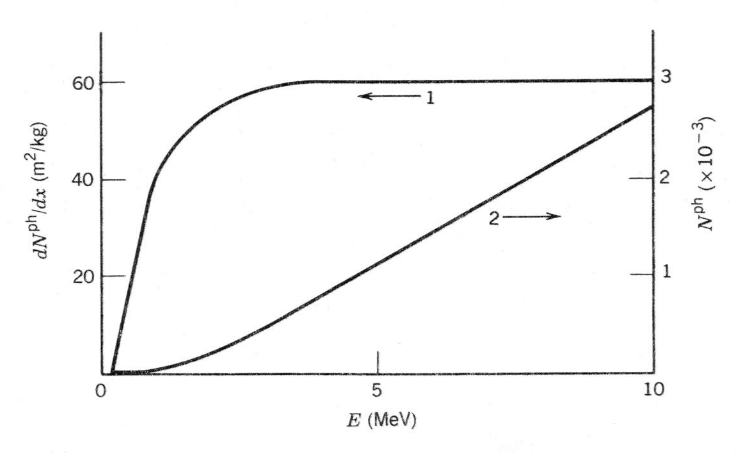

Fig. 18. Dependence on electron energy of the VCR photon yield in the spectral range (200–600 nm) per unit path length (1), and of the total number of photons in the spectral range (200–600 nm) emitted along the track (2) in water.

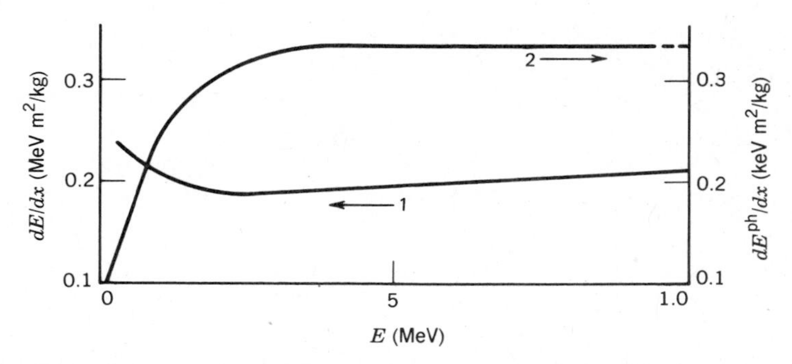

Fig. 19. Dependence on electron energy of energy losses per unit path length in water: 1, total losses; 2, losses on VCR in the spectral range (200–600 nm).

VCR photons is comparable in magnitude with light fluxes given by lamps used in experiments.

In the case of irradiation by gamma sources, the VCR yield is many orders of magnitude smaller. In the general case, the VCR yield for gamma irradiation should be calculated taking into account the geometry of the irradiated sample using the Monte Carlo method. A simple analytical method of calculation for the cases where the energy of gamma quanta is absorbed completely has been developed in Ref. 328. For the usually employed Co gamma sources, the calculations give a VCR flux 10^5–10^7 times smaller than those obtained in accelerators. This is due to the much smaller power of gamma sources in comparison with electron accelerators, as well as to the small energy of Compton electrons for ^{60}Co ($\bar{E}_c \sim 0.59$ MeV). The energy of Compton electrons for cesium gamma sources is even smaller. So even at equal dose rates, the effect of accelerator irradiation will be greater owing to higher electron energy.

In radiation studies it is conventional to relate the magnitude of the effect to the energy absorbed by unit mass.[331] Usually one takes the stopping power $S_m^e = dE/\rho\, dx$. Introducing the mass emissive power with

TABLE XIII
Number of VCR Quanta Within the Spectral Range 200–600 nm Emitted in Water for Different Characteristics of the Electron Beam

E_e (MeV)	$I = 100\ \mu$A	$I = 10$ mA
1	7.8×10^{14}	7.8×10^{18}
5	7.8×10^{17}	7.8×10^{19}
10	1.4×10^{18}	1.4×10^{20}

respect to VCR,

$$S^{\mathrm{ph}} = \frac{1}{\rho}\frac{dN^{\mathrm{ph}}}{dx}, \tag{8.8}$$

we can define the relative VCR yield as the ratio

$$C_e^{\mathrm{ph}} = S_m^{\mathrm{ph}}/S_m^e \tag{8.9}$$

which is the number of photons emitted within a certain spectral range per unit of absorbed energy. Since C_e^{ph} is expressed in terms of differentials, it corresponds to an infinitely thin layer; therefore, the electron energy is constant and equals E.

If the energy of an electron passing through an irradiated volume changes from E to E_1, the average relative yield of VCR for this volume is given by the formula

$$\bar{C}_e^{\mathrm{ph}} = \frac{N^{\mathrm{ph}}(E) - N^{\mathrm{ph}}(E_1)}{E - E_1}. \tag{8.10}$$

In Table XIV we present the results of calculation of C_e^{ph} and $\bar{C}_e^{\mathrm{ph}}$ in the spectral range 200–600 nm for electrons with energy $E_e = 0.5$–10 MeV in water. Let us note the characteristic initial increase and the following saturation that is clearly seen for $\bar{C}_e^{\mathrm{ph}}$.

The simplest way to distinguish the VCR-induced chemical processes from those induced by ionizing radiation itself is to compare the dependence of these processes on electron energy. First of all, the VCR has a threshold; for water $E_0(200\,\mathrm{nm}) = 260$ keV and $E_0(600\,\mathrm{nm}) = 300$ keV. The next important feature is the behavior of the yield of reaction products in the region from the threshold to $E_e = 3$ MeV. In the energy region between 0.5 and 10 MeV. In the energy region between 0.5 and 10 MeV, the yield per unit absorbed energy is practically independent of the electron beam energy, whereas, according to Table XIV, this is not so for the VCR-induced reactions.

TABLE XIV

Relative VCR Yield Within the Spectral Range 200–600 nm at Different Energies of a Monoenergetic Electron Beam in Water

E_e (MeV)	0.5	1	2	3	4	5	6	7	8	9	10
C_e^{ph} (10^{-15} J^{-1})	1.1	1.8	2.05	2.07	2.04	2.01	1.98	1.95	1.93	1.88	1.85
$\bar{C}_e^{\mathrm{ph}}$ (10^{-15} J^{-1})	0.29	0.81	1.20	1.40	1.50	1.56	1.60	1.64	1.66	1.68	1.70

In Refs. 328 and 329 still another effect characteristic of VCR has been pointed out—the so-called cumulative effect. This effect is observed at small concentrations of the VCR-absorbing substances, and manifests itself in an increase in the total number of VCR quanta with increase of the depth of irradiation. On the other hand, the absorbed electron energy depends on the depth of penetration in an opposite way: first, it reaches a maximum and then decreases.

Thus, for irradiation in accelerators, the effects produced in a medium always have the photoradiation character, which should be taken into account for photosensitive systems. Photosensitive processes are much more efficient in accelerators than in gamma installations.

The most suitable system for studying the VCR effects are transparent solutions with small concentrations of photosensitive additives. While the energy of fast electrons is distributed in proportion to the electronic fraction of each component, the absorption of VCR quanta is selective with respect to additives. As a result, the VCR energy absorbed by an additive can exceed the energy transferred to it by a fast electron, although the relative fraction of energy spent on VCR is small.

One can expect that further studies of the role of VCR in radiation chemistry will reveal many new, yet unknown features of radiolysis of photosensitive systems.

D. The Structure of Tracks of Heavy Charged Particles

1. *Specific Features of the Way Heavy Charged Particles Transfer Their Energy to a Medium*

As in the case of fast electrons, the main contribution to energy losses of heavy charged particles (ions) comes from ionization losses (see Section V), which are responsible for most of the radiation-chemical effects. Therefore, we consider here only the structure of that part of an ion's track where the ionization losses are dominating. The role of elastic interaction between ions and atoms of the medium, which becomes essential only at the end of the ion's track, is not considered in this section.

The efficiency of inelastic interaction of a charged particle with a medium is mainly determined by the ratio z^2/v^2 and does not depend explicitly on the particle's mass (see formulas for the cross sections in Section IV and formulas for the stopping power in Section V). The size of the region where the interaction is efficient (b_{eff}) and the dimensions of the area of energy localization (b_{pl} and Δx) are determined only by velocity of the charged particle; therefore, in general, a heavy charged particle generates the same track structures as does a fast electron with same velocity. However, since the mass of a heavy particle is more than 10^3 times greater than the electron mass and $z \neq 1$, there are certain specific features in the way its energy is transferred and distributed throughout the medium.

First of all, let us note that the interaction of an ion with electrons has little effect on the curvature of its trajectory, since the energy it loses in each such event and the corresponding change in its momentum amount to only a small fraction of its total energy and momentum ($\Delta E_{max}/E_0 \approx 4m/M < 2 \times 10^{-3}$). So in the region of the track where the ionization collisions are dominating, the trajectory of a heavy particle can be considered a straight line.

A heavy charged particle can knock out an electron from a molecule with maximum energy $E_{max} \simeq 2mv^2$ (at $v > v_0$), whereas for a fast electron with the same velocity the knocked out electron has $E_{max} \simeq mv^2/2$. Consequently, while an electron can knock out electrons with velocity no greater than its own, a heavy particle, in head-on collisions, produces delta electrons with velocities which can be twice as high as that of the ion. As a result, the energy of such delta electrons can be distributed to the regions of the medium far more remote from the point of initial ionization than in the case of electron irradiation.

The probability of interaction between an ion and a molecule of a medium is proportional to the square of the ion's charge, z^2. Therefore, at a given velocity, the ion with a greater charge looses more energy per unit path length. Since σ_i is proportional to z^2 [formula (4.37)], at greater z the number of secondary electrons produced per unit path length is greater. Since σ_{tot} is also proportional to z^2, with increase of z the average free path length of an ion $\bar{l}_{in}$ becomes shorter [see formula (8.1)], and track structures overlap each other at lower energy.

Finally, let us mention a feature that concerns multicharged ions. Since such ions have inner electron shells, they can produce Auger electrons the energy of which can be much higher than the maximum energy of secondary electrons. Although the energy of Auger electrons gives a smaller contribution to total energy losses, they can be absorbed a considerable distance away from the ion's trajectory and, thereby, can smooth out the steep decrease of energy losses with increase of the impact parameter.

2. *The Track Structure of Ions of Different Nature*

At high velocities the track of an ion can be considered as consisting of separate track structures similar to those of a fast electron. However, the ions used in radiation chemistry usually have initial energies not exceeding 10 MeV per nucleon. At such energies, even in the case of protons, the track structures overlap each other, forming a track similar to short tracks of slow electrons. Indeed, a proton with $E_p \simeq 10$ MeV has the velocity $v \simeq 4 \times 10^9$ cm/s; at this velocity the energy of an electron is about 5.5 keV, that is, still above the energy at which short tracks are formed.

Thus, an ion with energy $E \lesssim 10$ MeV per nucleon produces a region consisting of overlapping spurs along its straight-line path. The inner part of this region adjoining the ion's trajectory is called a *core*.[23,332,333] Inside the core the density of ionizations and excitations produced both by the primary ion itself and by the secondary electrons it has generated is very high. Since the probability of excitation $P_{0m}(b)$ is independent of the azimuthal angle [see formula (4.44)], the core has the shape of a cylindrical column with radius r_{core}.

When estimating the initial radius of the core, most authors use the adiabatic criterion of Bohr. However, the values they obtain are different (see, e.g., Refs. 332 and 333). The radius of the core can be taken to be the effective value of the impact parameter at which it is still possible for a molecule to be excited into the state with the lowest transition energy $\hbar\omega_{01}$ by the primary ion. Hence, in general, the radius of the core in a dilute medium can be found using formulas (4.45) or (4.54). But in condensed media the estimation of r_{core} is not so trivial.[334] However, at nonrelativistic velocities, when $\beta^2\epsilon(\omega) \ll 1$, the screening effect the medium has on the field of a charged particle is weak, so the radius of the core is determined by formula (4.45), namely,

$$r_{\text{core}} = \pi v / \omega_{01} \ . \tag{8.11}$$

For small velocities of heavy particles the size of the core in water is usually taken to be 1 nm.[334] This is the path length of a secondary electron with energy ~100 eV in water. Taking $\hbar\omega_{01} = 8.4$ eV, we find that the value $r_{\text{core}} = 1$ nm given by formula (8.11) corresponds to the ion velocity $v = 4.5 \times 10^8$ cm/s.

The distribution of primary excited and ionized states produced directly by the primary ion inside the core of the track can be found knowing the probability $P_{0m}(b)$. If n is the concentration of molecules in the ground state, the distribution of molecules in the mth excited state over the distance from the axis of the track is described by the function

$$N_{0m}(b) = nP_{0m}(b) \ .$$

In dilute media $P_{0m}(b)$ is given by formulas (4.44) and (4.53). As was shown in Ref. 335, a sufficiently good approximation for $N_{0m}(b)$ for nonrelativistic particles ($\gamma = 1$) is

$$N_{0m}(b) = \frac{2z^2e^4f_{0m}n}{mv^2\hbar\omega_{0m}b^2}\exp\left(\frac{-\omega_{0m}^2b^2}{2v^2}\right), \tag{8.12}$$

which after F_{0m} is substituted for f_{0m} (F_{0m} is defined in Section V.B.2) is also valid for condensed media. According to this formula, $N_{0m}(b)$ decreases very rapidly with increase of the impact parameter b, and the more steeply, the smaller the ion's velocity. Since multicharged ions have $z = z_{\text{eff}} \sim v/v_0$, in their case the decrease of $N_{0m}(b)$ with lowering of v is sharper than in the case of protons and alpha particles, because the factor preceding the exponent becomes independent of v.

The probability of formation of a secondary electron is approximately determined by the Rutherford cross section (4.20), from which it follows that the majority of secondary electrons have small energy. On the other hand, the cross section $d\sigma(\theta)$ of a secondary electron being ejected at an angle between θ and $\theta + d\theta$ with respect to the ion's direction is proportional to $d\sigma(\theta) \sim \sin\theta\cos^{-3}\theta\, d\theta$ (for $\theta \neq \pi/2$).[5] Therefore, it is most probable that a secondary electron be ejected at an angle close to $\pi/2$ to the axis of the track.

Part of the energy of secondary electrons is absorbed in the core of the track. Only the high-energy delta electrons are capable of leaving the core and of forming individual track structures. The region of the track outside the core that consists of track structures produced by delta electrons is called either penumbra[336] or an ultratrack.[337] If the tracks of delta electrons overlap each other, this region can become a continuous shell surrounding the core.

The distribution of ionization and excitation events produced by secondary electrons has been studied in many papers. In most of them the authors studied the radial distribution $Q(r)$ of energy losses of secondary electrons, using either analytical methods (see, e.g., Refs. 335 and 338–345) or the Monte Carlo method.[346,347] With the Monte Carlo method one is able to obtain a more detailed distribution of energy in an ion track. For instance, the results of calculations in Refs. 346 and 347 are in good agreement with experimental data on the radial energy distribution in tissue-equivalent gases.[348,349] On the other hand, the simple analytical formulas describing the radial distribution of secondary electron energy obtained in Refs. 335 and 339 are more convenient in a number of practical applications, for instance, in radiobiology.

However, in radiation chemistry the function $Q(r)$ alone in obviously not enough, for with it one can, at the best, obtain only the spatial distribution of ionized states [$N_i(r) = Q(r)/W_i$, W_i is the average energy required to form an ion pair]. A detailed distribution of excited and ionized states in ion tracks has been found in Ref. 13 using the Monte Carlo method, where the authors have also studied the way the energy distribution of delta electrons changes with distance from the axis of the track and along the track.

Using the results of Ref. 13, we have made a comparative study[15] of track structures for ions of different nature with specific energy ranging from 82 to 420 keV per nucleon. At such energies, the ion tracks consist of a core surrounded by a shell that has the shape of a cone pointed toward the end of the track. The value of its radius r_{sh} is determined by the path length of a delta electron with the maximum energy $E_{max} \simeq 2mv^2$, and, therefore, like the size of the core, depends on the ion's velocity. Consequently, at a given velocity, the geometric characteristics of tracks (r_{core} and r_{sh}) of different ions are the same. However, the qualitative composition of the core and the dependence of track characteristics on the ion velocity and on LET for multicharged ions are different than for alpha particles and protons (see Table XV and Fig. 20).

With retardation of an alpha particle, the characteristics of its track

TABLE XV
Characteristics of Tracks of an Alpha Particle and of a Multicharged Ion ^{127}I in Water[a]

Track Characteristic	Velocity of the Ion (10^8 cm/s)				
	9	7	6	5	4
E (MeV)	1.7	1.0	0.75	0.5	0.3
	53	32	24	16.5	10
S_e (eV/nm)	168	212	230	240	252
	6000	5540	5240	3600	3210
z_{eff} (e)	1.85	1.68	1.58	1.47	1.38
	11.0	8.6	7.4	6.0	5.0
N_δ (nm^{-1})	3.3	4.5	5.0	5.7	6.7
	118	118	117	87	84.5
N_i (nm^{-1})	5.2	6.8	7.4	7.9	8.4
	190	177	168.5	118	107
$\bar{n}_i^c$ (10^{21} cm^{-3})	1.46	2.0	2.18	2.4	2.66
	52.6	52.3	51	36.6	33.7
$\bar{n}_i$ (10^{19} cm^{-3})	0.27	0.54	1.0	3.8	7.4
	9.7	14.0	23.8	58.7	94
r_{sh} (nm)	25.0	20.0	15.0	8.0	6.0

[a] For each parameter of the track the top number corresponds to the alpha particle and the bottom number – to the multicharged ion ^{127}I. N_i and N_δ are, respectively, the number of ions and the number of delta electrons the particle produces per 1 nm of its path length; $\bar{n}_i^c$ is the average concentration of ions in the core of the track; $\bar{n}_i$ is the average concentration of ions in the track; and r_{sh} is the radius of the track shell.

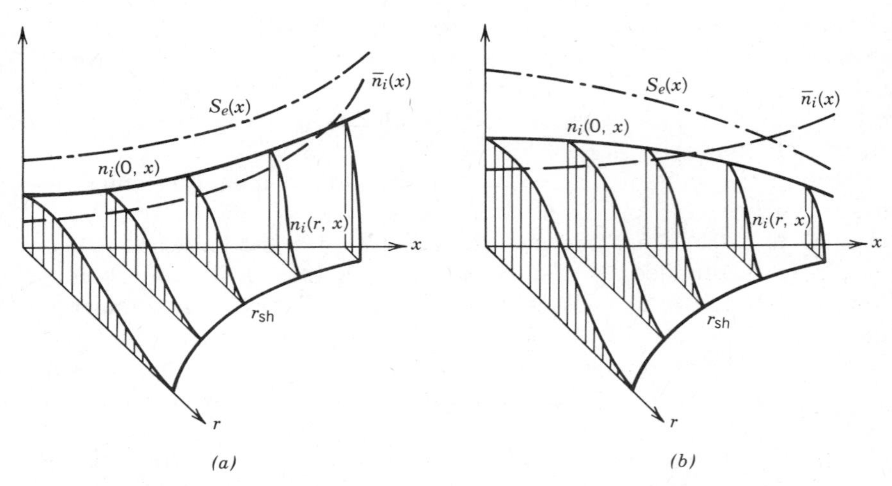

Fig. 20. Schematic diagram of the changes in the characteristics of the track of heavy charged particles with retardation: (*a*) the alpha particle; (*b*) a multicharged ion.

change in the following way. The ionization losses, the number of primary ions, and their concentration averaged over the radius of the track increase, while the average energy W_i the alpha particle spends on formation of an ion pair slightly decreases. Since at the considered velocities the retardation of an alpha particle is mainly due to ionization losses, the data presented in Table XV characterizes the track depending on the LET.

As we have mentioned previously, owing to effective neutralization of the charge of a multicharged ion with its retardation in a medium, in the region where the ionization losses are dominating its LET diminishes with energy. At the same time, the radius of the track contracts more rapidly than the LET decreases, and so the average concentration and the yield of primary ions behave the same way as they do in tracks of alpha particles, that is increase with lowering of ion's energy. This increase continues until the energy decreases to a level at which the losses in elastic collisions become comparable in magnitude with ionization loss (i.e., until $v \sim 4 \times 10^6$ m/s).

The primary effect of a multicharged ion is the ionization of practically all the molecules in a cylinder with diameter 0.7 nm (see estimates made in Ref. 335). Most of the energy of secondary electrons is also localized inside the core, and so the number of ionization events inside the core of the track of a multicharged ion exceeds the number of molecules in it. Consequently, practically every molecule of the core is ionized and excited many times, where, in the end, all these molecules are not only ionized but have dissociated into fragments. Thus, the region adjoining the axis of the track of a multicharged ion (with a radius about 1 nm) is a cylindrical ionic core filled mainly with ionic fragments. As was shown in Ref. 350, by the end of the physical stage the core of the track of a multicharged ion consists of low-temperature, nonisothermal and spatially nonhomogeneous nonideal plasma. Later, Ritchie and Claussen came to a similar conclusion.[351]

Thus, despite a certain similarity of tracks of different heavy ions, the increase of their charge leads not only to proportional increase of concentration of active particles, including the charged ones, but also to a qualitative change in the structure of the track. A partially ionized track transforms into a low-temperature plasma column whose components are in localized quasi-equilibrium states (for details see Ref. 350). The processes in such tracks take place under essentially nonequilibrium conditions, under which we cannot introduce a single temperature for all degrees of freedom of a molecule. As a result, the usual ideas concerning the reaction constants no longer have any meaning. In order to describe the processes in tracks of multicharged ions, one should use the methods of plasmochemistry.[231]

IX. SPECIFIC FEATURES OF RADIATION-CHEMICAL REACTIONS IN TRACKS OF PARTICLES OF DIFFERENT NATURE

A. The Influence of the Spatial Structure of a Track on the Features of Radiation-Chemical Reactions

1. The Influence of the State of Aggregation

As we have already discussed in Section VIII, at the physical stage of radiolysis the primary active particles (ions, excited molecules, and electrons) are localized in separate microregions—in the track structures. The dimensions of track structures, the concentration of active particles in them, and the subsequent transformations of these particles depend on the density of the medium.

In gases the track structures usually contain a small number of active particles and are separated from each other by considerable distance. Owing to efficient diffusion, the initial inhomogeneity in the distribution of active particles rapidly smoothes out, and by the time the chemical reactions begin, the intermediate chemically active particles are distributed practically homogeneously in the irradiated volume. For this reason the influence of tracks on radiation-chemical processes in gaseous media is negligibly small. Only when a gaseous medium is irradiated by multi-charged ions, such as fission fragments, can the spatial inhomogeneity affect the course of chemical reactions.[352,353]

The density of a liquid is hundreds of times greater than that of a gas. The spatial inhomogeneity in the distribution of reagents is more pronounced here, the remains for quite a long time, therefore, the diffusion processes in condensed media proceed more slowly. As a result, the kinetics of reactions in a substance in the condensed phase is different than it is in gases, and gives a different final radiation effect.

However, the spatial inhomogeneity in the distribution of reagents is not the only reason why the radiolysis of substances in the condensed state is different from that of gases. As we have already mentioned in Section VIII, as we pass from the gaseous state to the condensed one, at the primary stage of radiolysis we already observe a redistribution of yields of primary active particles (resulting in the increase of the yield of ionized states). Also different are the subsequent relaxation processes, as well as the processes of decay of excited and ionized states.[354] Another specific feature of processes in a condensed medium is the cage effect, which slows down the decay of a molecule into radicals.[355] Finally, the formation of solvated electrons is also a characteristic feature of radiation-chemical processes in liquids.[356]

Thus, the difference in radiation effects in gases and in condensed media is due to many factors. Therefore, by comparing the radiation effects in gases and in liquids, we do not get a pure contribution of track effects. A more fruitful way is to study the radiation effects depending on the characteristics of radiation at fixed physical and chemical properties of a medium.

One of the characteristics of radiation considered in radiation chemistry and in radiobiology is the linear energy transfer (LET). For fast charged particles the LET practically equals the ionization losses (or polarization losses, in condensed media) and is given by the formulas for the stopping power presented in Section V.A.

In early studies the track effect was attributed to the dependence of product yields on the LET. It was believed[357] that the higher the LET, the greater is the concentration of chemically active particles in the track, and, consequently, the more probable are the bimolecular reactions between them and the smaller part of them is spent in reactions with the solute. For instance, it is a well-known fact that in the case of water radiolysis the increase of LET leads to the increase of the yield of molecular hydrogen and lowers the yield of stabilized electrons and free radicals.[358,359] The studies of radiolysis of a number of organic liquids give data indicating that the increase of LET leads to increase of yields of molecular products the production of which can be attributed to recombination of chemically active particles (such as radicals) in the track.[360]

Considering the track structures as spherical or cylindrical formations and using the methods of diffusion kinetics, it proved to be possible to explain many experimental facts concerning the radiolysis of water solutions, in particular, the dependence of yields on LET.[361] It is owing to this that the LET was considered to be a universal qualitative characteristic of radiation, and the concentration of active particles was considered to be in direct dependence on the LET with no regard for the type of charged particle.

However, many experimental results concerning the dependence of yields of radiolysis products for organic liquids on LET in a wide range of values of the latter do not have a proper theoretical explanation within the conception relating the structure of a track to LET. These are (1) the sharp increase of yields of low-molecular products of radiolysis at high LET[362,363]; (2) in the case of fission fragments, the increase of yields of some products of methanol radiolysis, the production of which is attributed to recombination of radicals in the track, with decrease of LET of a fragment[364]; (3) the difference between the yields $G(H_2)$ of hydrogen for radiolysis of benzene by protons and by alpha particles with same LET[365] (see also the results of a more comprehensive study in Ref. 367). In order to explain all these experimental facts, it proved to be necessary to consider the detailed structure of a track and its dependence on the type of the charged particle.[15]

2. *The Influence of the Track Structure on Track Reactions and the Insufficiency of the LET Conception*

The role of the track structure is most clearly illustrated by the example of radiolysis of liquids by heavy ions. In this case it is possible to vary broadly the geometric dimension of tracks and the concentrations of active particles in them. The dependence of track effects on the track structure has been studied in Refs. 365 and 366 The qualitative relation between the structure of a track and the features of radiation-chemical processes has been analysed.[15,18]

Using the detailed information about the structure of ion tracks, one can explain the dependence of yields on the LET in the following way. The low values of LET are observed for radiolysis of liquids induced by high-energy protons. The track of a proton in this case consists of a core and branches, the latter being the tracks of delta electrons. The higher the proton energy, the greater fraction of it is carried outside the core by delta electrons. For protons with energy above 10 MeV, the core transforms into separate nonoverlapping spurs, that is, except for some peculiarities, the track of a proton is similar to an electron track. As a result, the features of the radiolysis in this case are close to those of the radiolysis induced by fast electrons.

As a proton slows down, its LET, as well as the number of delta electrons, increase, while the size of the track, on the contrary, becomes smaller since the maximum energy of delta electrons lowers. As a result, one observers a sharp increase both of the local concentrations near the track's axis and of average concentrations. This favors the increase of yields of products of recombination of active particles with increase of LET, that is, there is a direct correspondence between the density of active particles in the track and the LET. A similar picture is observed in tracks of alpha particles.

The high values of LET (more than 1 keV/nm) are observed when a liquid is irradiated by multicharged ions, such as fission fragments. The structure of tracks of such ions has a number of specific features. The first one has to do with the increase of the average concentration of active particles with lowering of LET, that is, these two quantities behave in an opposite manner (see Fig. 20). This leads to the fact that the yields of certain radiolysis products first increase as the LET of a fragment lowers, then reach a maximum, and after than (as with retardation of the fragment, the energy losses in elastic collisions become essential) decrease. It is with this that one can explain the experimental results on the radiolysis of methanol induced by fission fragments with different initial energies.[364]

The second feature of the track structure of multicharged ions—the high concentration of fission fragments in the track's core—manifests itself in the preferential production of low-molecular products in the core of the track. The yields of products with more complicated structure, on the contrary, become smaller, which explains the high yields of H_2, CO, and CH_4 observed in radiolysis of liquid hydrocarbons induced by multicharged ions.[362,364,365]

Intermediate values of LET correspond to radiolysis of liquids by different ions. It is not difficult to find the conditions under which two or more ions have equal LET (this was done in Ref. 367). Except for the region of small energies, the LET is determined by ionization losses, that is, by formula (5.2). If the charges of two ions are different, their S_e have the same values at different velocities (we consider the region before the Bragg's peak): the ion with a greater charge must have a higher velocity. Consequently, both the core [see formula (8.11)] and the adjoining region of the track (formed by tracks of delta electrons) have larger dimensions for an ion with a larger charge. As a result, the corresponding local concentrations, and quite naturally, the concentrations averaged over the volume of the track, are lower. A good illustration of what we have just said are the data in Table XVI, where we present the average characteristics of proton and alpha-particle tracks in water with the same LET (for two different values of the latter). The effective radius of the track we have introduced in Table XVI is the path length of a delta electron with the energy equal to the average energy of those delta electrons whose path length exceeds the radius of the core.

Thus, the higher the ion's charge, the larger is the volume of the track throughout which its linear energy losses are distributed, that is, the lower are the local concentrations of active particles. As a result, the radiolysis in the track of an ion with a large z is actually very close to the radiolysis in the track of an ion with a smaller charge at a lower LET. Let us illustrate this point with the following example. Consider a product of radiolysis formed as a result of recombination of radicals: for instance, the H_2 molecule formed via the reaction $H + H \rightarrow H_2$. And there is also a competing reaction in which the radical H is captured but which does not lead to production of H_2. In this case the increase of the concentration of

TABLE XVI
Characteristics of Proton and Alpha-Particle Tracks at Equal LET in Water

	$S_e = 98$ eV/nm		$S_e = 50$ eV/nm	
Characteristic	Proton	Alpha Particle	Proton	Alpha Particle
Energy (MeV)	0.082	4.5	0.4	10
Velocity (10^8 cm/s)	4	15	8.7	22
Effective charge	0.69	2	0.93	2
Average number of delta electrons ejected over 1 nm of the ion track	1.56	1.7	0.86	1
Maximum energy of a delta electron (keV)	0.181	2.5	0.86	5.45
Radius of the core $r_{core} = \pi v / \omega_{01}$ (nm)	1.2	4.5	2.65	6.7
Effective radius of the track (nm)	2.5	21	6.0	35
Energy carried away by delta electrons outside the core of the track (eV/nm)	4.3	16.3	7	12.4
Average density of ionization events in the core of the track (10^{20} cm^{-3})	5.3	0.45	0.69	0.1

H radicals will enhance the first reaction and, consequently, will increase the yield $G(H_2)$. This is apprently the case with radiolysis of benzene[367] (see Fig. 21). In favor of this is the fact that as an ion slows down, and the size of its track becomes smaller, while the LET and the local and average concentrations of active particles grow, the yield $G(H_2)$ monotonically increases. So when we replace an ion by another ion with a greater charge and the same LET, the local and average concentrations become smaller, and, consequently, the yield $G(H_2)$ lowers.

Thus, in the general case, there is no direct correspondence between the magnitude of LET and the local and average concentrations of active particles in the track. Therefore, the LET cannot serve as a universal characteristic of the quality of radiation without any connection to the specific type of particle. Attempts have been made to introduce in place of LET other comparative characteristics of radiation, such as the ratio z^2/v^2,[368] or an average density of absorbed energy inside the core of the track.[367] However, at a fixed value of each of these parameters, the radiation effect is still not independent of the type of ion.

It seems that all the attempts to introduce a single parameter that would characterize the effect of radiation are unsuccessful. The only characteristic that can serve this purpose is the spatial distribution of

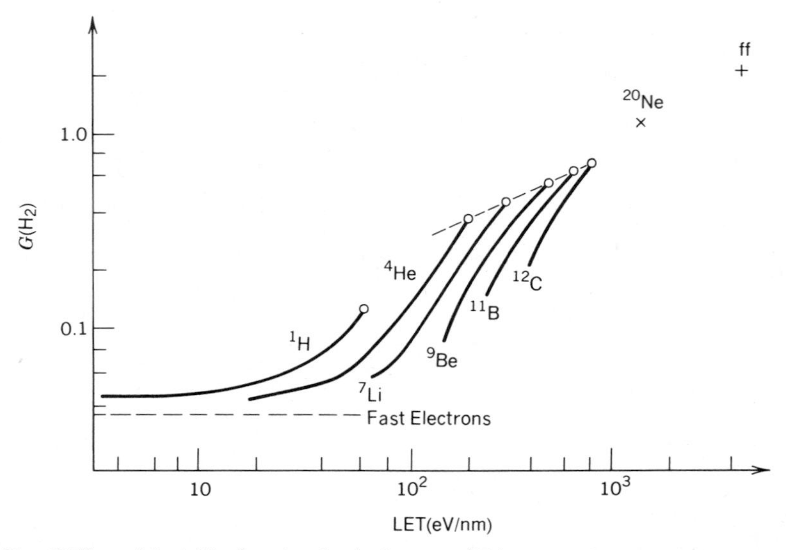

Fig. 21. Differential yield of molecular hydrogen $g(H_2)$ versus the LET in benzene for different ions (Reproduced by permission from Ref. 367). Data for ^{20}Ne and for fission fragments (ff) correspond to integral yields. Circles represent the maximum value of LET for each ion.

active particles in the track. For the greatest part of the track, the latter is determined both by the charge and by the velocity, and not by their ratio alone. Therefore, it is these two parameters one should use for determining the qualitative characteristics of charged particles.

B. The Thermochemical Effect Produced by Charged Particles

In early theories of the chemical effect of radiation it was supposed that the changes one observes in an irradiated medium are due to formation of local heated regions, which subsequently became known as thermal spikes. Later it was shown that for a particle with low LET the heating is neither considerable nor is sufficiently long to have an essential effect on chemical transformations. For instance, according to estimates made by Mozumder,[23] the rise of the temperature inside a spur is only about 30 K, while the time $\tau_{1/2}$ by which the temperature in the center of the spur lowers to a half of its initial value is about 6×10^{-12} s.

In the case of radiation with a high LET, the density of absorbed energy in the volume of the track is very high. Most of this energy is localized in the core, the radius of which in water is 1–2 nm. The density of absorbed energy is particularly high in the core of the track of a multicharged ion (in water $Q_{\text{core}} = 2.6 \times 10^{11}\ \text{J/m}^3$). If we assume that the volume of the track is heated adiabatically by this internal energy, the rise of the temperature inside the core will, naturally, be considerable (at $\rho = 10^3\ \text{kg/m}^3$ and $c_v = 4 \times 10^3\ \text{J/kg K}$, $\Delta T \sim Q_{\text{core}}/c_v\rho = 6.5 \times 10^4\ \text{K}$). At such heating the initiation of endothermal reactions is quite possible.

Based on such an approach, Goldanskii and Kagan[369] have considered the general characteristics of the thermochemical effect of ionizing radiation. In a number of studies[23,362,363] the model of thermal spikes was used to explain the effects one observes in the radiolysis of liquids induced by particles with high LET, for instance, the increase of the yields of low-molecular products. Goldanskii et al.[370] have noted that the energy released inside the track leads to creation of a shock wave. The shock contraction of the substance results in a sharp rise in the concentration of electrons in the track, which can stimulate the radiation-chemical processes (this is the so-called hydrodynamical effect).

However, as is speculated in Ref. 360, it is doubtful that the local heating can have an essential effect on the kinetics of radiation-chemical reactions. According to the thermal spike model, all the energy lost by the ionizing particle in a very short time transforms into heat. In reality, this energy first exists in the form of electron excitation of molecules and ions, as well as in the form of kinetic energy of subexcitation electrons. These intermediate active particles, with an excess of energy, live longer than a thermal spike.

A detailed analysis of the relaxation processes occurring in tracks of multicharged ions is presented in Ref. 350. It is shown that by the end of the physical stage, the track of an ion is a nonisothermal and spatially nonhomogeneous plasma column. The rate of heating of the medium is determined by the efficiency of internal conversion and of the decay of molecules, and, therefore, the heating is a *consequence* and not the cause of the decay (which is contrary to the usual ideas concerning the influence of local heating on the yield of low-molecular radiation products).

In Ref. 350 it was also noted that, with increase of the degree of ionization in the medium, for a short time states of an ion system with negative energy may be formed in the track (i.e., the bound states) and there may even appear a certain order in the arrangement of particles. The pressure in this case can become negative. The transfer of electron excitation energy to other degrees of freedom tends to destroy the bound states and leads to heating of the medium and to the rise of pressure. However, the rate with which the pressure rises is practically comparable with that for the hydrodynamical mechanism of energy transfer. Therefore, it is unlikely that owing to shock contraction the concentration of electrons would increase much in comparison with the value it acquires owing to primary degradation of the energy of a heavy ion, as is stated in Ref. 370.

As far as we know, only LaVerne and Meisels[371] have made an attempt to study the effect of local heating experimentally. For this purpose they irradiated gaseous propane, as well as the gaseous mixture of methane (90%), ammonia (6%), and propane (4%), by fission fragments at different densities of the gas. As temperature probes they used the relative yields of isopropyl radicals and of normal propyl radicals. According to the results of their measurements, the effect of heating is much smaller than predicted by the thermal spike model, which, in general, is in agreement with the conclusions of Ref. 350.

C. On the Problem of Equivalence of the Effects Produced by Different Types of Radiation

Scientists are often faced with problems directly related to the problem of equivalence of the radiation effects produced by different types of radiation. The most typical are the two following problems. The first has to do with prediction of the radiation effect produced by a given type of radiation based on the data of radiative transformations produced by another type of radiation. This problem is very closely related to the problem of determining the limits of applicability of dosimetric systems (especially, of liquid chemical dosimeters). The second problem concerns the choice of equivalent radiation that can be substituted for a difficult-to-study type of radiation we might be interested in.

As a rule the radiation effect produced by any type of emission is a superposition of direct effects of the primary radiation and of the secondary (and even tertiary) radiation the latter induces. Consequently, if the radiation effect is mostly due to the effect produced by secondary (or tertiary) emission, the latter can be used instead of the primary radiation. As concerns the structure of tracks, such a simulation will be correct if the spatial distribution of chemically active particles in the irradiated volume remains close to the one produced by the primary source.

On general grounds, the X-ray and gamma irradiation can be replaced by irradiation with electrons with a corresponding energy spectrum, while neutron irradiation can be replaced by irradiation with heavy ions that would simulate the recoil nuclei of the medium.

The radiation effect produced by heavy ions can be simulated with electrons if the ions have high velocities. It is in this case that the spatial distribution of active particles in the track is close to the one in tracks of fast electrons. At small velocities of heavy ions the tracks of delta electrons overlap each other to a considerable extent, which results in the concentration of charged particles in a microvolume of the ion's track being very high. With development of powerful pulsed electron accelerators it became possible to create high concentrations of active particles in a medium. According to Ref. 372, with such accelerators one is able to reproduce and study the processes occurring in tracks of heavy ions.

However, as regards the structure of the track, generally speaking, such simulation is problematic, since we do not reproduce the corresponding spatial distribution. In the case of irradiation with ions, each of the latter forms it own track, which does not overlap the neighboring ones, and the concentrations of active particles are high only in these separate tracks, the diameter of which is several nanometers. In high-intensity electron beams the tracks of individual electrons overlap each other. As a result, the concentration of active particles is high in macroregions. This increase in the size of the region with high concentration of charged particles worsens the conditions for energy transfer, owing to which the temperature in an irradiated volume can be higher than in individual tracks. At high concentration of charged particles in macroregions there appear a number of new effects connected with collective properties of a dense electron–ion system.[373] For this reason the kinetics of radiation-chemical reactions will also be different.

The problem of equivalence of the radiation effects produced by different types of radiation is very important in dosimetry of ionizing radiation (especially, for liquid chemical dosimeters). As we have shown in Section IX.A, in the general case, the radiation effect in a condensed medium depends not only on the LET but also on the type of radiation. So only those dependences of the radiation effect in a dosimeter on the LET are of any real meaning that correspond to only one type of radiation. A dependence satisfying this requirement must have the form of curves similar to those featured in Fig. 21 for each type of radiation. On the other hand, the use of a dependence such as that of $G(Fe^{3+})$ on LET presented in Ref. 374 for a Fricke dosimeter (see also Fig. 7.4 in Ref. 375), in which different parts of a single curve correspond to different types of radiation, can lead to errors in measurements of the dose absorbed for a given type of ion.

References

1. I. G. Kaplan, in *Elementary Processes of High Energy Chemistry*. Nauka, Moscow, 1965, p. 253 (in Russian).
2. R. Platzman, *Rad. Res.* **17**, 419 (1962).
3. J. M. Heller, R. N. Hamm, R. D. Birkhoff, and L. R. Painter, *J. Chem. Phys.* **60**, 3483 (1974).
4. C. J. Kutcher and E. S. Green, *Rad. Res.* **67**, 408 (1976).
5. H. A. Bethe and J. Ashkin, in *Experimental Nuclear Physics* E. Segré (ed.). Wiley New York, 1953, Vol. 1, Part 2.
6. R. L. Fleischer, P. B. Price, and R. M. Wolker, *Nuclear Tracks in Solids: Principles and Applications*. University of California Press, Berkeley, 1975.
7. G. Girija and G. Gopinathan, *Radiat. Phys. Chem.* **16**, 245 (1980); **19**, 107 (1982).
8. J. W. Hunt, R. R. Wolff, M. J. Bronskill, C. D. Johah, E. J. Hart, and M. S. Matheson, *J. Phys. Chem.* **77**, 425 (1973).
9. J. W. Hunt and W. J. Chase, *Canad. J. Chem.* **55**, 2080 (1977).
10. A. Mozumder and J. L. Magee, *Int. J. Radiat. Phys. Chem.* **7**, 83 (1975).
11. M. J. Berger, *Methods in Computational Physics*. Academic Press, New York, 1963, Vol. 1, p. 135.
12. H. J. Paretzke, G. Leuthold, G. Burger, and W. Jacobi, in *Proc. IV Symposium on Microdosimetry*, Commission of the European Communities, Luxembourg, 1974, Vol. 1, p. 123.
13. I. G. Kaplan, A. M. Miterev, and L. M. Khadzhibekova, *High Energy Chemistry* **11**, 432 (1977).
14. H. J. Paretzke, J. E. Turner, R. N. Hamm, H. A. Wright, and R. H. Ritchie, *Rad. Res.* **92**, 47 (1982).
15. I. G. Kaplan and A. M. Miterev, in *Proc. 5th Symp. on Radiat. Chem.* J. Dobo, P. Hedvig, and R. Schiller (eds.). Akademiai Kiadó, Budapest, 1983, p. 33.
16. J. E. Turner, J. L. Magee, H. A. Wright, A. Chatterjee, R. N. Hamm, and R. H. Ritchie, *Rad. Res.* **96**, 437 (1983).
17. R. L. Platzman, in *Radiation Research*, J. Silini (ed.). North-Holland, Amsterdam, 1967, p. 20.
18. I. G. Kaplan and A. M. Miterev, *Khimiya Vysokikh Energii* **19**, 208 (1985).

19. I. G. Kaplan, A. M. Miterev, and V. Ya. Sukhonosov, *Khimiya Vysokikh Energii* **20**, 495 (1986).
20. R. L. Platzman, in *Radiation Biology and Medicine*, W. D. Claus (ed.). Addison-Wesley, Reading MA, 1958, p. 15.
21. E. G. Hart and R. L. Platzman, in *Mechanisms in Radiobiology*, A. Forssberg and M. Errera (eds.). Academic Press, New York, 1961, p. 93.
22. U. Fano, *Ann. Rev. Nucl. Sci.* **13**, 1 (1963).
23. A. Mozumder, in *Advances in Radiation Chemistry*, M. Burton and J. Magee (eds.). Wiley-Interscience, New York, 1969, Vol. 1, p. 1.
24. M. Burton, K. Funabashi, R. R. Hentz, P. K. Ludwig, J. L. Magee, and A. Mozumder, in *Transfer and Storage of Energy by Molecules*, G. M. Barnett and A. M. North (eds.). Wiley-Interscience, London 1969, Vol. 1, p. 161.
25. I. G. Kaplan and V. E. Skurat, *High Energy Chemistry* **6**, 224 (1972).
26. J. H. Baxendale and F. Busi (eds.), *The Study of Fast Processes and Transient Species by Electron Pulse Radiolysis*. D. Reidel, Dordrecht, 1982.
27. V. A. Fok, *Principles of Quantum Mechanics*, 2nd ed. Nauka, Moscow, 1976 (in Russian).
28. A. Messiah, *Quantum Mechanics*. Wiley, New York, 1966.
29. B. M. Garin and V. M. Byakov, Preprint ITEF No. 26, Moscow, 1983 (in Russian).
30. V. V. Konovalov, A. M. Raytzimring, and Yu. D. Tsvetkov, *Khimiya Vysokikh Energii* **18**, 5 (1984).
31. E. J. Hart and M. Anbar, *Hydrated Electron*. Wiley-Interscience, New York 1970.
32. A. K. Pikaev, *Solvated Electron in Radiation Chemistry*. Nauka, Moscow, 1969 (in Russian).
33. G. A. Kenney-Wallace and C. D. Jonah, *J. Phys. Chem.* **86**, 2572 (1982).
34. G. A. Kenney-Wallace, *Canad. J. Chem.* **55**, 2009 (1977).
35. T. A. Karlson, *Photoelectron and Auger Spectroscopy*. Plenum, New York, 1976.
36. T. Förster, in *Comparative Effects of Radiation*, M. Burton, J. S. Kirby-Smith, and J. L. Magee (eds.). Wiley, London, 1960, p. 300.
37. I. G. Kaplan and V. G. Plotnikov, *Khimiya Vysokikh Energii* **1**, 507 (1967).
38. J. A. Barltrop and J. D. Coyle, *Excited States in Organic Chemistry*. Wiley, London, 1975.
39. Th. Förster, *Fluoreszenz Organischer Ferbindunger*. Vandenhoeck and Ruprecht, Göttingen, 1951.
40. D. L. Dexter, *J. Chem. Phys.* **21**, 836 (1953).
41. V. L. Ermolaev, E. N. Bodunov, E. B. Sveshnikova, and T. A. Shakhverdov, *Nonradiative Transfer of Electron Excitation Energy*. Nauka, Moscow, 1977 (in Russian).
42. G. R. Freeman, in Ref. 26, p. 19.
43. E. S. Parilis, *Auger Effect*. Izd. FAN Uzbek. SSR, Tashkent, 1969 (in Russian).
44. D. Chattarji, *The Theory of Auger Transition*. Academic Press, New York, 1976.
45. E. S. Parilis and V. I. Matveev, *Usp. Fiz. Nauk* **138**, 573 (1982).
46. H. Siegbahn, L. Asplund, and P. Kelve, *Chem. Phys. Lett.* **35**, 330 (1975).
47. R. R. Rye, T. E. Madey, J. E. Houston, and P. H. Holloway, *J. Chem. Phys.* **69**, 1504 (1978).

48. A. Aberg and G. Howat, *Handbuch der Physik*. Springer Verlag, Berlin, 1982, Vol. 31, p. 469.
49. R. Manne and H. Agren, *Chem. Phys.* **93**, 201 (1985).
50. M. Higashi, E. Hiroike, and T. Nakajima, *Chem. Phys.* **68**, 377 (1982); **85**, 133 (1984).
51. G. E. Laramore, *Phys. Rev.* **A29**, 23 (1984).
52. C. M. Liegener, *Chem. Phys.* **92**, 97 (1985).
53. R. Arneberg, J. Müller, and R. Manne, *Chem. Phys.* **64**, 249 (1982).
54. D. Moncrieff, I. H. Hiller, and S. A. Pope, *Chem. Phys.* **82**, 139 (1983).
55. T. A. Karlson and M. O. Krause, *Phys. Rev.* **140A**, 1057 (1965).
56. T. A. Karlson and M. O. Krause, *Phys. Rev.* **137A**, 1653 (1965).
57. J. Bednář, *Jaderna Energie* **30**, 211 (1984).
58. M. J. van der Wiel, in *Radiation Research-Biomedical, Chemical and Physical Perspectives*, O. F. Nygaard, H. I. Adler, and W. K. Sinclair (eds). Academic Press, New York, 1975, p. 205.
59. J. Berkowitz, in *Radiation Research-Biomedical, Chemical and Physical Perspectives*, O. F. Nygaard, H. I. Adler, and W. K. Sinclair (eds). Academic Press, New York, 1975, p. 188.
60. G. L. Weissler, *Handbuck der Physik*. Springer Verlag, Berlin, 1956, Vol. 21.
61. W. P. Jesse and R. L. Platzman, *Nature* **195**, 790 (1962).
62. M. Inokuti, *Rad. Res.* **59**, 343 (1974).
63. V. I. Makarov and L. S. Polak, *High Energy Chemistry* **4**, 1 (1970).
64. M. Kasha, *Disc. Farad. Soc.* **9**, 14 (1950).
65. Y. Hatano, *Bull. Chem. Soc. Japan* **41**, 1126 (1968).
66. P. M. Dehmer and W. A. Chupka, *J. Chem. Phys.* **65**, 2243 (1976).
67. U. Fano, *J. Phys.* **B7**(14), 1401 (1974).
68. S. M. Tarr, J. A. Schavone, and R. S. Freund, *Phys. Rev. Lett.* **44**, 25 (1980); **44**, 1660 (1980).
69. S. E. Kupriyanov, A. A. Perov, A. Yu. Zayats, and A. N. Stepanov, Pis'ma ZhTF **14**, 861 (1981).
70. A. Yu. Zayats, A. A. Perov, and A. P. Simonov, *Khimicheskaya Fizika*, No. 3, 333 (1983).
71. S. Nishikawa and T. Watanabe, *Chem. Phys. Lett.* **22**, 590 (1973).
72. J. Bednář, *Jaderna Energie* **28**, 8 (1982).
73. I. G. Kaplan, *High Energy Chemistry* **17**, 159 (1983).
74. G. V. Golubkov and G. K. Ivanov, *ZhETF* **80**, 1321 (1981); *Chem. Phys. Lett.* **81**, 110 (1981).
75. G. V. Golubkov and G. K. Ivanov, *Khimicheskaya Fizika*, No. 9, 1179 (1982).
76. E. M. Balashov, G. V. Golubkov, and G. K. Ivanov, *ZhETF* **86**, 2044 (1984).
77. J. E. Aarts, C. I. Beenakker, and F. J. de Heer, *Physica* **53**, 32 (1971).
78. C. I. Beenakker and F. J. de Heer, *Chem. Phys.* **6**, 291 (1974); **7**, 130 (1975).
79. F. J. de Heer, *Int. J. Rad. Chem.* **7**, 137 (1975).
80. E. Palke and W. N. Lipscomb, *J. Am. Chem. Soc.* **88**, 2384 (1966).
81. R. S. Knox, *Theory of Excitons*. Academic Press, New York, 1963.

82. A. S. Davydov, *Theory of Molecular Excitons*. Plenum, New York, 1971.
83. N. Swanson and C. J. Powell, *Phys. Rev.* **145**, 195 (1966).
84. R. Voltz, in *Progress and Problems in Contemporary Radiation Chemistry*, J. Teplý (ed.). Prague, 1971, Vol. 1, p. 139.
85. J. Daniels, *Opt. Commun.* **3**, 240 (1971).
86. L. R. Painter, E. T. Arakawa, M. W. Williams, and J. C. Ashley, *Radiat. Res.* **83**, 1 (1980).
87. J. J. Ritsko, L. J. Brillson, R. W. Bigelow, and T. J. Fabish, *J. Chem. Phys.* **69**, 3931 (1978).
88. R. A. McRae, M. W. Williams, and E. T. Arakawa, *J. Chem. Phys.* **61**, 861 (1974).
89. M. W. Williams, R. N. Hamm, E. T. Arakawa, L. R. Painter, and R. D. Birkhoff, *Int. Rad. Phys. Chem.* **7**, 95 (1975).
90. D. Pines, *Elementary Excitations in Solids*. Benjamin, New York, 1963.
91. L. D. Landau and E. M. Lifshitz, *Electrodynamics of Continuous Media*. Addison-Wesley, Reading, MA, 1960.
92. U. Fano, *Phys. Rev.* **118**, 451 (1960).
93. W. Brandt and R. H. Ritchie, in *Physical Mechanism in Radiation Biology*, R. D. Cooper and R. W. Wood (eds.). U.S. Atomic Energy Commission, Washington, DC, 1974, p. 20.
94. H. Raether, *Excitation of Plasmons and Interband Transitions by Electrons*. Springer Tracts in Modern Physics. Springer-Verlag, Berlin 1980, Vol. 88, p. 1.
95. J. C. Slater, *Insulators, Semiconductors and Medals*. McGraw-Hill, New York, 1967.
96. R. D. Birkhoff, L. R. Painter, and J. M. Heller, Jr., *J. Chem. Phys.* **69**, 4185 (1978).
97. U. Killat, *Z. Phys.* **263**, 83 (1973).
98. J. J. Ritsko and R. W. Bigelow, *J. Chem. Phys.* **69**, 4162 (1978).
99. A. S. Davydov, *Quantum Mechanics*. Pergamon Press, Oxford, 1965.
100. R. N. Hamm, H. A. Wright, R. H. Ritchie, J. E. Turner, and T. P. Turner, in *Proc. 5th Symp. Microdosimetry*, J. Booz, H. G. Ebert, and B. G. R. Smith (eds.). Commission of the European Communities, Brussels, 1976, p. 1037.
101. D. Pines, *Rev. Mod. Phys.* **28**, 184 (1956).
102. P. Wölff, *Phys. Rev.* **92**, 18 (1953).
103. P. Nozieres and D. Pines, *Phys. Rev.* **109**, 762 (1958).
104. N. F. Mott and H. S. Massey, *The Theory of Atomic Collisions*. Clarendon Press, Oxford, 1965.
105. N. F. Lane, *Rev. Mod. Phys.* **52**, 29 (1980).
106. L. D. Landau and E. M. Lifshitz, *Quantum Mechanics, Nonrelativistic Theory*. Fizmatgiz, Moscow, 1963 (in Russian).
107. G. F. Drukarev, *Collisions of Electrons with Atoms and Molecules*. Nauka, Moscow, 1978 (in Russian).
108. A. Temkin, in *Electronic and Atomic Collisions*, N. Oda and K. Takayanagi (eds.). *Proc. 11th Int Conf.*, Kyoto, 1979. North-Holland, Amsterdam, 1980, p. 95.
109. J. Kistemaker, in Ref. 108, p. 3.
110. P. G. Burke, in *Atomic and Molecular Collisions Theory, Proc. NATO, Adv. Study Inst.*, F. A. Gianturco (ed.). Plenum, New York, 1982, Vol. B71, p. 69.
111. C. J. Joachain, in *Atomic and Molecular Physics of Controlled Thermonuclear*

Fusion, *Proc. NATO*, *Adv. Study Inst.*, Ch. J. Joachain and Douglass E. Post (eds.). Plenum, New York, 1983, Vol. B101, p. 139.

112. J. Hinze (ed.). *Electron-Atom and Electron-Molecule Collisions*. Plenum, New York, 1983.
113. M. Inokuti, *Rev. Mod. Phys.* **43**, 297 (1971).
114. M. Inokuti, G. Itikawa, and J. E. Turner, *Rev. Mod. Phys.* **50**, 23 (1978).
115. F. H. Read and G. L. Whiterod, *Proc. Phys. Soc. London* **82**, 434 (1963).
116. F. H. Read, in *Atoms, Molecules and Lasers*, Lect. Inst. Winter Coll., Trieste, 1973, Wienne, 1974, p. 567.
117. C. E. Brion and A. Hamnett, in *Advances in Chemical Physics*, I. Prigogine and S. Rice (eds.). Wiley, New York, 1981, Vol. 45, p. 2.
118. G. D. Zeiss, W. J. Meath, J. C. F. MacDonald, and D. J. Dawson, *Radiat. Res.* **70**, 284 (1977); *Can. J. Phys.* **55**, 2080 (1977); *Molec. Phys.* **39**, 1055 (1980).
119. J. Berkowitz, *Photoabsorption*, *Photoionization*, *and Photoelectron Spectroscopy*. Academic Press, New York, 1979.
120. V. S. Asoskov, V. M. Grishin, V. K. Ermilova, L. P. Kotenko, G. I. Merzon, and V. A. Chechin, *Trudy Fiz*, *Instit. An SSSR* **140**, 3 (1982).
121. C. E. Brion and J. P. Thomson, *J. Elect. Spectroscopy Relat. Phenom.* **33**, 301 (1984).
122. K. J. Miller, S. R. Mielczarek, and M. Krauss, *J. Chem. Phys.* **51**, 26 (1969).
123. K. J. Miller, *J. Chem. Phys.* **51**, 5235 (1969).
124. M. Krauss and S. R. Mielczarek, *J. Chem. Phys.* **51**, 5241 (1969).
125. S. R. Mielczarek and K. J. Miller, *Chem. Phys. Lett.* **10**, 369 (1971).
126. J. R. Oppenheimer, *Phys. Rev.* **32**, 361 (1928).
127. D. R. Bates, A. Fundaminsky, J. W. Leech, and H. S. Massey, *Phil. Trans. Roy. Soc. A* **243**, 93 (1950).
128. V. I. Ochkur, *Sov. Phys. JETP* **18**, 503 (1964).
129. G. F. Drukarev, in *The Theory of Electron–Atomic Collisions*, J. B. Husted (ed.). Academic Press, New York, 1967.
130. I. G. Kaplan, *Symmetry of Many-Electron Systems*. Academic Press, New York, 1975.
131. M. R. H. Rudge, *Proc. Phys. Soc. London*, **85**, 607 (1965); **86**, 763 (1965).
132. M. R. H. Rudge, *Adv. Atom. Mol. Phys.* **9**, 47 (1973).
133. M. Matsurawa, *J. Chem. Phys.* **51**, 4705 (1969).
134. D. C. Cartwright, *Phys. Rev.* **A2**, 1331 (1970).
135. Y.-K. Kim, in *Physics of Ion-Ion and Electron-Ion Collisions*, Proc. NATO Adv. Study Summer Inst., F. Brouillard and J. W. McGowan (eds.). Plenum, New York, 1983, Vol. B83, p. 101.
136. M. Inokuti, R. P. Saxon, and J. L. Dehmer, *Int. J. Radiat. Phys. Chem.* **7**, 109 (1975).
137. J. Schutten, E. J. de Heer, H. R. Moustafa, A. J. H. Boerboom, and J. Kistemaker, *J. Chem. Phys.* **44**, 3924 (1966).
138. B. L. Schram, M. J. Van der Wiel, F. J. de Heer, and H. R. Moustafa, *J. Chem. Phys.* **44**, 49 (1966).
139. H. J. Grosse and H. K. Bothe, *Z. Naturforschung* **B23a**, 1583 (1968); **25a**, 1970 (1970).

140. H. Bethe, *Z. Phys.* **135**, 325 (1930).

141. G. D. Zeiss, W. J. Meath, J. C. F. MacDonald, and D. J. Dawson, *Radiat. Res.* **63**, 64 (1975).

142. F. F. Rieke and W. Prepejehal, *Phys. Rev.* **A6**, 1507 (1972).

143. I. G. Kaplan, A. M. Miterev, and V. Ya. Sukhonosov, *Rad. Phys. Chem.* **27**, 83 (1986).

144. H. G. Partetzke and M. J. Berger, in *Proc. 6th Symp. on Microdosimetry*, J. Booz and H. G. Ebert (eds.). Commission of the European Communities, Brussels, 1978, p. 749.

145. D. K. Jain and S. P. Khare, *J. Phys.* **B9**, 1429 (1976).

146. L. Vriens, in *Case Studies in Atomic Collision Physics*, E. W. McDaniel and M. R. C. McDowel (eds.). North-Holland, Amsterdam, 1969, Vol. 1, p. 335.

147. V. I. Ochkur, in *Atomic Collision Problems*, Yu. N. Demkov (ed.). Leningrad State University, 1975, p. 42 (in Russian).

148. J. D. Jackson, *Classical Electrodynamics*. Wiley, New York, 1962.

149. M. Abramovitz and I. A. Stegun (eds.). *Handbook of Mathematical Functions with Formulas, Graphs and Mathematical Tables*, NBS, Washington DC, 1964.

150. U. Fano, in *Charged Particle Track in Solid and Liquids*, G. E. Adams, D. K. Bewley, and J. W. Boag (eds.). The Phys. Soc. Conf. Series, No. 2, Cambridge, 1970, p. 1.

151. E. Fermi, *Z. Phys.* **29**, 315 (1924).

152. L. D. Landau and E. M. Lifshitz, *Theory of Field*. Addison-Wesley, Reading, MA, 1959.

153. G. L. Yudin, *ZhETF* **83**, 908 (1982) (in Russian).

154. Chr. Lehman, *Interaction of Radiations with Solid and Elementary Defect Production*. North-Holland, Amsterdam, 1977.

155. S. V. Starodubtsev, *Complete Scientific Works*. FAN Uzb. SSR, Tashkent, 1970, t.2, kn.2 (in Russian).

156. M. A. Kumakhov and F. F. Komarov, *Energy Losses and Path-Lengths of Ions in Solids*. Izd. BGU, Minsk, 1979 (in Russian).

157. S. P. Ahlen, *Rev. Mod. Phys.* **52**, 121 (1980).

158. R. L. Platzman, in *Symposium on Radiobiology*, J. J. Nickson (ed.). Wiley-Interscience, New York, 1952, p. 139.

159. J. F. Janni, *Atom. Data Nucl. Data Tabl.* **27**, 147 (1982).

160. L. Pages, E. Bertel, H. Joffre, and L. Sklavenitis, *At. Data*, **4**, 1 (1972).

161. J. C. Ashley, *Radiat. Res.* **89**, 25 (1982).

162. J. Neufeld, *Proc. Phys. Soc. London*, **60**, 590 (1953).

163. A. Mozumder, in *Proc. 3d Tihany Symp. Rad. Chem.*, J. Dobo and P. Hedvig (ed.). Akademiai Kiado, Budapest, 1972, Vol. 2, p. 1132.

164. H. Sugiyama, *Radiat. Eff.* **66**, 205 (1981).

165. I. G. Kaplan and A. P. Markin, *Optika Spektrosk.* **19**, 856 (1965).

166. D. Powers, *Accounts Chem. Res.* **13**, 433 (1980).

167. D. Powers, H. G. Olson, and R. Gowda, *J. Appl. Phys.* **55**, 1274 (1984).

168. E. Kamaratos, *Nucl. Instr. Meth.* **215**, 337 (1983).

169. Y. J. Xu and G. S. Khandelwal, *Phys. Rev.* **A29**, 3419 (1984).

170. Yu. V. Gott, *Interaction of Particles with Matter in Plasma Studies*. Atomizdat, Moscow, 1978 (in Russian).
171. O. B. Firsov, *Sov. Phys. JETP* **9**, 1076 (1959).
172. J. Lindhard, M. Scharff, and H. E. Schiott, *K. Dan Vidensk. Selsk. Mat.-Fys. Medd.* **33**, No. 14 (1963).
173. M. D. Brown and C. D. Moak, *Phys. Rev.* **B6**, 90 (1972).
174. I. A. Akhiezer, L. N. Davydov, and Z. A. Spol'nik, *Ukr. Fiz. Zhurn.* **23**, 601 (1978).
175. V. S. Nikolaev, *Usp. Fiz. Nauk* **85**, 679 (1965).
176. H. D. Betz, *Rev. Mod. Phys.* **44**, 465 (1972).
177. N. Bohr, *K. Dan. Vidensk. Selsk. Mat.-Fys. Medd.* **18**, *No.* 8 (1948).
178. A. M. Miterev and E. A. Borisov, *Atomn. Energiya* **36**, 320 (1974).
179. J. Jortner and A. Gaathon, *Can. J. Chem.* **55**, 1801 (1977).
180. B. Baron, D. Hoover, and F. Williams, *J. Chem. Phys.* **64**, 1997 (1979).
181. J. Casanovas, R. Grob, D. Delaeroix, J. P. Guellfucci, and D. Blanc, in *Proc. 7th Symposium on Microdosimetry*, H. G. Ebert and H. D. Hartfiel (eds.). Harwood Academic Publishing for the Commission of the European Communities, Brussels and Luxembourg, 1981, p. 157.
182. R. Schiller, Sz, Vass, and J. Mandics, *Int. J. Radiat. Phys. Chem.* **5**, 491 (1973).
183. R. A. Holroyd, B. K. Dietrich, and H. A. Schwarz, *J. Phys. Chem.* **76**, 3794 (1972).
184. R. A. Holroyd, *J. Chem. Phys.* **57**, 3007 (1972).
185. R. A. Holroyd, S. Tames, and A. Kennedy, *J. Phys., Chem.* **79**, 2857 (1975).
186. R. A. Holroyd and R. L. Russel, *J. Phys. Chem.* **78**, 2128 (1974).
187. R. A. Holroyd and M. Allen, *J. Chem. Phys.* **54**, 5014 (1971).
188. V. A. Benderskii and A. M. Brodsky, *Photoemission from a Metal Into Electrolytic Solution*. Nauka, Moscow, 1977 (in Russian).
189. P. Delahay and K. von Burg, *Chem. Phys. Lett.* **83**, 250 (1981).
190. P. Delahay, *Acc. Chem. Res.* **15**, 40 (1982).
191. B. Raz and J. Jortner, *Chem. Phys. Lett.* **4**, 155 (1969).
192. I. Messing and J. Jortner, *Chem. Phys.* **24**, 183 (1977).
193. R. R. Dogonadze and A. M. Kuznetsov, in *Itogi Nauki i Tekh., ser. Phys. Chem., Kinetics*. VINITI, Moscow, 1973, Vol. 2.
194. R. R. Dogonadze, A. M. Kuznetsov, and T. A. Marsagishvili, *Electrochim. Acta.* **25**, 1 (1980).
195. V. M. Byakov, Preprint ITEF, No. 165, Moscow, 1983 (in Russian).
196. A. G. Khrapak and I. T. Yakubov, *Electrons in Dense Gases and in Plasma*. Nauka, Moscow, 1981.
197. A. Henglein and M. Gratsel, in *Solar Power and Fuels*, J. R. Bolton (ed.). Academic Press, New York, 1977, p. 53.
198. D. Grand, A. Bernas, and E. Amouyal, *Chem. Phys.* **44**, 73 (1979).
199. J. W. Boyle, J. A. Ghormley, C. J. Hochanadel, and J. F. Riley, *J. Phys. Chem.* **73**, 2886 (1969).
200. D. N. Nikogosyan, A. A. Oraevsky, and V. I. Rupasov, *Chem. Phys.* **77**, 131 (1983).
201. E. N. Lassettre and A. Skerbele, in *Methods of Experimental Physics*, D. Williams (ed.). Academic Press, New York, vol. 3, p. 868.

202. C. J. Powell, *Health Physics* **13**, 1265 (1967).
203. U. Fano, *Phys. Rev.* **103**, 1202 (1956).
204. R. H. Ritchie, R. N. Hamm, J. E. Turner, and H. A. Wright, in *Proceedings 6th Symposium on Microdosimetry*, J. Booz and H. G. Ebert (eds.). Commission of the European Communities, Harwood, Brussels, 1978, p. 345.
205. W. F. G. Swann, *J. Franklin Inst.* **226**, 598 (1938).
206. E. Fermi, *Phys. Rev.* **57**, 485 (1940).
207. R. M. Sternheimer, M. J. Berger, and S. M. Seltzer, *At. Data Nucl. Data Tables* **30**, 261 (1984).
208. R. B. J. Palmer and A. Akhavan-Rezayat, *J. Phys.* **D11**, 605 (1978).
209. D. I. Thwaites and D. E. Watt, in *Proceedings 6th Symposium on Microdosimetry*, J. Booz and H. G. Ebert (eds.). Commission of the European Communities, Harwood, Brussels, 1978, p. 777.
210. I. Frank and Ig. Tamm, *Comptes rendus de l'Acad. Sci. URSS* **14**, 109 (1937).
211. B. M. Bolotovsky, *Usp. Fiz. Nauk* **62**, 201 (1957).
212. H. Fröhlich and R. L. Platzman, *Phys. Rev.* **92**, 1152 (1953).
213. R. L. Platzman, *Radiat. Res.* **2**, 1 (1955).
214. C. E. Klots, in *Progress and Problems in Contemporary Radiation Chemistry*, J. Teplý (ed.). Prague, 1971, Vol. 1, p. 120.
215. J. Bednář, *Radiochem. Radioanal. Lett.* **55**, 131 (1982).
216. W. P. Jesse and J. Sadauskis, *Phys. Rev.* **88**, 417 (1952).
217. J. L. Magee and M. Burton, *J. Am. Chem. Soc.* **73**, 523 (1951).
218. G. El Komoss and J. L. Magee, *J. Chem. Phys.* **36**, 256 (1962).
219. D. A. Douthat, *Radiat. Res.* **61**, 1 (1975).
220. K. Kowari and S. Sato, *Bull. Chem. Soc. Jpn.* **51**, 741 (1978).
221. G. J. Shulz, in *Principles of Laser Plasmas*, B. Bekefi (ed.). Wiley, New York, 1976.
222. A. K. Kazansky and I. I. Fabrikant, *Usp. Fiz. Nauk* **143**, 601 (1984).
223. J. A. D. Stockdale, in *Radiation Research*. Jpn. Assoc. Radiat. Res., Tokyo, 1979, p. 100.
224. A. Herzenberg, *J. Phys.* **B1**, 548 (1968).
225. L. Dube and A. Herzenberg, *Phys. Rev.* **A20**, 194 (1979).
226. J. Bardsley, *J. Phys.* **B1**, 349 (1968).
227. L. G. Christophorou, *Radiat. Phys. Chem.* **12**, 19 (1978).
228. D. E. Golden, N. F. Lane, A. Temkin, and E. Gerjuoy, *Rev. Mod. Phys.* **43**, 642 (1971).
229. E. Gerjuoy and S. Stein, *Phys. Rev.* **97**, 1761 (1955).
230. Y. Itikawa, *J. Phys. Soc. Jpn.* **30**, 835 (1971); **32**, 217 (1972).
231. L. S. Polak, A. A. Ovsyannikov, D. I. Slovetsky, and F. B. Vurzel, *Theoretical and Applied Plasmochemistry*. Nauka, Moscow, 1975 (in Russian).
232. H. S. W. Massey and E. H. S. Burhop, *Electronic and Ion Impact Phenomenon*. Oxford University Press, London, 1952.
233. E. W. McDaniel, *Collision Phenomena in Ionized Gases*. Wiley, New York, 1964.
234. Y. Itikawa, *Atom. Data Nucl. Data Tabl* **14**, 1 (1974); **21**, 69 (1978).
235. W. Sohn, R. Jung, H. Ehrhardt, *J. Phys.* **B16**, 891 (1983).

236. H. Tanaka, T. Okada, L. Boesten, T. Sazuki, T. Jamamoto, and M. Kubo, *J. Phys. B* **15**, 3305 (1982).
237. V. F. Sokolov and Yu. A. Sokolova, *Pis'ma v ZhTF* **7**, 627 (1981).
238. M. G. Lynch and D. Dill, *J. Chem. Phys.* **71**, 4249 (1979).
239. D. Thirumalai, K. Onda, and D. G. Truhlar, *J. Chem. Phys.* **74**, 526 (1981).
240. A. V. Eletsky and B. M. Smirnov, *Usp. Fiz. Nauk* **136**, 25 (1982).
241. M. A. Biondi, in *Principles of Laser Plasmas*, B. Bekefi (ed.). Wiley, New York, 1976.
242. M. A. Biondi, in *Applied Atomic Collision Physics*. Academic Press, New York, 1982, p. 173.
243. L. G. Christophorou and R. N. Compton, *Health Physics* **13**, 1277 (1967).
244. D. L. McCorkle, I. Szamrej, and L. G. Christophorou, *J. Chem. Phys.* **77**, 5542 (1982).
245. R. N. Compton, in *Electronic and Atomic Collisions, Proc. 11th Int. Conf. Kyoto*, 1979, N. Oda and K. Takayanagi (eds.). North-Holland, Amsterdam, 1980, p. 251.
246. D. L. McCorkle, L. G. Christophorou, and S. R. Hunter, in *Electron and Ion Swarms*, L. G. Christophorou (ed.). Pergamon Press, New York, 1981, p. 21.
247. A. V. Eletsky and B. M. Smirnov, *Usp. Fiz. Nauk* **147**, 459 (1985).
248. J. A. La Verne and A. Mozumder, *Radiat. Phys. Chem.* **23**, 637 (1984).
249. K. Koura, *J. Chem. Phys.* **79**, 3367 (1983); **80**, 5800 (1984); **81**, 303 (1984); **81**, 4180 (1984).
250. A. M. Samuel and J. L. Magee, *J. Chem. Phys.* **21**, 1080 (1953).
251. W. J. Chase and J. W. Hunt, *J. Phys. Chem.* **79**, 2835 (1975).
252. J. L. Magee and W. P. Helman, *J. Chem. Phys.* **66**, 310 (1977).
253. J. L. Magee, *Canad. J. Chem.* **55**, 1847 (1977).
254. M. Tachiya and H. Sano, *J. Chem. Phys.* **67**, 5111 (1977).
255. V. M. Byakov and V. L. Grishkin, Preprint ITEF, No. 41, 1977.
256. L. M. Hunter, D. Lewis, and W. H. Hamill, *J. Chem. Phys.* **52**, 1733 (1970).
257. P. B. Merkel and W. H. Hamill, *J. Chem. Phys.* **55**, 1409 (1971).
258. K. Hiraoka and W. H. Hamill, *J. Chem. Phys.* **57**, 3870 (1972).
259. K. Hiraoka and W. H. Hamill, *J. Chem. Phys.* **59**, 574 (1973).
260. K. Hiraoka and M. Nara, *Bull. Chem. Soc. Jpn.* **54**, 1589 (1981).
261. A. M. Raitsimring, R. I. Samoilova, and Y. D. Tsvetkov, *Radiochem. Radioanal. Letters* **38**, 75 (1979).
262. A. Mozumder and J. L. Magee, *J. Chem. Phys.* **47**, 939 (1967).
263. W. Schmidt and A. O. Allen, *J. Phys. Chem.* **72**, 3730 (1968); *J. Chem. Phys.* **52**, 2345 (1970).
264. A. Hummel and W. F. Schmidt, *Radiat. Res. Rev.* **5**, 199 (1974).
265. L. Onsager, *Phys. Rev.* **54**, 554 (1938).
266. G. C. Abell and K. Funabashi, *J. Chem. Phys.* **58**, 1079 (1973).
267. A. Mozumder, *J. Chem. Phys.* **60**, 4300 (1974); **60**, 4305 (1974).
268. J. P. Dodelet, K. Shinsaka, U. Kortsch, and G. R. Freeman, *J. Chem. Phys.* **59**, 2376 (1973).

269. J. Casanovas, R. Grob, D. Blanc, G. Brunet, and J. Mathieu, J. Chem. Phys. **63**, 3673 (1975).
270. S. S. S. Huang and G. R. Freeman, *Can. J. Chem.* **55**, 1838 (1977).
271. J. P. Dodelet, *Can. J. Chem.* **55**, 2050 (1977).
272. J. P. Dodelet, P. G. Fuochi, and G. R. Freeman, *Can. J. Chem.* **50**, 1617 (1972).
273. M. G. Robinson and G. R. Freeman, *Can. J. Chem.* **51**, 1010 (1973).
274. J. M. Warman and S. J. Rzad, *J. Chem. Phys.* **52**, 485 (1970).
275. J. P. Dodelet and G. R. Freeman, *Can. J. Chem.* **50**, 2729 (1972).
276. A. Mozumder, *J. Chem. Phys.* **50**, 3153 (1969).
277. Yu. V. Pleskov and Z. A. Rotenberg, *Khimiya Vysokikh Energii* **8**, 99 (1974).
278. A. M. Brodsky, Yu. Ya. Gurevich, Yu. V. Pleskov, and Z. A. Rotenberg, *Modern Photoelectrochemistry. Photoemission Phenomena.* Nauka, Moscow, 1974 (in Russian).
279. Yu. Ya. Gurevich, Yu. V. Pleskov, and Z. A. Rotenberg, *Itogi Nauki i Tekh., ser. Rad. Chem., Photochemistry.* VINITI, Moscow, 1978, Vol. 1.
280. A. M. Brodsky and Yu. Ya. Gurevich, *ZhETF* **54**, 213 (1968).
281. W. Tauchert, H. Junglut, and W. F. Schmidt, *Can. J. Chem.* **55**, 1860 (1977).
282. G. C. Barker, A. W. Gardner, and D. C. Sammon, *J. Electrochem. Soc.* **113**, 1182 (1966).
283. Yu. Ya. Gurevich and Z. A. Rotenberg, *Elektrokhimiya* **4**, 529 (1968).
284. Z. A. Rotenberg, *Elektrokhimiya* **10**, 1031 (1974).
285. L. I. Korshunov, Ya. M. Zolotovitskii, V. A. Benderskii, and V. I. Gol'danskii, *Khimiya Vysokikh Energii* **4**, 461 (1970).
286. V. A. Benderskii, S. D. Babenko, Ya. M. Zolotovitskii, A. G. Krivenko, and T. S. Rudenko, *J. Electroanalyt. Chem.* **56**, 325 (1974).
287. S. D. Babenko, V. A. Benderskii, Ya. M. Zolotovitskii, A. G. Krivenko, and T. S. Rudenko, *Elektrokhimiya* **12**, 693 (1976).
288. A. M. Raitsimring, S. D. Babenko, G. I. Velichko, A. G. Krivenko, V. A. Benderskii, A. A. Ovchinnikov, and Yu. D. Tsvetkov, *Dokl. AN SSSR* **247**, 627 (1979).
289. A. M. Raitsimring and Yu. D. Tsvetkov, *Khimiya Vysokikh Energii* **14**, 229 (1980).
290. I. V. Kreytus, V. A. Benderskii, V. A. Beskrovnii, and Yu. E. Tiliks, *Khimiya Vysokikh Energii* **16**, 112 (1982).
291. V. V. Konovalov, V. V. Tregub, and A. M. Raitsimring, *Elektrokhimiya* **20**, 470 (1984).
292. V. V. Konovalov, A. M. Raitsimring, and Yu D. Tsvetkov, *Chem. Phys.* **93**, 163 (1985).
293. L. I. Korshunov, Ya. M. Zolotovitskii, and V. A. Benderskii, *Uspekhi Khimii* **40**, 1511 (1971).
294. G. C. Barker, A. W. Gardner, and G. Bottura, *J. Electroanalyt. Chem.* **45**, 21 (1973).
295. Cz. Stradowski and W. H. Hamill, *J. Phys. Chem.* **80**, 1431 (1976).
296. H. T. Choi, D. S. Sethi, and C. L. Braun, *J. Chem. Phys.* **77**, 6027 (1982).
297. A. A. Balakin, L. V. Lukin, A. V. Tolmachev, and B. S. Yakovlev, *Khimiya Vysokikh Energii* **15**, 123 (1981).

298. B. S. Yakovlev, *Khimiya Vysokikh Energii* **15**, 435 (1981).

299. U. Fano, in *Comparative Effects of Radiation*, M. Burton, J. S. Kirby-Smith, and J. L. Magee (eds.). Wiley, London, 1960, p. 14.

300. I. G. Kaplan and A. M. Miterev, *Radiat. Phys. Chem.* **26**, 53 (1985).

301. A. I. Burshtein, *Usp. Fiz. Nauk* **143**, 553 (1984).

302. V. V. Tregub, A. M. Raitsimring, and V. M. Moralev, Preprint No. 7, 1980; Preprint No. 8, 1981, Novosibirsk.

303. V. A. Pitkevich and V. V. Duba, *Radiobiologiya* **21**, 829 (1981).

304. V. A. Pitkevich and V. G. Vidensky, *Medits. Radiologiya* **7**, 24 (1975).

305. H. G. Paretzke and F. Schindel, in *Proc. 7th Symposium on Microdosimetry*, J. Booz, H. G. Ebert, and H. D. Hartfiel (eds.). Harwood Academic Publishing for the Commission of the European Communities, Brussels and Luxembourg, 1981, p. 387.

306. D. E. Lea, *Action of Radiations on Living Cells*. Macmillan, New York, 1947.

307. L. H. Gray, *J. Chem. Phys.* **48**, 172 (1951).

308. J. L. Magee and A. Chatterjee, *Radiat. Phys. Chem.* **15**, 125 (1980).

309. A. Mozumder and J. L. Magee, *Radiat. Res.* **28**, 203 (1966); *J. Chem. Phys.* **45**, 3332 (1966).

310. J. L. Magee and A. Chatterjee, *J. Phys. Chem.* **82**, 2219 (1978).

311. I. Santar and J. Bednář, *Int. J. Radiat. Phys. Chem.* ***1***, 133 (1969).

312. *Average Energy Required to Produce an Ion Pair*, ICRU Report, 31, Washington, DC, 1979.

313. A. Singh, W. Chase, and J. W. Hunt, *Faraday Discus. Chem. Soc.* **63**, 28 (1977).

314. C. D. Jonah and J. R. Miller, *J. Phys. Chem.* **81**, 1974 (1977).

315. C. D. Jonah, M. S. Matheson, J. R. Miller, and E. J. Hart, *J. Phys. Chem.* **80**, 1267 (1976).

316. L. T. Bugaenko, V. M. Byakov, and S. A. Kabakchi, *Khimiya Vysokikh Energii* **19**, 291 (1985).

317. P. A. Cherenkov, *Dokl. AN SSSR* **2**, 451 (1934).

318. S. I. Vavilov, *Dokl. AN SSSR* **2**, 457 (1934).

319. V. P. Zrelov, *Valilov-Čherenkov Radiation and Its Application in High Energy Physics*. Atomizdat, Moscow, 1968, Vols. 1 and 2 (in Russian).

320. M. J. Bronskill and J. Hunt, *J. Phys. Chem.* **72**, 3762 (1968).

321. R. K. Wolf, M. J. Bronskill, and J. W. Hunt, *J. Chem. Phys.* **53**, 4211 (1970).

322. Z. P. Zagorski and Z. Zimek, *Int. J. Rad. Phys. Chem.* **7**, 529 (1975).

323. H. B. Steen, *Int. J. Rad. Phys. Chem.* **7**, 489 (1975).

324. T. S. Zhuravleva, N. A. Bach, and L. T. Bugaenko, *Izv. AN SSSR, ser. Fiz.* **39**, 2428 (1975).

325. T. S. Zhuravleva, *Khimiya Vysokikh Energii* **11**, 279 (1977).

326. Y. Katsumura, S. Tagawa, and Y. Tabata, *Rad. Phys. Chem.* **19**, 243 (1982).

327. V. K. Milinchuk, E. R. Klinshpont, and S. Ya. Pshezhetsky, *Macroradicals*. Khimiya, Moscow, 1980 (in Russian).

328. I. G. Kaplan and N. V. Polyansky, *Dokl. AN SSSR* **267**, 1110 (1982); *Khimiya Vysokikh Energii* **16**, 387 (1982).

329. I. G. Kaplan and N. V. Polyansky, in *Proc. 5th Symp. on Radiation Chemistry*, J. Dobo, P. Hedvig, and R. Schiller (eds.). Akademiai Kiado, Budapest, 1983, p. 3.
330. G. D. Zeiss and W. I. Meath, *Mol. Phys.* **30**, 161 (1975).
331. V. I. Ivanov, *Course on Dosimetry*. Atomizdat, Moscow, 1978 (in Russian).
332. A. Mozumder, A. Chatterjee, and J. L. Magee, in *Advances in Chemistry Series*. Amer. Chem. Soc., Washington, DC, 1968, Vol. 81, p. 27.
333. J. L. Magee and A. Chatterjee, *J. Phys. Chem.* **84**, 3529 (1980).
334. A. Mozumder, *J. Chem. Phys.* **60**, 1145 (1974).
335. A. M. Miterev, I. G. Kaplan, and E. A. Borisov, *High Energy Chem.* **8**, 461 (1974).
336. A. Chatterjee and J. L. Magee, in *Proc. 6th Symp. on Microdosimetry*, J. Booz and H. G. Ebert (eds.). Comission of the European Communities, Brussels, 1978, p. 283.
337. R. H. Ritchie and W. Brandt, *Radiation Research, Biomedical, Chemical and Physical Perspectives*, O. Nygaard, H. I. Alder, and W. K. Sinclair (eds.). Academic Press, New York, 1975, p. 205.
338. Yu. K. Kagan, *Dokl. AN SSSR* **119**, 247 (1958).
339. J. J. Butts and R. Katz, *Radiat. Res.* **30**, 855 (1967).
340. D. I. Vaisburd, Yu. V. Volkov, and A. M. Kol'chuzhkin, in *Radiation Disturbance in Solids and Liquids*. FAN Uzb. SSR, Tashkent, 1967, p. 83 (in Russian).
341. J. W. Baum, S. L. Stone, and A. V. Kuehner, in *Proc. of Symp. on Microdosimetry*, H.G. Ebert (ed.). Euratom, Brussels, 1968, p. 269.
342. E. I. Kudryashov, A. M. Marenny, V. I. Popov, and O. M. Mesheryakova, *Cosmic Biology and Medicine* **4**, 35 (1970) (in Russian).
343. A. Chatterjee, H. Maccabee, and C. A. Tobias, *Radiat. Res.* **54**, 479 (1973).
344. J. H. Miller and A. E. S. Green, *Radiat. Res.* **57**, 9 (1974).
345. J. Fain, M. Monnin, and M. Montret, *Radiat. Res.* **57**, 379 (1974).
346. H. G. Paretzke, in *Proc 4th Symp. on Microdosimetry*, J. Booz, H. G. Ebert, R. Eickel, and A. Waker (eds.). Luxembourg, 1974, Vol. I, p. 141.
347. M. Zaider, D. J. Brenner, and W. E. Wilson, *Radiat. Res.* **95**, 231 (1983).
348. M. N. Varma, J. W. Baum, and A. V. Kuchner, *Radiat. Res.* **62**, 1 (1975).
349. C. L. Wingate and J. W. Baum, *Radiat. Res.* **65**, 1 (1976).
350. A. M. Miterev, *Khimiya Vysokikh Energii* **14**, 483 (1980).
351. R. H. Ritchie and C. Claussen, *Nucl. Instr. Meth.* **198**, 133 (1982).
352. G. G. Meisels, J. P. Gregory, A. A. Siddiqi, J. P. Freeman, and W. C. Richardson, *J. Am. Chem. Soc.* **97**, 987 (1975).
353. G. G. Meisels, J. A. LaVerne, W. B. Richardson, and T. C. Hsieh, *J. Phys. Chem.* **82**, 2231 (1978).
354. S. Lipsky, *Chem. Education* **58**, 93 (1981).
355. G. R. Freeman, in *Advances in Radiation Research Physics and Chemistry*, J. F. Duplan and A. Chapiro (eds.). Gordon and Breach, New York, 1973, Vol. 2, p. 351.
356. S. Ya. Pshezhetskii, *Mechanism and Kinetics of Radiation-Chemical Reactions*. Khimiya, Moscow, 1968 (in Russian).
357. W. Burns and R. Barker, in *Aspects of Hydrocarbon Radiolysis*, T. Gaumann and J. Hoigne (eds.). Academic Press, New York, 1968, p. 33.

358. M. Anbar, in *Fundamental Processes in Radiation Chemistry*, P. Ausloos (ed.). Interscience, New York, 1968, p. 651.
359. J. K. Tomas, in *Advances in Radiation Chemistry*, M. Burton and J. Magee (eds.). Wiley-Interscience, New York, 1969, Vol. I, p. 103.
360. R. A. Holroyd, in *Fundamental Processes in Radiation Chemistry*, P. Ausloos (ed.). Interscience, New York, 1968, p. 413.
361. A. Kupperman and G. G. Belford, *J. Chem. Phys.* **36**, 1412 (1962).
362. M. Matsui and M. Imamura, *IPCR Cyclotron Prog. Rept.* **5**, 93 (1971).
363. S. P. Kil'chitskaya, E. P. Petryaev, and E. P. Kalyazin, *Izv. AN BSSR, ser. Fiz.-Energ. Nauk* No. 4, 26 (1979).
364. L. A. Bulanov, E. V. Starodubtseva, and E. A. Borisov, *Khimiya Vysokikh Energii* **6**, 476 (1972).
365. W. G. Burns, in *Charged Particle Track in Solids and Liquids*, G. E. Adams and J. W. Boag (eds.). The Phys. Soc. Conf. Series No. 2, Cambridge, 1970, p. 143.
366. W. G. Burns, in *Proc. 7th Symp. on Microdosimetry*, J. Booz, H. G. Ebert, and H. D. Hartfiel eds.). Harwood Academic Publishers for the Comiss. European Communities, Brussels, 1982, p. 471.
367. J. A. LaVerne and R. H. Schuler, *J. Phys. Chem.* **88**, 1200 (1984).
368. R. Katz, in *Proc. 7th Symp. on Microdosimetry*, J. Booz, H. G. Ebert, and H. D. Hartfiel (eds.). Harwood Academic Publishers for the Commiss. European Communities, Brussels, 1982, p. 583.
369. V. I. Goldanskii and Yu. M. Kagan, *Int. J. Appl. Radiat. Isotopes* **11**, 1 (1961).
370. V. I. Goldanskii, E. Ya. Lantsburg, and P. A. Yampol'skii, *Pis'ma v ZhETF* **21**, 365 (1975).
371. J. A. LaVerne and G. G. Meisels, *Radiat. Phys. Chem.* **21**, 329 (1983).
372. D. I. Vaisburd, V. P. Kuznetsov, V. A. Moskalev, and M. M. Shafir, *Atomnaya Energiya* **39**, 366 (1975).
373. D. I. Vaisburd, B. N. Semin, E. G. Tavanov, S. B. Matlis, I. N. Balychev, and G. I. Gering, *High-Energy Electronics of a Solid*. Nauka, Novosibirsk, 1982 (in Russian).
374. N. E. Bibler, *J. Phys. Chem.* **79**, 1991 (1975).
375. A. K. Pikaev, *Modern Radiation Chemistry. Principles. Experimental Technique and Methods*. Nauka, Moscow, 1985 (in Russian).

AUTHOR INDEX

Numbers in parentheses are reference numbers and indicate that the author's work is referred to although his name is not mentioned in the text. Numbers in italic show the pages on which the complete references are listed.

SUBJECT INDEX